C0-AWS-213

CONTAINERS OF CLASSICAL GREECE

Pelike, Apulian polychrome, fourth century B.C. 24 cm.
Woman at laver (louterion). An alabastron stands in front of her foot.
Courtesy of Los Angeles County Museum of Art, The William Randolph Hearst Collection.

CONTAINERS OF CLASSICAL GREECE

A Handbook of Shapes

M.G. KANOWSKI

University of Queensland Press
ST LUCIA • LONDON • NEW YORK

First published 1984 by University of Queensland Press
P.O. Box 42, St Lucia, Queensland, Australia

Typeset by University of Queensland Press
Designed by Paul Rendle
Printed in Hong Kong by Silex Enterprise & Printing Co.

Distributed in U.K., Europe, the Middle East, Africa, and the Carribbean by
Prentice Hall International, International Book Distributors Ltd,
66 Wood Lane End, Hemel Hempstead, Herts. England

Distributed in the U.S.A. and Canada by
Technical Impex Corporation
5 South Union Street
Lawrence, Mass. 01843 U.S.A.

National Library of Australia
Cataloguing-in-Publication data

Kanowski, M.G. (Max G.), 1932-
Containers in classical Greece.

Includes index.
ISBN 0 7022 1744 1.

1. Pottery, Greek. 2. Art, Greek. 3. Art
metal-work, Greek. I. Title.

730'.0938

Library of Congress Cataloguing in Publication Data

Kanowski, M.G. (Maxwell George), 1932-
Containers of classical Greece.

1. Vases, Greek. 2. Ceramic tableware – Greece.
3. Drinking vessels – Greece. 4. Containers – Greece.
I. Title.
NK4645.K36 1984 738.3'0938 83'14726
ISBN 0-7022-1744-1

Contents

Illustrations

PHOTOGRAPHS

Preface

This book deals with the shapes of ancient Greek containers. Most of our information comes from surviving pottery and to an extent from metal vessels. I have not taken many examples from glassware. The restrictions imposed by ancient glass-making processes produced a rather different set of shapes in that medium, especially for balsamaria, unguentaria and other small vessels.

One of the useful features of Gisela Richter's and Marjorie Milne's *Shapes and Names of Athenian Vaes* was that it dealt separately with each of the important shapes. I have decided on a similar approach. There are simple descriptions for each important shape (and a few less important ones), supplemented by line drawings. More detailed and technical information may be gained from the additional references provided for each shape, and from general indices. A selection of photographs has also been included, since the exterior decoration of containers is an important source of information about the ways in which they were used.

It has been my intention to provide a useful handbook for students of Greek art. But the main section will also be comprehensible to potters and non-specialists who may have more general interests.

A book that deals with ancient Greek containers must be greatly indebted to the painstaking work of such experts as Sir John Beazley, Robert Cook, Marjorie Milne, Gisela Richter, Brian Sparkes, Lucy Talcott and many others. I gratefully acknowledge this debt.

My thanks go also to Danny Sheehy of the University of Queensland for the preparation of line drawings.

General Introduction

Materials used for containers

Containers in the ancient Mediterranean world were made from different materials including clay (pottery), glass, metal (copper, bronze, and in more affluent areas, gold and silver), wood and other plant material (reeds for basketry, dried gourds for drinking vessels), substances of animal origin (skins and horn for wineskins and drinking horns), and stone.

We are apt to get a false impression of the relative popularity of these substances because of their different survival rates. Gold and silver objects have hardly any chance of survival except within graves, hoards, or from shipwrecks and sites such as Pompeii. Ancient writers mention wooden vats, tubs, troughs and chests. Though this sort of thing does not survive, quite often the fastenings, handles, and decoration such as inlay that were made of more permanent materials do survive. Wood was also regularly used by poorer people for platters, plates and bowls. The name *pyxis* is clearly derived from *pyxos* (boxwood) and *ollix* is described as a wooden drinking bowl. The *kissybion* was a cup either made of ivy-wood (*kissos*) or with ivy decoration. A *chalkion* or *chalkōma* must have been a bronze or copper vessel. Stone vessels were difficult to make, but, once made, were likely to have a long life. A surprisingly large number appear from certain places and periods (for example, Egyptian, Minoan, and Mycenaean burials).

But it is from pottery that most of our knowledge of Greek containers has come, because it is pottery containers that survive in the largest numbers.

Pottery in Greece

The art of pottery making in Greece goes back to the Neolithic period. Some extensively studied sites such as Dimini and Sesklo in northern Greece have been shown to have acquired the essential characteristics of Neolithic settlements before they acquired the art of pottery making, but pottery manufacture was introduced before long. There is evidence for the widespread use of the pottery wheel in the Aegean early in the second millennium B.C. Pottery of good quality from the Greek classical period (and most poor quality pottery as well) was wheel-made.

Types of pottery

Greek pottery of the Classical period is assigned to broad classifications according to type of decoration and function of the pot, for example, figured pottery, plain black wares, domestic pottery, and trade amphorae. Grades of pottery may vary a great deal within classifications, but in general the finest pottery is found amongst the figured wares. It is for this reason, and for the great interest of much of the subject-matter of the decoration, that this type of pottery has attracted most attention. Black pottery (commonly called black-painted or black-glaze) was often as good technically as the figured wares. The coarser domestic pottery has been little studied until recently, and the many local styles are still imperfectly understood. Export amphorae are of interest mainly in the areas of trade, economics, and chronology.

Chronology of painting styles

Type and style of decoration, rather than shape, have regularly been taken as the most important evidence for the date of a pot. In some cases shape is just as reliable a criterion, and if shapes were more closely studied, this evidence could be more widely used.

The following table refers mainly to the figured pottery of two of the most important Greek pottery centres (Athens and Corinth). The dates have been rounded off and are very approximate.

Geometric

The name comes from the predominantly geometric motifs

of the decoration. The dates given for this period apply to Athens. At Corinth the end of the geometric comes a couple of decades earlier. There are other less important areas of production where in some cases geometric motifs survive longer.

Protogeometric	From between 1100 and 1050 B.C. to between 950 and 900 B.C.
Early Geometric	900–850 B.C.
Middle Geometric	850–760 or 750 B.C.

(This period is sometimes divided into two half centuries – Geometric severe followed by Geometric ripe.)

Late Geometric	760 or 750 B.C.–700 B.C.

Orientalizing and Black-figure (Athens and Corinth)

(a) *Athens*

Early Protoattic	700–675 B.C.
Black and White	675–650 B.C.
Later Protoattic	650–610 B.C.
Attic Black-figure (Corinthian influence)	610–550 B.C.
Attic Black-figure (Mature)	570–530 B.C.
Late Attic Black-figure	530–450 B.C.

(b) *Corinth*

Early Protocorinthian	720–690 B.C.
Middle Protocorinthian	690–650 B.C.
Late Protocorinthian	650–640 B.C.
Transitional	640–625 B.C.
Mature Animal and Human Figure	625–550 B.C.

Later styles include those called "Black Pattern" and "White".

Red-figure (Athens)

The following gives the Attic sequence. During the Classical period and later, red-figure vase painting became important in South Italy and Sicily also. Red-figure was never very important at Corinth.

Early Attic Red-figure	530–500 B.C.
Late Archaic Red-figure	500–480 B.C.
Early Classical Red-figure	480–450 B.C.
Classical Red-figure	450–425 B.C.
Late Classical Red-figure	425–400 B.C.

Fourth Century Red-figure lasts to about 320 B.C.

White-Ground (Athens)

Painting on a white ground was a moderately important style, current at Athens during much of the fifth century B.C.

Black (Black-glazed or Black-painted)

The plain black style became important at the height of the black-figure period, and it was produced in quantity well into Hellenistic times. Though much of the best quality ware was made at Athens early in the period, it was also widely imitated.

Hellenistic pottery

A number of different styles appeared including black with painted decoration, and decoration consisting of impressed and relief work.

Names and Shapes

A search through Liddell and Scott's Greek dictionary will produce many hundreds of Greek words for containers. As most of these are not much more than names, which we can attach only vaguely to shapes, modern books on Greek pottery usually highlight only a small number of the more important ones. These are names that are generally accepted in modern usage.

In the search for shapes for these names the ones that hint at the shape's function are important clues. *Arytēr* or *arytaina* should be ladles of some sort (*aryō*, "I ladle" and compare *kolokyntharytaina*, "gourd-dipper"). The name *halia* indicates a salt box (*hals*, "salt"), *milteion* was clearly a box for keeping *miltos* (red ochre) in, *phykion* a rouge pot (*phykos*, "rouge"). The *ichthya* ought to have something to do with fish (*ichthys*, "fish") – a container for preserved fish? Or perhaps it was shaped like a fish.

But the information that we can extract from such names is really minimal. By modern analogy we might guess that an *oxis* (vinegar cruet) was a small vessel with a narrow mouth. The *ekpōma* ("drinking out of") must have been a cup, but what sort of cup? The function of a *thermantēr* is fairly obvious (the word means heater) but many shapes could come into that category. The same applies to such words as *christērion* and *exaleiptron* (unguent containers), *phrygetron* (vessel for roasting barley), *potēr* and *potērion*, related to the Greek word "to drink", and *enourēthra* (chamber-pot).

A physical attribute of a vessel may be present in the name: *deirokypellon*, necked cup; *kampsa*, *kamptra* and related words, of plaited baskets? (*kamptō*, "I bend"); *lepastē*, limpet-shaped vessel? (*lepas*, "a limpet"); *mastos*, a breast-shaped cup; *aspis* (shield) used of a round, flat bowl;

gastra and *gastris*, pot-bellied containers (*gastēr*, "belly"); *kōnis*, apparently a cone-shaped container; *ōon*, an egg-shaped vessel.

Epithets can give additional information, for example, *aōtos* (without handles), *monouatos* (single-handled), *triōton* (a three-handled vessel), *apythmenos* (without a base), *amblycheiles* (with rounded rim), *bathys* (deep), *brachystomos* (with short mouth), *pachystomos* (with thick mouth), *kōthōnocheilos* (with lip like a kothon).

There are also clues to capacity, but mostly referring to particular vessels, and not really very helpful when it comes to judging the range of possible sizes of a class of vessels. One might mention *dikotylos* (two cups); *heptakotylos* (seven cups capacity – used of a lekythos); *dekamphoros* (ten amphorae capacity – used of a krater and a pithos); compounds using *hēmi* (half) – *hēmichōon, hēmikadion, hēmikotylion, hēmikyathos*; and such words as *mikrokeramon* (small pot).

There are certain classes of words with endings like *-thēkē* ("repository for"), *-ouchos* ("holder"), *-docheion, -dokē* ("receptacle for"), *-phorion* ("carrier"). These are suffixes attached to words denoting objects put or kept in these containers. The variety of sizes of objects associated with *-thēkē* ("repository for") shows that the containers must range in size from small boxes to large chests or cupboards. Things that appear in combination with *-thēkē* include bread, meat, vegetables and other food, musical instruments, weapons, needles, finger-rings, alabaster, incense, myrrh, spices, and many others. Liddell and Scott's dictionary has several dozen such words combined with *-thēkē*. Probably the smallest of these containers are what the pottery books call *pyxides*.

A number of Greek vessels have *chous* or *choē* ("pourer") as part of the name – sometimes a helpful clue, sometimes enigmatic: *prochoē, prochous, oinochoē* (wine-pourer), *spendochoē* (libation-pourer), *plemochoē* (?).

A few names are applied to both containers and boats, for example, *kantharos, kymbē, skaphis*.

But many names of containers are names alone as far as we are concerned. Let us take as an example the shapes used for drinking cups. In the modern literature names have been conventionally assigned to the more important shapes: *kylix* to the broad, shallow type, *skyphos* or *kotylē* to the deeper type that looks more like our cup-shape, *kantharos* to the chalice-like tall-handled type. But the little evidence that we have tends to show that ancient usage was more promiscuous, and with the more obscure shapes we are completely at a loss. The following can really only be described as "some sort of cup": *batiakē, bēssa, hystiakon, kapēlikon, kratanion, manēs, mathalis, melē, pōma, sabrias, skalis, sōroteros, troulla, tryblion. Kobathos* is presumably a word for a cup, as it was found in a list of cups, and there are many others.

Some names of vessels are local and regional, and probably ephemeral. The local nature of some is vouched for by ancient commentators. A few, including some export wine-containers, are named after their place of origin, like *Chion* or *Memphition.* A *kandaulos* was a Lydian shape; *kondy* was supposed to be a Persian word for drinking vessel. We hear of a Rhodian *skyphos* and a Corinthian *kratēr* – apparently referring to special shapes. *Aryballis* was stated by Hesychius to be a Dorian word. Modern archaeologists have applied the name *lydion* to a small krater-shaped vessel, common in the East Greek area in the sixth century B.C., but we do not know what its ancient name was.

As there were regional variations in nomenclature, so there were certainly changes in usage over the centuries. Earlier Attic names for the small box now conventionally called *pyxis* might have been *kylichnis* and *libanōtis.*

The important shapes

When modern archaeologists and art historians became interested in Greek pottery very little was known about it. Names that were applied to shapes early in the modern era proved difficult to detach, even when later shown to be patently wrong.

This book deals with about thirty different shapes. The names currently used for these shapes are certainly not in every case those regularly used by the ancients. Although we can be fairly confident that in the majority of cases modern usage corresponds partly with ancient practice, it is likely that the ancient lines of demarcation for various shapes were different from ours. To take one of the commonest shapes, the amphora: although our use of the name *amphora* corresponds roughly with the ancient use of the word *amphoreus*, the rather arbitrary distinctions that we have tended to make between *amphora* and such shapes as *kados, pelikē* and *stamnos* produce categories that might bear little relation to the ancient. Certainly the ancient *pelikē* was not our *pelikē*.

Other names that are used differently now from ancient times include *askos, dīnos, kōthōn, loutrophoros, onos* and *phormiskos.* Ancient and modern usage seem to be close for *alabastron, hydria, kantharos, lebēs* (especially the *gamikos* variety), *lēkythos, kratēr mastos, oinochoē* and *phialē.*

Words such as *aryballos, pyxis,* and some names for cups are probably partly right, the ancient usages perhaps being closer to ours at certain periods. Fuller details are given under the individual names.

Shapes in non-figured pottery

Though the general trends are usually much the same in both plain black and figured wares there are also a few important differences, especially in the popularity of different shapes in the different media. Occasionally, for example, a shape is common in plain black, but hardly appears at all in figured pottery. (See *Agora*, pp. 10 and 88.)

Domestic pottery shapes were not influenced to any great extent by the wider world of fashion. Here the main criteria were utility, sturdiness, durability and cheapness, and though not much work has been done on such pottery apart from a few centres like Athens and Corinth, it does appear that local types developed fairly independently of other areas. Sometimes one area specialized in a particular type and sold it to its neighbours. The sturdy, thick-walled Corinthian mould-made mortars and lekanai (exported to Athens in the fifth century B.C.) are a case in point.

Trade amphorae

Of the ordinary pottery shapes the amphora is the most important and most studied. Variation in shape is important evidence for date and provenance, and it is sometimes possible to get very exact information from the stamps on the handles. Many of these stamps were concerned with quality control of a commodity (usually wine). The good reputation of such commodities could be damaged by adulteration and inferior imitations.

Calendar dates (including the month of the year), references to officials in office at the time, statements of provenance, contents, and maker's name are other useful data appearing on amphorae from time to time. This information can be supplemented by a number of clay analysis techniques to discover provenance, and other techniques such as thermoluminescence which relate to dating.

Amphora studies have been concerned particularly with the Hellenistic and Roman periods when amphorae were very prolific. An indication of the abundance of Roman amphorae can be gauged from a glance at some of the more remarkable finds, for example a deposit of 1,350,000 amphorae at Turin, their contents consisting of prepared potters' clay and the Monte Testaccio at Rome – a "mountain" consisting largely of potsherds from Spanish amphorae. (See *Bulletino della Commissione archaeologica* and Tenney Frank in Amphora References.)

Literature and amphora inscriptions and graffiti bear witness to the wide variety of commodities stored and exported in these containers. Apart from those usually mentioned (wine, olive oil, and olives) there are references to grain, meal and flour, fish such as tunny and mackerel and fish products, dried and preserved fruits, vegetables and nuts, honey, vinegar, unguents and medicines.

Literature and archaeology supply evidence for secondary use of storage vessels after they have been emptied of their original contents. Amphora burials are widespread, especially for children. Amphorae were commonly used in the building industry, sometimes whole, sometimes broken into sherds, especially in Roman times when concrete work became important. They were used for drainage and toilet purposes, and the tops of the larger ones as well-heads.

It was necessary to treat unglazed amphorae (and other similar coarse pottery) to make it water-tight. Pitch and pine resin were used for this purpose. Resin imparted the typical "retsina" flavour to wine stored in these treated containers, a demand for which persisted in Greece long after such treatment of containers was necessary. It was then customary to introduce the flavour into some wines artificially.

Other vases designed for the export trade

Several of the Attic pottery shapes were especially popular in the West Greek area and elsewhere overseas and were exported in large numbers. These included amphora types ("Nikosthenic" and "Tyrrhenian") and other shapes such as kyathos and stamnos. In some cases styles that were exported in quantity were apparently not particularly popular at home, and must have been designed with the export market in view.

Lids and stoppers

A variety of methods was used to seal the mouth of a container. Clay disks and lids are often found on archaeological sites, and wooden disks were also made. Where it was necessary to make a vessel air-tight, various waxes and plasters such as gypsum were used to seal the crack between lid and lip of the vessel. Sometimes bungs of clay or cloth were used as stoppers, and Pliny (*N.H.* 16, 13, 14) refers to cork.

Bits of linen and traces of a pitch-like substance used for

sealing have survived on some of the vessels from Tutenkhamen's tomb, and this kind of stopper was no doubt used in the Greek world too. Geometric pots often had very elaborate lids. In the archaic Greek period (black-figure and early red-figure), amphorae and sometimes other vessels had finely painted lids.

Literature

The attractiveness of the decoration has encouraged the production of many books dealing mainly with the painting of fine pottery, and treating shapes in less detail. The most spectacular advances in the study of Greek pottery this century have been in the area of vase painting, especially in the work of Sir John Beazley and his successors in identifying and analyzing styles of individual artists and schools and their influence on one another.

General references for shapes

The following books are useful for the study of pottery shapes. A more comprehensive list of references is included in the sections for each specific shape.

Arias, P., Hirmer, M. and Shefton, B.B. *History of Greek Vase Painting.* London: Thames & Hudson, 1962. Excellent illustrations, and good, concise accounts.

Beazley, John D. *Attic Black-figure Vase-Painters.* Oxford: Clarendon, 1956.

______. *Attic Red-figure Vase-Painters.* 2nd ed. Oxford: Clarendon, 1963. This and *Attic Black-figure Vase-Painters* contain short introductions on shapes. Artists' works are then listed and categorized according to shapes.

______. *Paralipomena.* Oxford: Clarendon, 1971.

Boardman, John. *Athenian Black Figure Vases.* London: Thames & Hudson, 1974.

______. *Athenian Red Figure Vases.* London: Thames & Hudson, 1975. Each of Boardman's books has a chapter relating to shapes, *BF* Chapter 9 and *RF* Chapter 5, and also deals with them incidentally. Good bibliography.

Cook, R.M. *Greek Painted Pottery.* 2nd ed. London: Methuen, 1972. This has been used as a standard text book for the last decade. The emphasis is particularly on decoration which is dealt with systematically, region by region, under the headings of the various styles. There are also chapters dealing with the major shapes, technique, inscriptions, chronology, the pottery industry, uses of pottery for other studies, practical comments for handlers and collectors of pottery, a history of the study of vase painting, and an excellent bibliography under various categories with a critical comment on each entry.

Corpus Vasorum Antiquorum (CVA). This is a series of fascicules covering the holdings of ancient Greek vases in museums and other collections throughout the world. Each fascicule has approximately fifty plates of photographs, together with descriptions of the vases, and the more recent fascicules often have line drawings and profiles. Some fascicules also contain non-Greek material as well as Greek, for example Egyptian or Danube Valley. Although the size and general format of the fascicules are the same (under the direction of a coordinating committee), there are some local variations and idiosyncracies in such matters as numbering, indexing and terminology. Text and illustrations vary greatly in quality, but the recent fascicules generally have very good illustrations.

The number of published fascicules of the *CVA* is approaching 230, and it is now possible for students to do much of their basic research from these fascicules alone. A comprehensive index is sorely needed. The *CVA* index included as an appendix to this book is intended to point students to those fascicules which contain provenances and large numbers of a particular shape that interests them.

Gericke, Helga. *Gefässdarstellungen auf Griechischen Vasen.* Berlin: Verlag Bruno Hessling, 1970. Discussion of the important shapes, together with a table listing representations of vessels portrayed on other vases.

Richter, Gisela M.A. and Milne, Marjorie J. *Shapes and Names of Athenian Vases.* New York: Metropolitan Museum of Art, 1935. Useful basic introduction to the study of Athenian vase shapes of the classical period. A useful set of illustrations, chosen mainly from objects in the Metropolitan Museum.

Simon, Erika. *Die Griechischen Vasen.* Munich: Hirmer Verlag, 1976. Lavishly illustrated and concise descriptions.

Technical studies

Caskey, L.D. *Geometry of Greek Vases*: Attic vases in the Museum of Fine Arts analysed according to the principles of proportion discovered by Jay Hambridge. Boston: Museum of Fine Arts, 1922.

Noble, Joseph V. *The Techniques of Painted Attic Pottery.* London: Faber & Faber, 1966.

Schiering, Wolfgang. *Griechische Tongefässe: Gestalt, Bestimmung und Formenwandel.* Berlin: Gebr. Mann Verlag, 1967. Deals with the significance of various shapes, comparisons with sculpture, architecture and principles behind decoration.

Jucker, Hans. "Blätter zur Geschichte der Griechischen Vasenformen". *HASB* 1 (1975): 36-44. Bibliography on vase shapes.

Studies of coarser pottery, domestic shapes

Amyx, D.A. "The Attic Stelai: Vases and Other Containers". *Hesperia* 27 (1958): 169-310.

Sparkes, B.A. "The Greek Kitchen". *Journal of Hellenic Studies* 82 (1962): 121-137.

______ and Talcott, Lucy. *The Athenian Agora XII: Black and Plain Pottery*. Princeton: American School of Classical Studies at Athens, 1970.

General classical references

Hammond, N.G.L. and Scullard, H.H. *Oxford Classical Dictionary*. 2nd ed. Oxford: Clarendon Press, 1970.

Harvey, P. *Oxford Companion to Classical Literature*. Oxford: Clarendon Press, 1937.

CONTAINER SHAPES

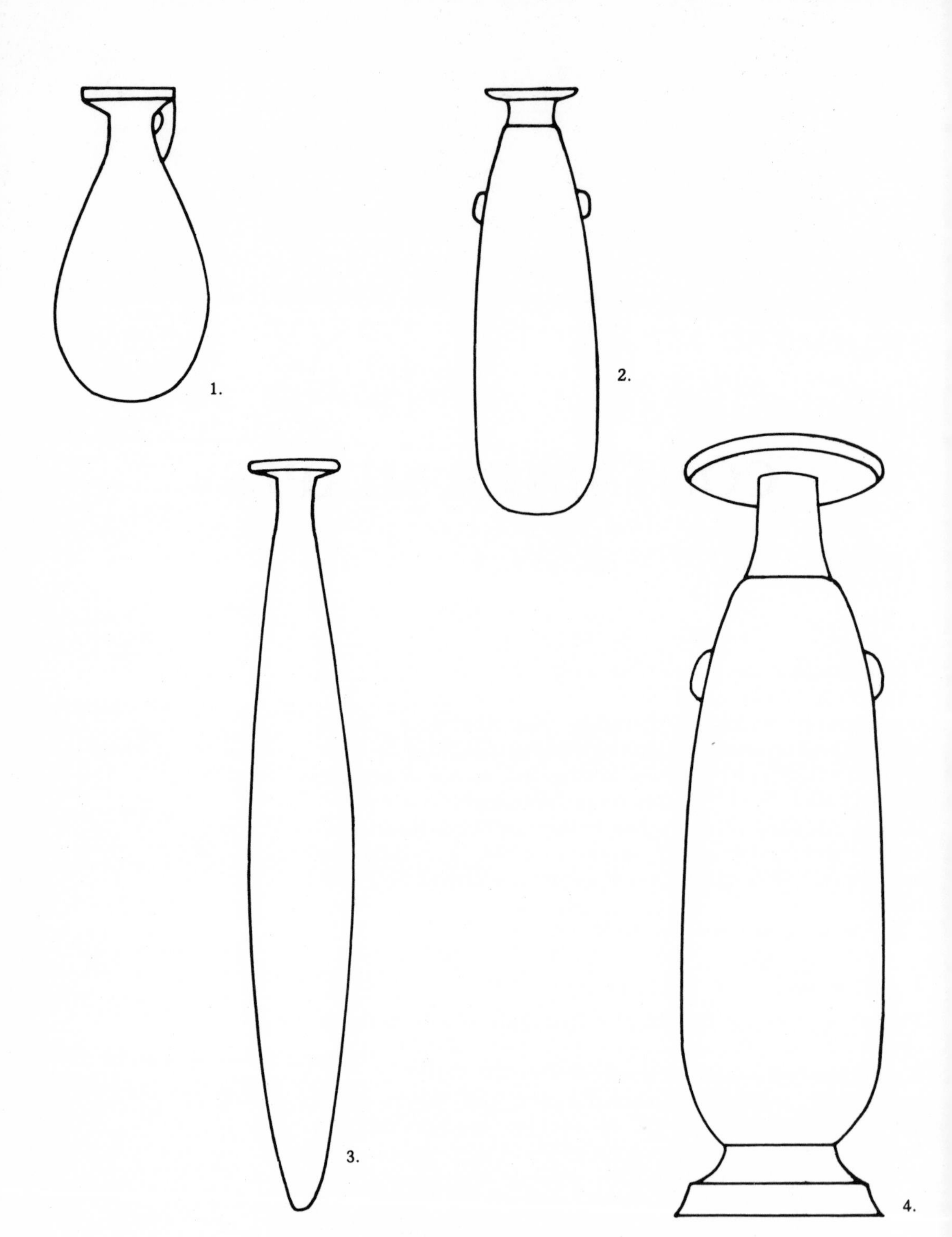

1. Alabastron, Corinthian, first half of sixth century B.C. 11 cm.
2. Alabastron, Attic r.f., second half of fifth century B.C. 15 cm.
3. Alabastron, East Greek, c. 600 B.C. 24 cm.
4. Alabastron, Apulian r.f. fourth century B.C. 24 cm.

Alabastron [ἀλάβαστρον]

Variants ALABASTOS (early) and ALABASTROS (later)

The alabastron has a thin neck, and a longish body shaped like a drop or elongated pear. The lip is flat and disk-like, considerably wider in diameter than the neck. Some alabastra have small lugs or handles at shoulder or neck level to which a string could be attached, as was done with the aryballos. Such handles might also have been used to facilitate sealing (Webb). The bottom of the alabastron is usually round, though sometimes it is flat enough for the vessel to stand. South Italian alabastra are often fitted with bases.

The main types of alabastron are as follows.

1. Corinthian

This type appears late in the first half of the seventh century B.C. and is very common by the end of the century. It continues well into the sixth century B.C. Fairly squat to begin with, it later becomes slimmer, and the smallish bevelled mouth is replaced by a wider, flat lip which is regular thereafter. Usually a single perforated lug joins lip to neck. The Italo-Corinthian version often has a flat bottom.

2. Etruscan, Italocorinthian and East Greek type

Slim, pointed, handleless. Roughly contemporary with Type 1.

3. Attic

The usual Attic type has a rounded base and sometimes lugs or handles. Usually ten to twenty centimetres long. Common for a little over a century, from about 500 B.C. to a little after 400 B.C.

There are examples of Types 2 and 3 made of alabaster as well as clay, which would explain the name.

Suidas would have us believe that etymologically the name alabastron means "not having handles", as if from *a-labē a*, "not, without"; *labē*, "a handle"), which is certainly not a correct description of all alabastra.

Another ingenious etymology connects it with Ebaste, a goddess of Bubastos in Egypt. According to this suggestion the name would mean "Vase of Ebaste" (Sethe).

The name *bombylios* was formerly applied to some alabastra (*CVA Great Britain* 12, p. 6).

The alabastron was used mainly by women for perfume. Illustrations on vases show a liquid being decanted from vessels such as pelike to alabastron, and the application of an alabastron's contents to a woman's hair and body after a bath. Illustrations on white-ground lekythoi show alabastra being carried to the grave.

Some of the types of perfume used can be seen from occasional inscriptions on alabastra – *hirinon, kinnamōmon* and *kypros* (iris, cinnamon and cyprus).

References

Amyx, pp. 213 ff.

H.E. Angermeier, "Das Alabastron" (Diss. Giessen, 1936).

ARV2, chap. 39 etc.

Athenaeus 6. 268a (quoting Crates, *Wild Animals*). "And the alabastos of perfume will come at once automatically, and sponge and sandals as well."

Cook, pp. 230, 232.

CVA France 12 (Corinthian); *Germany* 4 (Corinthian), 11 (Pl. 32 Illustration showing fetching of oil in an alabastron), 14 (Corinthian), 20 (Pl. 247 ff. Women using alabastron at the bath); *Great Britain* 3 (Pl. 41 Attic r.f.); *Greece* 1 (III Jc Pl. 3 ff. Illustrations showing alabastron, suspended by a string, being carried to a grave); *Italy* 52 (Protocorinthian and Corinthian), 53 (Corinthian), *USA* 1 (Corinthian, Attic b.f., r.f., w.g.), 5 (passim Corinthian and Boeotian, Pl. 12, 2 "Andrian" shape), 8 (Fogg Pl. 21, 2 Attic b.f.; Gallatin Pl. 42, 1 Attic b.f. of unusual shape).

Daremberg and Saglio, s.v. *alabaster*.

Gericke, pp. 72 ff. and Tab. 1. Gives names of perfumes inscribed on alabastra (p. 73).

Herodotus 3. 20. 1. "An alabastron (full) of myrrh." Similar expressions to this, which assume that the alabastron is a perfume container, appear in many other authors (see Amyx) including Alexis Comicus, Callimachus, Theocritus and the New Testament; and compare *unguenti alabastrum* in Roman writers.

Ursula Knigge, "Ein rotfiguriges Alabastron aus dem Kerameikos", *AM* 79 (1964): 105-113.

Payne, passim.

Pliny *NH* 9. 13.

RE, s.v. *alabastron* 2.

Richter and Milne, p. 18.

Konrad Schauenburg, "Unteritalische Alabastra", *JdI* 87 (1972): 258-298. Very good bibliography, p. 259.

K. Sethe, *Sitzungsberichte der preuss. Akad. der Wissenschaften* (1933), pp. 887 ff. Possible Egyptian origin.

Sparkes and Talcott *Pots and Pans of Classical Athens.* Agora Picture Books no. 1. Princeton: American School of Classical Studies at Athens, 1976. no. 47. Filling alabastron from a pelike.

Suidas. "Alabastron. A vessel for myrrh without handles. A myrrh-container made of stone." "Lekythos. The Athenians call the perfume lekythos an alabastron."

Webb, pp. 45-46 and n. 29b.

1. Neck-amphora, Attic Protogeometric, tenth century B.C. 46 cm.
2. Neck-amphora, Attic Geometric, ninth century B.C. 77 cm.
3. Neck-amphora, Attic Geometric, c. 850 B.C. 69 cm.
4. Neck-amphora, Attic b.f., 540–530 B.C. 33 cm.
5. Neck-amphora, Fikellura (East Greek), 540–520 B.C. 30 cm.
6. Amphora, Attic b.f. (Nikosthenic), c. 530 B.C. 31 cm.
7. Amphora (Type B), Attic b.f., c. 520 B.C. 43 cm.
8. Amphora (Type A), Attic r.f., c. 500 B.C. 60 cm.
9. Amphora, Panathenaic, c. 500 B.C. 66 cm.
10. Amphora (Type C), Attic r.f., c. 490 B.C. 40 cm.
11. Neck-amphora, Campanian, 330–310 B.C. 67 cm.
12. Amphora (pointed, trade type), found at Athens but provenance uncertain, second century B.C. 84 cm.

Amphora [ἀμφορεύς]

The Greek words are *amphoreus* and *amphiphoreus*, Latin *amphora*, the Latin form being conventionally used in English. A similar word with an amphora ideogram appears on the Linear B tablets.

The name is generally assumed to refer to the fact that the vessel can be grasped and carried by the handles on either side (*amphi*, "on both sides"; *phoros*, "carrying"; that is, two-handled. Compare *amphithetos, amphōtis*). The possession of two handles is one of the criteria for the modern definition of the shape, which excludes similar-looking but three-handled vessels like the hydria. The other requirement is that the vessel be essentially "closed", with a mouth of suitable size to be stoppered or fitted with a lid if necessary (though many amphorae were not designed with stepped lips convenient for receiving such lids). Neck and mouth are narrower than the body of the vessel (contrast such "open" vessels as cups and kraters).

Apparently the name was applied in ancient times to roughly the same shapes as in modern convention, but possibly encompassing others as well, including the shape that modern archaeologists usually call *pelikē*.

It is convenient to divide the amphora shape into two classes, neck-amphorae and one-piece amphorae. In neck-amphorae neck meets body at an angle. The one-piece amphora is not so articulated; instead neck merges into shoulder and body in a curve.

The neck-amphora is the earlier shape, descended from a Mycenaean prototype, and appearing in Attic Protogeometric with either horizontal or vertical handles. These Protogeometric amphorae are often plump, almost globular. By the Geometric period proper the shape is on the whole slimmer, especially in the lower section, which narrows down towards the base from high shoulders. The neck too is taller, and the vertical handles are attached to the upper part of the shoulder and upper part of the neck. Horizontal handles still appear on some of the large funerary types.

Throughout the Greek world during the Orientalizing and early black-figure period slim varieties of neck-amphora are found for example Cycladic "Heraldic" Group, and also much broader types such as Fikellura and Cycladic Ad Groups. In mature black-figure there is a wide-shouldered variety at Athens in which the top part of the shoulder, where it meets the neck, is almost horizontal.

In red-figure the smaller Nolan amphora (about thirty centimetres) was made during most of the fifth century B.C. Its name comes from the site (Nola) near Naples where a number of the type were found.

An amphora type deserving special mention is the Panathenaic. It was designed to be filled with olive oil as a prize for the Panathenaic Games. Although the shape changes over the centuries it retains its narrow neck and mouth and narrow foot. There is a standard form of decoration too. Athena usually appears on one side, while an athletic event is portrayed on the other. There is often an inscription which advertises the fact that the vase is "from the Games at Athens". The series is a long one, beginning at about 560 B.C. and carrying through to at least the second century B.C. The conservative black-figure technique was used for Panathenaic prize amphorae even in the red-figure period. But in the fifth and fourth centuries B.C., especially in southern Italy, the shape attracted some red-figure artists as well, who did not restrict their painting to the traditional Panathenaic themes.

Overall the neck-amphora was not one of the most important Hellenistic shapes.

The one-piece amphora, also called "belly-amphora", or simply *amphora* (without a qualifying word, in contrast to its cousin the "neck-amphora") appears sporadically throughout the early part of the seventh century B.C., and is quite common by 600 B.C. The usual form of decoration at this time was a panel front and back. Very common in early red-figure, it has just about disappeared by 400 B.C.

One-piece amphorae are conventionally divided into three classes, largely according to differences of treatment of extremities (mouth, foot, and handles). As far as chronology is concerned the system of nomenclature usually adopted is a little confusing because Type A begins later than Type B. (Note that the Richter and Milne classification is a little different.) The usual classification is as follows.

Type A (Richter and Milne Type 1b)

Flaring (concave) lip, complex "stepped" foot, flat strap handles with flanges, decorated with ivy-pattern. Approximately 550–450 B.C., occasionally outside these limits.

Type B (Richter and Milne Type 1a)

Begins earlier, but then exists alongside Type A. Lip straight or slightly concave, simple roundel or echinus foot, handles round in cross section. Approximately 625–425 B.C.

Type C (Richter and Milne Type 1c)

Convex lip, round and protruding to begin with, then of echinus profile; simple foot, usually of torus (round) or echinus profile, handles of various types. Approximately 570–470 B.C.

In the late sixth century Nikosthenes, a potter or proprietor of a pottery establishment at Athens, produced a large number of amphorae, perhaps especially for the Etruscan market since they copy a shape that was in vogue there. These Nikosthenic amphorae were mass produced and were often painted by inferior artists. This type of amphora has a broad band around the shoulder, bordered by a ridge which encloses the band, top and bottom. Sometimes this band is decorated as a unit, at other times the borders are disregarded and the decoration overlaps the border. The neck, broad at the bottom, slopes inwards towards the top, before finally reversing direction and spreading out into a broad lip. Two strap handles join lip to shoulder.

Pointed amphorae, with pointed or knobbed bottom instead of a foot, made of coarse clay and undecorated, were made by peoples in the Near East and Egypt for trading purposes at a very early period. Callender (p. 4) lists some of the oldest known (Malta – Neolithic period, and Anatolia – about 3000 B.C.). They appear in the Greek world from about the seventh century on, where they were soon manufactured locally, becoming very common later, especially in Hellenistic and Roman times. As these amphorae had no bases they could not stand upright without support. If it was necessary for them to stand when opened, they had to lean against something or be placed in a stand made for the purpose. Coarse amphorae were the all-

Amphora, (Type A) Attic red-figure, painted by the Deepdene Painter, 470–460 B.C. 47 cm.
Athena pours wine from an oinochoe into a kantharos held by Herakles.
Courtesy of Los Angeles County Museum of Art, The William Randolph Hearst Collection.

purpose containers of antiquity, having a similar function to barrels and casks of later civilizations, and used to store many different commodities.

The better-quality amphorae were used especially as wine containers and decanters. Panathenaic and other amphorae were used to store olive oil. At certain times, as in the Geometric period, amphorae were used for funerary purposes, as grave-markers. Sometimes they were also used for the bones or ashes of the dead. It is possible that some of the amphorae designed to cater for overseas fashions might have been exported empty, as desirable in their own right, and not primarily as containers.

The amphora as a liquid measure was equal to about eight and a half gallons.

Midget amphorae (*amphoriskoi*) like other small vessels, were used as unguent and perfume containers.

References

Agora 12, pp. 47 ff.

Amyx, pp. 174 ff. (esp. Chian, Eretrian amphorae, Panathenaics).

Athenaeus 11. 501a.

John D. Beazley, *ABV* 94 Tyrrhenian; *BSA* 18 (1911–12): 217 ff. Neck-amphora; *BSA* 19 (1912–13): 239-40. Panathenaic; *Der Berliner Maler* (Berlin, 1930) 11 Nolan; "The Master of the Achilles Amphora in the Vatican", *JHS* 34 (1914): esp. 185 ff. Deals with several amphora shapes, "Citharoedus", *JHS* 42 (1922): 70 ff. Defines one-piece amphora types, *V. Amer.* 37-38. Nolan amphorae.

Hansjörg Bloesch, "Stout and slender in the late archaic period", *JHS* 71 (1951): 29 ff.

Boardman, *BF* pp. 185-86; *RF* p. 208; *Boston* II, pp. 39-40; III, 1 (Type A); III, 16 (Type C).

D. von Bothmer, "The Painters of 'Tyrrhenian' Vases", *AJA* 48 (1944): 161 ff.

Georg von Brauchitsch, *Die Panathenaischen Preisamphoren* Berlin: Teubner, 1910.

P. Bruneau, *Exploration Archéologique de Delos* 27 (1968) chap. 14 (Grace and Petropoulakou). Amphora stamps.

Bulletino della Commissione archaeologica communale di Roma (1879): 193.

M.H. Callender, *Roman Amphorae.* London: Oxford University Press, 1965. Useful for general information and for comparisons, especially for coarse amphora types.

P.J. Connor, "Attic neck amphorae decorated with black bands", *AA* (1978): 273 ff.

Cook, pp. 217 ff. An excellent account of the development of the amphora shape.

CVA Belgium 1 (Attic b.f.); *Canada* (Attic b.f.); *Denmark* 3 (b.f.); *France* 1 (Attic b.f.), 4 (b.f. belly), 5 (Nikosthenic and other b.f. neck), 7 (b.f.), 8 (b.f., r.f., Panathenaic), 9 (r.f. Nolan), 18 (Attic b.f.); *Germany* 2 (early Attic), 3 (b.f. mostly Type B), 6 (r.f. Nolan), 12 (r.f., Type A, Panathenaic, pointed), 17 (Fikellura and Attic b.f.), 18 (Attic r.f. neck), 20 (Attic r.f. neck), 25 (Attic b.f. neck), 32 (Attic b.f. neck), 35 (Attic b.f. neck, panel), 37 (Attic b.f. neck), 41 (Attic b.f. neck), 44

(Attic b.f. panel and neck), 45 (Attic b.f.); *Great Britain* 1 (Panathenaic), 4 (Attic b.f. and r.f.), 5 (Attic b.f. neck), 7 (Nolan), 14 (Attic b.f. esp. neck), 15 (esp. b.f.); *Italy* 1 (Attic b.f.), 3 (Attic b.f. mainly neck), 7 (Attic b.f. belly, neck, Panathenaic), 17 (Attic b.f.), 20 (Attic b.f.), 25 (Attic b.f.), 26 (Attic b.f.), 32 (South Italian), 36 (Attic b.f. neck), 49 (Apulian r.f.), 57 (Attic b.f.); *Netherlands* 3 (Attic b.f. incl. Panathenaics); *New Zealand* 1 (Attic b.f.); *Poland* 4 (Attic Protogeometric and b.f.); *Switzerland* 2 (Attic b.f. and Apulian), 3 (Attic b.f.); *U.S.A.* 2 (Pl. 29 Bail amphora), 4 (Attic b.f.), 5 (Attic b.f.), 5 (Attic b.f.); 8 (Attic b.f. and r.f.), 10 (Attic b.f.), 12 (Attic b.f. belly, panel, Panathenaic), 14 (Attic b.f. Types A and B. Pl. 33 ff. "Psykter" neck-amphora with double walls), 15 (Attic b.f.), 16 (Attic b.f. neck), 17 (Attic b.f. neck), 17 (Attic b.f.), 18 (Attic b.f. and r.f.).

Daremberg and Saglio, s.v. *amphora.*

Tenney Frank, *An Economic Survey of Ancient Rome* (Baltimore: Johns Hopkins, 1937), p. 184.

Manuel F. Galiano, "Les noms Mycéniens en -e-u", *Acta Mycenea* II (1972). The name amphora in Mycenaean Greek.

E.N. Gardiner, "Panathenaic Amphorae", *JHS* 32 (1912): 179-193.

Gericke, pp. 64 ff.

Virginia Grace, *AM* 89 (1974): 193 ff.; "Timbres amphoriques trouvés à Délos", *BCH* 76 (1952): 515-540; "Standard Pottery Containers of the ancient world", *Hesp. Suppl.* 7, 175 ff. Commercial amphora shapes and capacities. Late Classical and Hellenistic.

V.R. Grace and A. Frantz, *Amphoras and the ancient Wine Trade.* Princeton: American School of Classical Studies at Athens, 1961.

E. Langlotz, *Frühgriechische Bildhauerschulen* (Nuremberg: Frommann, 1927), pp. 14 ff. and Pl. 13. One-piece amphora.

S.B. Luce, *AJA* 20 (1916): 439 ff. Nolan.

R. Lullies, *AK* 7 85 ff.

Moeris, s.v. *amphorea. Amphoreus* given as the Attic name for the two-handled vessel that was elsewhere known as a stamnos.

G.E. Mylonas, *The Protoattic Amphora of Eleusis* (Athens: Athenian Archaeological Society, 1957). Written in Greek with summary in English.

Dimitrios Papastamos, *Melische Amphoren* (Münster: Aschendorf, 1970).

RE, s.v. *amphora.*

Richter and Milne, pp. 3-4.

E. Schmidt, *Archaistische Kunst in Griechenland und Rom* (Munich: Heller, 1922) pp. 70-91. Panathenaic.

H. Thiersch, *Tyrrhenische Amphoren* (Leipzig: Seemann, 1899).

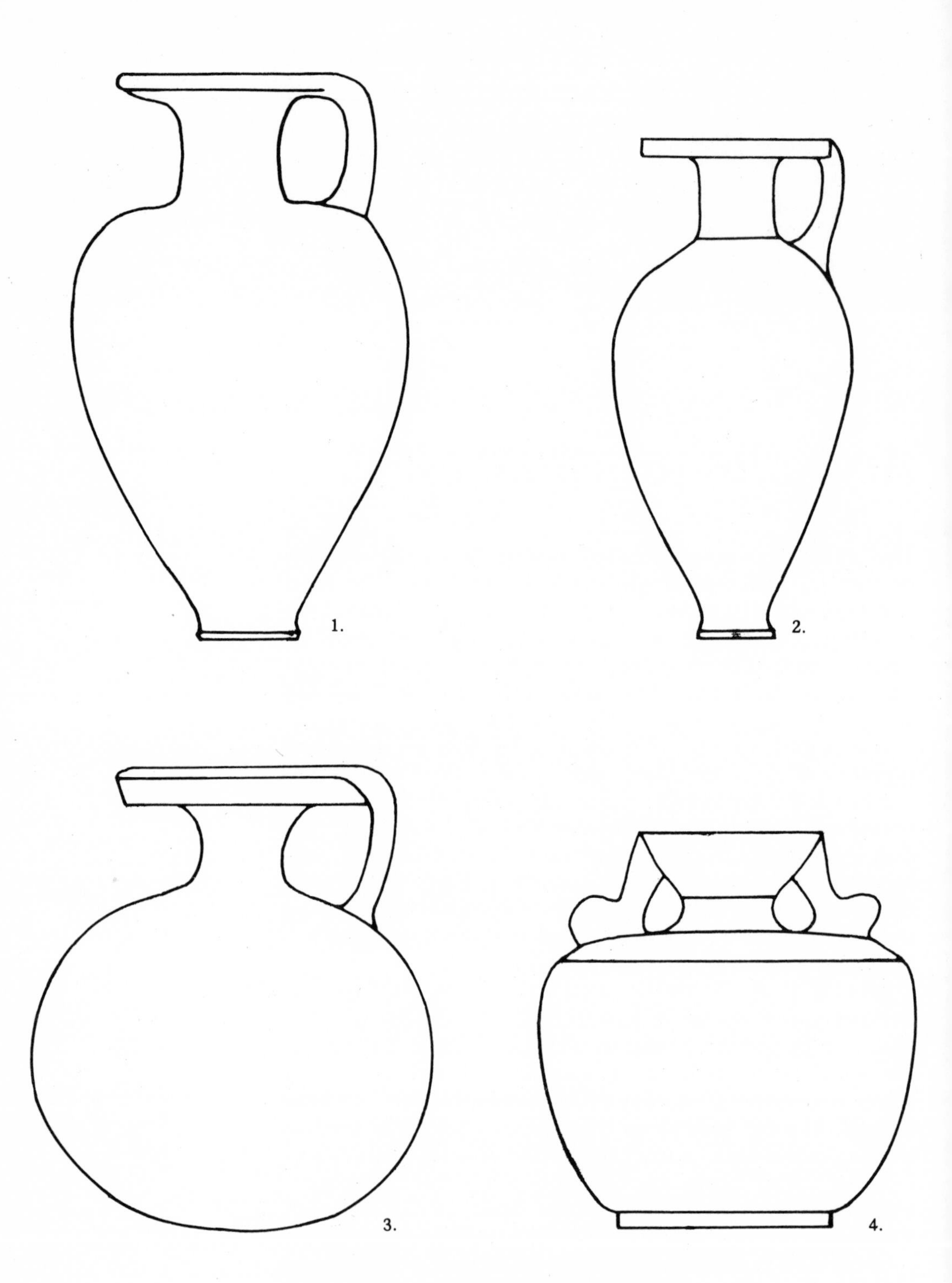

1. Aryballos (ovoid), Corinthian, 680–670 B.C. 8 cm.
2. Aryballos (piriform), Corinthian, c. 630 B.C. 7 cm.
3. Aryballos (round), Corinthian, 525–500 B.C. 7 cm.
4. Aryballos (round), Attic r.f., 490–480 B.C. 6 cm.

Aryballos [ἀρύβαλλος]

The name is conventionally applied to small, narrow-necked oil flasks used especially by athletes. The most common type is round, but they may be elongated, or sometimes fashioned into less regular shapes such as human heads. The round ones rarely exceed seven or eight centimetres in height. The ovoid and piriform types are sometimes called *lēkythoi* (*CVA Great Britain* 12, p. 5).

The round variety was divided by Beazley into two types.

Type I ("Corinthian"). Globular body with rounded bottom and single handle joining shoulder to broad, flat, disk-shaped mouth. Common from the mid seventh century (though there are some as early as Early Protocorinthian) to the fifth century B.C. Widely imitated, the shape was produced in Laconian, East Greek, South Italian, Boeotian, and for a while, Attic workshops.

Type II ("Attic"). In the early examples two narrow handles join shoulder to mouth. Each handle has a short, spur-like projection acting as a sort of handle buttress at the shoulder junction. Later, handles and their associated spurs disappear. The mouth is "hemispherical" or bell-shaped, like the normal lekythos type. Made from the last quarter of the sixth century to the early fourth century B.C.

Some "round" aryballoi are made with a flattish bottom, and sometimes there is a simple base. Other aryballoi shapes (ovoid, piriform, "Attic") regularly have bases.

Ovoid and pointed aryballoi are fairly common in areas of Corinthian influence in the Middle and Late Protocorinthian period.

The word *aryballos* and its relatives were believed to be Dorian (see Hesychius). On the rare occasions where *aryballos* was used in Attic Greek it meant firstly a purse, and secondly a vessel of some sort. It seems that the pot that we call a round aryballos was given the same name as the purse because it looks like a draw-purse with the string tightly drawn. Athenaeus believed the borrowing of the name came in the reverse direction, from pot to purse, but this seems less likely.

Aristophanes uses the word once (*Knights* 1094), apparently for the play on the word *arytaina* which he mentions a few lines before. Probably the regular Attic name for small vessels of this type was *lēkythos* or *lēkythion,* at least at Athens in the latter part of the fifth century B.C.

The name *lēkythos,* written in antiquity, is extant on a few aryballoi – a piriform aryballos from Cumae (first half of seventh century B.C.), as well as an Apulian lekythos of the fourth century B.C. A Corinthian round aryballos of the last quarter of the sixth century B.C. has the name *olpē* on it. So it appears that the shape that we call *aryballos* was known by various names in ancient times, including *aryballos, lēkythos* and *olpē.*

Many illustrations on vases show aryballoi in use. Most of these are from the early red-figure period, tapering off in the later Classical period. But there are some illustrations by black-figure artists also. These illustrations on vases show that the aryballos was, with few exceptions, a shape used by men, especially in the palaestra (Gericke, p. 97). Athletes are shown pouring oil from aryballoi into the hand (for massage of the limbs), and aryballoi are also shown suspended by a string attached to the vessel's neck, hanging from the wrist or a wall. Aryballoi often appear in the company of sponges and strigils.

Compare also *aryballis, arytaina, lēkythos, lēkythion, olpē.*

References

Antiphanes (in Pollux 10. 152). Draw-purse.

Aristophanes *Frogs* 1200 ff. (e.g. 1208, "he lost his lekythion"; Aristophanes' lekythion was possibly thought of as being the shape that we call *aryballos.*) *Knights* 1090 ff. "Paphlagon: 'I had a dream. I dreamt the goddess herself was pouring health and wealth on Demos out of a ladle (arytaina). Sausage seller: 'By heavens I had a dream too. I dreamt the goddess herself came from the city with her owl perched upon her. Then she poured ambrosia from an aryballos on your head, garlic sauce on his."

Athenaeus 11. 783-84. "A cup (poterion), wider at the bottom, drawn in at the top like draw-purses. These also, because of their similarity, some people call *aryballoi* . . . The aryballos is not much different from the *arystichos.* It derives from *arytein* (to draw) and *ballein* (to throw)."

John D. Beazley, "Aryballos", *BSA* 29 (1927–8): 187 ff. Discussion of the name, types of aryballoi, materials from which they were made. "Charinus: Attic Vases in the form of Human Heads", *JHS* 49 (1929): 39 ff. Descriptions of a number of aryballoi in the shape of heads, including Janiform types.

Cook, pp. 47, 230-232.

CVA France 12 (Corinthian); *Germany* 14 (Corinthian), 15 (Variously shaped Etrusco-Corinthian aryballoi), 36 (Corinthian ovoid, pointed and round aryballoi), 39 (Pl. 33, 4-8. Ring aryballos, MC, first quarter of sixth century B.C.); *Great Britain* 9 (Pl. 64, 1-7. Attic round aryballos, 490–480 B.C.), 12 (Pl. 3 ff. Corinthian and Boeotian); *Italy* 52 (Corinthian); *New Zealand* 1 (Corinthian).

Daremberg and Saglio, s.v. *aryballos.*

Gericke, p. 75 and Tab. 6.

C.H. Emilie Haspels, "How the Aryballos was suspended", *BSA* 29 (1927–8): 216 ff.

Hesychius. *Aryballis* the Doric word for *lēkythos.*

Lazzarini, 360 ff.

Pollux 7. 166, "Aryballos and arytaina are vessels of the baths. Aristophanes mentions both of them"; 10. 63: Arytaina and aryballos are included in a list of objects from the baths.

RE, s.v. *aryballos.*

Richter and Milne, p. 16.

P.N. Ure, *Aryballoi and Figurines from Rhitsona in Boeotia*, Reading University Studies (Cambridge: Cambridge University Press, 1934). Section IV describes the common types of aryballoi.

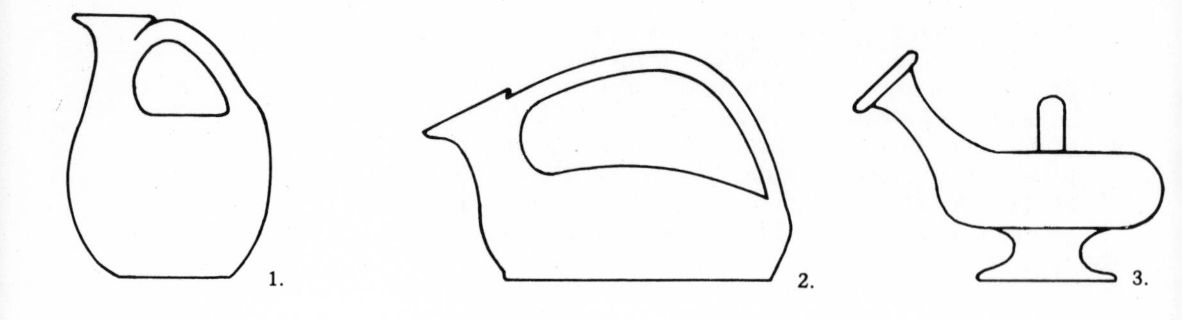

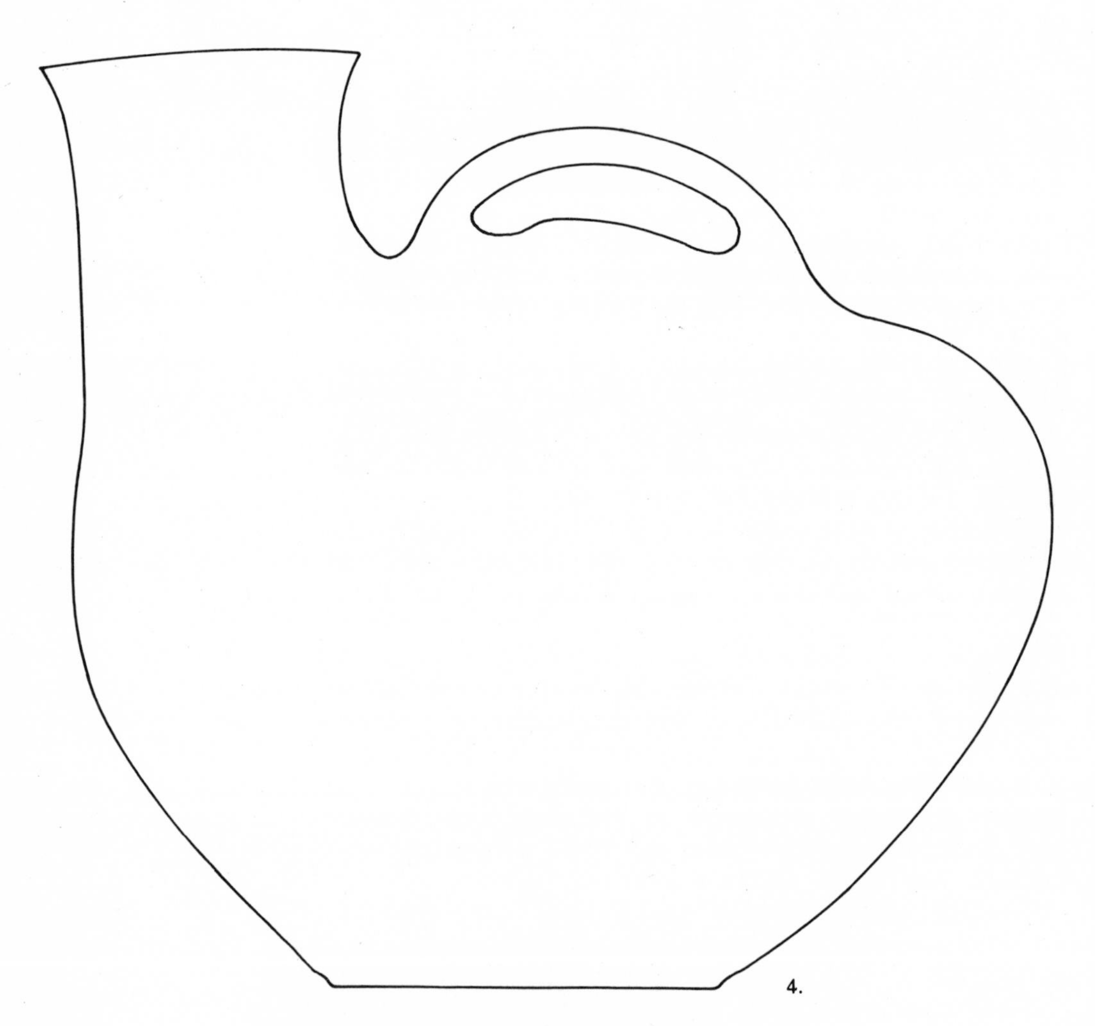

1. Askos, East Greek, sixth century B.C. 8 cm.
2. Askos, Attic, c. 430 B.C. 7 cm.
3. Guttus, South Italian, second half of third century B.C. 7 cm.
4. Askos, Canosa Ware, fourth/third century B.C. 31 cm.

Askos [ἀσκός]

See also Rhyton, Figure and Plastic Vases

The ancient *askos* was a skin made into a bag, usually a wine-container. Ancient references must be taken to apply to this type of wine-skin. The application of the word to a number of Greek pottery shapes is modern, and originates through the fancied resemblance of some of them to wine-skins.

Classical Athenian askoi are often small vessels; the Eastern "sway-backed" types and some of the South Italian varieties are much larger.

The bodies of Attic askoi vary from flattish to quite deep. They have a short neck or spout rising obliquely from one end, and usually the mouth is small. A handle joins neck to body basket-fashion, usually sweeping right over the top of the pot from just below the mouth to the opposite end of the body, with usually a low foot. In some varieties of askoi (for example, South Italian) the handle may be independent of the neck and attached to the top of the body. Later examples often have a small ring handle.

Spouts are sometimes made to look like lions' heads. Some wider mouthed varieties are made with a strainer, and some are designed to take a lid.

Vessels similarly shaped to askoi but with vertical necks and small ring handles are often classified as *gutti* (*Agora*, pp. 12, 160).

Ascoid vases are found in Greek pottery from Geometric times on, and there are Mycenaean vases reminiscent of the shape. The askos starts to appear in high quality pottery at Athens in the first quarter of the fifth century

B.C., possibly imitating East Greek varieties, and it continues to be made for a century or so. Its popularity persisted longer in the West, where, in fact, the shape had existed in the Geometric period. The askoi of this area look more like real wine-skins than do some of the other types.

Shape and design of askoi make them very efficient unguent vessels. The larger ones must have been used for other liquids as well.

A very common decoration for Attic askoi consists of two figures on the body, one on either side.

References

Agora, pp. 12, 157 ff., 210-11.

John D. Beazley, "An Askos by Macron", *AJA* 25 (1921): 326 ff. This article refers to several varieties of the shape.

Cook, pp. 231-32. Cook gives the main subdivisions of the type. 1. East Greek, often fairly large, flat-bottomed. 2. Low, smallish askos of Attic r.f. and black-painted wares. 3. Bulky, deep askos of the West Greek area. 4. Larger, round-bottomed askos of Daunian and Canosa Ware.

CVA Germany 6 (Attic r.f.), 23 (Pl. 85 Gnathian); *Great Britain* 6 (Pl. 39 Attic r.f.; Pl. 44, 45 Canosan and Apulian), 10 (IV D a passim, Daunian); *Switzerland* 1 (Pl. 28, 10 & 16 Attic black-glaze, contrasting shapes, fifth century B.C.); *U.S.A.* 2 (Pl. 30, 1 Etruscan, fourth century B.C.); 4 (Pl 1, 11 Askos or "duck vase" from Cyprus, Mid Bronze Age).

Daremberg and Saglio s.v. *guttus.*

Hesychius. *Askos* equivalent to *hydria.*

Herbert Hoffmann, "Sexual and Asexual Pursuit", Royal Anthropological Institute of Great Britain and Ireland, Occasional Paper 34 (1977). The decoration of Attic askoi.

M. Mayer, "Askoi", *JdI* 22 (1907): 207 ff.

Pollux 6. 14; 10. 71. Askoi for wine. (Hesychius and Pollux are both probably referring to askoi made of skin. and used as water and wine containers.)

RE, s.v. *guttus; Suppl. 3,* s.v. askos.

Richter and Milne, pp. 17 ff.

G.A.S. Snijder, "Guttus und Verwandtes", *Mnemosyne* 3 Ser. 1 (1934): 34 ff.

Exaleiptron [ἐξάλειπτρον]

The shapes listed in this group are also called: 1. Kothon, 2. Tripod Pyxis, 3. Plemochoe. (These shapes are also dealt with separately in the sections for Kothon, Pyxis, and Plemochoe). (See illustration p. 118.)

Exaleiptron, "unguent flask, unguent basin"; *aleiphō*, "I anoint"; *exaleiphō*, "I plaster, wash over".

The word *exaleiptron* is used by some archaeologists to include a variety of vessels whose shapes (and possibly functions) were similar. The main common characteristic of members of this group is the incurving lip, which is rolled over towards the inside, so that the opening of the mouth is only a third to a half the maximum diameter of the pot. The shape is like that of a short cylinder squashed flat, the squashing causing the sides to bulge. There may be a ring foot, or stem and disk, or other type of base, and from one to four handles (the later vessels are handleless).

Whether all these vessels were given the same name in antiquity, and if so, what this name was, is unknown. *Holkeion, kylichnis, plemochoē* and *smegmatothēkē* have all been suggested, as well as *exaleiptron*, and in the modern literature the names *kōthōn, tripod pyxis, plemochoē*, and *kōthōn plemochoē* have been used for various members of the group.

The name *kōthōn* for vessels of this sort (usually the earlier, smaller ones in the series) is purely a modern convention. The name *plemochoē* has been attached by modern writers mainly to the later development of the shape – larger, more elegant vessels on the whole. It is now generally believed that the ancient kothon was a simple cup or mug – at any rate a drinking vessel of some sort.

The shape appears in Greek pottery in the late seventh century B.C., and peters out at the beginning of the fourth. It was common in Corinthian pottery about 600 B.C. No plausible prototypes have been found outside Greece. A Mycenaean shape reminiscent of the exaleiptron is too remote to have exerted an influence (Scheibler 1964, p. 92). In any case these vessels are footless and the lip rolls outwards – the opposite way to the later Greek types. It might be an adaptation of a Geometric wide flat cup or pyxis, but a gradual evolution seems ruled out, as intermediate stages are lacking. Some of the earliest examples are Cretan.

1. Kōthōn

Convex-curved outer profile, low ring foot usually wider at the bottom. In some examples later in the series the bowl may have straight or even concave sides as in pyxides, angled or carinated top and bottom. Otherwise the main variations consist in different treatment of the handles. Early examples are usually lidless. The shape is a common Corinthian one from the early to late Corinthian period. There is a good deal of confusion in the nomenclature of such pieces in the literature.

2. Tripod Pyxis

Usually three feet as the name implies, exceptionally four. To begin with, vertical sides; a carination where sides join top and bottom; later the sides are curved. The wide, flat legs, vertical or curved outwards at the bottom, are attached to the outside bottom, or further up the side of the vessel, depending on its profile, and may end in the shape of animal feet. In some examples struts or stays join the legs to the bottom of the vessel. It is usually designed to have a lid, and normally there are no handles. The shape appears at Corinth before 600 B.C., and is common for half a century, persisting considerably longer in places such as Boeotia.

3. Plemochoē

Flat bowl of convex profile but sometimes carinated. The distinguishing feature is a central foot, wide at first, later replaced by a narrow stem on a flat disk base. Handles disappear later in the series. In Attica the shape is usually made to take a lid. It becomes common in Attica and Boeotia after sporadic appearances in the first half of the sixth century B.C. After about 550 B.C. figure decoration becomes rare, the usual surface being black glaze, with perhaps tongues or other minor decoration at the top.

Plemochoai are divided by Beazley into Types A and B, essentially according to the type of foot. Type A foot is the wider splayed variety; Type B (popular about 500 B.C. and reasonably common for a century or so) has a narrow stem on a flat disk base. This disk varies in profile from a complex stepped form to a simple concave type, whose surface is usually left reserved. There is often a convex ring at the junction of bowl and stem.

The use of the name *exaleiptron* for all these vessels with lip rolled inwards has the advantage of seeming to fit the ancient function of the shape (especially as portrayed on vases), though in ancient times the name might have been used for other unguent containers as well. The incurving lip does not make for an ideal pouring vessel, and the shape is quite unsuitable for drinking. There are no burn marks on surviving vessels which might be evidence for use as lamps or for burning incense (it has been suggested that floating wicks might have been used if they were lamps).

But the lip is of suitable design to prevent spillage while carrying the liquid. Illustrations of the shape on vases seem to show that it was most used (at least from the mid fifth century on) by women, especially while bathing and dressing. It probably contained scent, scented water, unguent, or perhaps powder, which could be sprinkled, smeared, or dabbed with fingers, sponge or cloth. There are many illustrations showing the vessel being carried, often together with a box or alabastron. It also appears in marriage, funerary, and possibly other rituals, and though it is nearly always associated with women, occasionally a man is involved, and it has been found in men's graves as well as women's. Part of its function could have been similar to that of the large, more public perirrhanterion. It might have been its more portable household counterpart.

Inscriptions show that exaleiptra could be included in a doctor's equipment, and that they were dedicated at the sanctuary of Asclepius at Athens and elsewhere.

Apart from pottery the shape was made in bronze, iron, precious metals and stone.

For the form of the word compare also *aleiptron*, *myraleiptron*.

References

ABV, pp. 348-49.

Agora, p. 180.

Antiphanes (ed. Meineke) III. 120. Exaleiptra included in a doctor's equipment.

Aristophanes *Acharnians* 1063. Exaleiptron in a wedding context. An *alabastos* is also mentioned (1. 1053). Possibly a similar association is seen in the two vessels held by a woman on a w.g. lekythos (Athens Nat. Mus. 1963) and the two vessels on *ARV*[2], p. 262, 30 and Richter and Milne p. 21. Some take the two names in Aristophanes (*exaleiptron* and *alabastos*) to refer to the same vessel, e.g. Amyx, p. 213.

Boston, I, pp. 49-50. Smegmatotheke probably used for liquid or semi-liquid perfume, but possibly as a lamp, censer, or for soap, paint. Marble smegmatothekai.

R.M. Burrows & P.N. Ure, "Excavations at Rhitsona in Boeotia", *BSA* 14 (1907–8): 268, 274, 305 ff.; "Kothons and Vases of allied Types", *JHS* 31 (1911): 72-99.

CVA Germany 1 (Pl. 39, 7 Attic, end of sixth century B.C.), 6 (Pl. 100 Attic, c. 500 B.C.), 10 (Pl. 19, 4-9 Corinthian, second half of sixth century B.C.; Pl. 31, 17 & 19, Pl. 32, 3 Attic tripod kothon, 580–570 B.C., with other references; Pl. 41, 11 Attic, c. 500 B.C.); 26 (Pl. 16, 7 & 8 Corinthian, 650–600 B.C.; Pl. 24, 12-15 Attic, late sixth century B.C.); *Great Britain* 6 (Pl. 4, 39 Corinthian, 650–600 B.C.); 9 (III C Pl. 2 Several Corinthian examples); *Greece* 1 (III Jc Pl. 3-7 and passim. Illustrations of plemochoai on w.g. lekythoi); *U.S.A.* 4 (Pl. 15, 3. Corinthian, 650–600 B.C. Good earlier bibliography).

Daremberg and Saglio, s.v. *exaleiptron.*

Gericke, p. 82 and Tab. 68.

IG IV 39 (Aegina); II/III^2 1534, 192 (Athens, Sanctuary of Asclepius); II/III^2 1517 B I 202. 223 (Brauron); XI^2 2, 161 B 125; 163 B 14; 198 B I 77 (Delos, Temple of Apollo). Exaleiptra dedications.

Pernice "Kothon und Räuchergerät", *JdI* 14 (1899): 60-72.

Payne, pp. 297-98.

E. Pfuhl, "Zur Geschichte der griechischen Lampen und Laternen", *JdI* 27 (1912): 52. "Kothons" from Aegina used as pigment containers.

Pollux 6. 105-6. "The vessel which held myrrh, similar to a phiale, was called an exaleiptron." Cf. also Pollux 10. 121.

RE, s.v. *kōthōn, plemochoē.*

Richter and Milne, pp. 21-22, s.v. *plemochoē.*

Ingeborg Scheibler, "Exaleiptra", *JdI* (1964): 72-108; "Kothon – Exaleiptra: Addenda", *AA* (1968): 389-397.

1. Hydria, Attic Geometric, eighth century B.C. 39 cm.
2. Hydria, Attic, early sixth century B.C. 15 cm.
3. Hydria, Attic b.f., c. 530 B.C. 40 cm.
4. Hydria, Caeretan, late sixth century B.C. 44 cm.
5. Hydria, Attic b.f., c. 500 B.C. 30 cm.
6. Hydria (Kalpis), Attic r.f., mid fifth century B.C. 34 cm.
7. Hydria, Attic coarse ware, late fifth century B.C. 38 cm.
8. Hydria, Apulian r.f., third quarter of fourth century B.C. 32 cm.

Hydria [ὑδρία]
Kalpis [κάλπις]

The name *hydria* was probably applied in antiquity to a number of shapes used as water-carriers, including the shape to which we apply the name. Nowadays the name is restricted to this one type of water container. Its most obvious characteristic is that it has three handles – two horizontal ones on the shoulder for lifting and carrying, and one vertical, extending from shoulder to neck or lip (like a jug handle) to make pouring easier.

The name *hydria* is written beside a hydria-shaped vessel on the Francois Vase, and as one would expect of vessels used to carry water, hydriai appear regularly in fountain-house scenes.

The hydria shape was made in Mycenaean times when it was apparently called *ka-ti*, a word that might be related etymologically to the Greek *kēthis* and *kathidos.* Several hydriai were found at the bottom of the Mycenaean well on the north slope of the Athenian Acropolis.

To begin with, the hydria is often a plump vessel. Some of the early hydriai, including these globular types, are similar in shape to the amphorae contemporary with them, the only difference being in the number of handles.

At the height of black-figure, towards the middle of the sixth century, there is an important change of shape. The body of the hydria becomes more triangular with a flatter shoulder. In Attic pottery during the last quarter of the sixth century a slimmer version becomes the rule, and continues for about half a century.

The name *kalpis* is often applied nowadays to the one-piece hydria whose neck meets body in a continuous curve. It appears before 500 B.C. and retains its popularity into the Hellenistic period, becoming slimmer as time goes

Hydria, Attic black-figure, second half of sixth century B.C. 49.6 cm.
Girls at the fountain house, some with pads on their heads to carry hydriai.
Courtesy of British Museum B330.

on. While *kalpis* was certainly applied to a water-carrier in ancient times, it is not clear whether it signified this shape. The name *krōssos* was also used in antiquity for some sort of water-container.

Hydriai are common in coarse pottery for a very long period, and in black-glaze from the fourth century on.

Hydria, Apulian red-figure, by the Painter of the Berlin Dancing Girl, 450–425 B.C. 29 cm.
Women working with wool. The one on the right makes a skein from wool in a kalathos. A lekythos stands on the table and a mirror hangs on the wall behind. Courtesy of Ashmolean Museum, 1974.343.

Bronze hydriai have survived from the seventh and sixth centuries B.C., and occasional silver ones from the fourth century B.C. Hydriai made of metal were favoured in cult practice. A bronze hydria is supposed to have stood in front of Phidias' statue of Zeus at Olympia.

Apart from their main use as everyday water-containers and in cult, hydriai had other functions as well. They were used as ballot boxes and as receptacles for judges' names in the law courts, and in some places, for example, Hadra near Alexandria in Hellenistic times, as funerary urns.

Compare *kalpē, kalpos, kathidos, kēthis, krōssos.*

References

Agora, pp. 53, 200.

ABV, p. 76, 1. Francois Vase. (See also Arias, Hirmer, and Shefton, Pl. 44; Minto, *Il vaso Francois.*)

Amyx, pp. 200-201.

Aristophanes *Lysistrata* 327, 358. Hydria/kalpis used for water-containers.

Brian F. Cook, *Inscribed Hadra Vases in the Metropolitan Museum of Art* (New York: Metrop. Mus. of Art, 1966).

Cook, pp. 225-26.

CVA Austria 3 (r.f.); *France* 9 (b.f. and r.f.), 14 (r.f.), 18 (b.f.); *Germany* 20 (r.f.), 41 (Attic b.f. neck and one-piece); *Great Britain* 7 (r.f.), 8 (b.f. and r.f.), 14 (b.f.); *Italy* 17 (r.f.), 20 (Attic b.f.), 42 (b.f.); *Netherlands* 3 (b.f.); *U.S.A.* 5, 17 (b.f.), 19 (b.f.).

Erika Diehl, *Die Hydria: Formgeschichte und Verwendung im Kult des Altertums* (Mainz: Philipp von Zabern, 1964). A very comprehensive study.

Elvira Fölzer, *Die Hydria.* (Leipzig: Seemann, 1906).

Gericke, pp. 48 ff.

Lucia Guerrini, *Vasi di Hadra.* (Rome: Bretschneider, 1961–2).

Hesychius. *Askos* equivalent to *hydria.* Also s.v. *kalpos* and *amphiphoreus.*

Plutarch *Demetrius* 53.

RE, s.v. *hydria.*

Richter and Milne, pp. 11-12. Includes ancient references to the word *kalpis.*

David M. Robinson, "New Greek Bronze Vases", *AJA* 46 (1942): 172 ff.

Michael Ventris and John Chadwick, *Documents in Mycenaean Greek* (Cambridge: Cambridge University Press, 1956), p. 338, n. 238. A tablet from Pylos. *Ka-ti* a possible name for hydria.

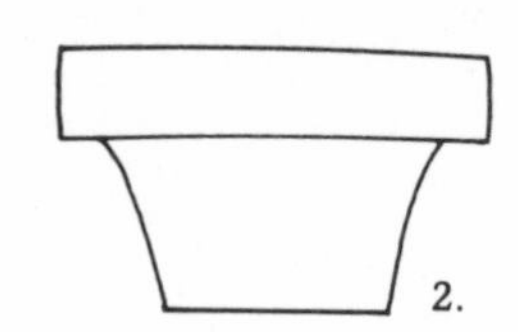

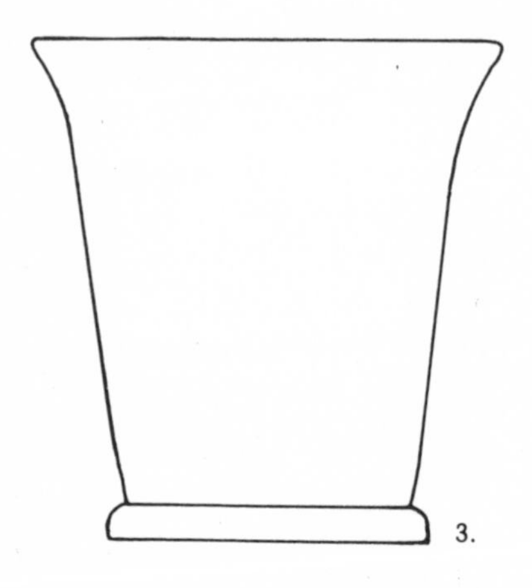

1. Kalathos, Attic geometric (lattice-work sides), c. 700 B.C. 17 cm.
2. Kalathos (miniature), Attic, c. 550 B.C. 4 cm.
3. Kalathos, Attic b.f., c. 540 B.C. 10 cm.
4. Kalathos, Attic r.f., c. 480 B.C. 25 cm.

Kalathos [κάλαθος] *Talaros* [τάλαρος]

Woman's basket, used especially for wool, spun and unspun (the name appears to be related to the Greek word *klōthō* ("I spin") and hence also to our word "cloth". It is not clear if there was any difference between the vessels to which the names *kalathos* and *talaros* are applied, but the words do appear separately in some lists (see Pollux 7. 173).

The name *kalathos* is applied nowadays to the container often illustrated in antiquity with women working with wool. It was wider at the top than at the bottom, calyx-shaped – rather like a calyx krater, but without the calyx krater's handles and base. Though illustrated on vases, surviving examples made of pottery are rare, which leads us to suppose that they were usually made of other materials. They were probably regularly woven out of reeds and canes like other types of baskets, as in the Pollux reference given above (and compare the basket painted on the Amasis Painter's belly-amphora – Basel, Antikenmuseum Kä. 420, Boardman *BF* Ill. 89). Phylo's talaros in the *Odyssey* was of silver. A pottery kalathos in Munich has a spout which shows that the shape was sometimes used as a container for liquids.

Very small pottery vessels of similar shape have also been called *kalathoi* by archaeologists. These have flaring walls and a projecting flat-sided lip. They are found in Athens from the seventh century on, but were also made elsewhere, including Corinth. The small size of these vessels and the context in which some of them were found have prompted the suggestion that they might have been used as stoppers for large vessels besides being containers in their own right.

In addition to use by women in their everyday work, the kalathos was also used for cult purposes.

The name *kalathos* was also used by the Greeks for other objects of basket shape, for example, the Corinthian column capital and a head-covering (see Hesychius).

Compare also *kalathion, kalathiskos, kophinos, talarion, talariskos, talaros;* and the Latin words *calathus, calathiscus, qualus, quasillus, quasillum.*

References

Agora, pp. 12, 56, 80. List of b.f. and r.f. kalathoi, and other references.

Aristophanes *Lysistrata* 579. "And then card into the kalathiskos".

ARV[2], p. 385 n. 228 (Kalathoid vase); p. 805 n. 89 & p. 1670 (Illustrations of a wool basket in use); p. 1034 n. 4 (Kalathoid vase).

CVA Germany 4 (Pl. 46, 14 Apulian, approx. third century B.C.) 15 (Pl. 2, no. 1 & 3 Protogeometric open-work kalathos. Full bibliography), 26 (Pl. 46, 1 & 3 Early S. Italian r.f. pelike whose decoration includes women with kalathos beside pillar), 27, (Pl. 114, 1 & 2 Geometric kalathos. Attic? Late eighth century B.C.), 33 (Pl. 185, 5 Boeotian kalathos. See also p. 56 where it is suggested that these early perforated "kalathoi" may actually be lids – the holes to allow circulation of air. Turned upside down they are very similar to the feet of Boeotian bird-bowls.); *Great Britain* 6, (Pl. 1, 17 South Italian, third century B.C.); *U.S.A.* 5 (Pl. 52 Attic r.f.), 17 (Pl. 33 Attic b.f. kalathos c. 540 B.C.).

Daremberg and Saglio, s.v. *calathus.*

V.R. d'A. Desborough, *Protogeometric Pottery* (Oxford: Clarendon, 1952), p. 113.

Arthur Evans, *Palace of Minos* II, Pt. 1, (London: Macmillan, 1928), pp. 134 ff.

Hesychius. "Kalathos. A cup, also a psykter. The parts above the face. A vessel in which iron is cast. A woman's vessel for storing wool."

Homer *Od.* 4. 125. "Phylo brought a silver talaros . . ."

Martial 8. 6. 15-16. Cup.

Pollux 7. 173 ". . . for weaving talaroi and kalathiskoi"; 10. 125.

RE, s.v. *kalathos.*

Richter and Milne, p. 13.

G. Rodenwaldt, "Spinnende Hetären", *AA* (1932): 7-21.

Williams, *AK* 4 (1961): 27-29.

1. Kantharos, Attic Geometric, third quarter of eighth century B.C. 8 cm. (incl. handles)
2. Kantharos, Etruscan Bucchero, c. 600 B.C. 12 cm.
3. Kantharos, Boetian black-painted, third quarter of fifth century B.C. 20 cm (incl. handles)
4. Kantharos, Attic r.f., 490–480 B.C. 19 cm. (incl. handles)
5. Kantharos (sessile), Boeotian black-painted, third quarter of fifth century B.C. 14 cm. (incl. handles)
6. Kantharos (St Valentin), Attic, 450–425 B.C. 11 cm.
7. Kantharos, Apulian, late fourth century B.C. 21 cm.
8. Cup kantharos, Attic, c. 320 B.C. 9 cm.

Kantharos [κάνθαρος]

The name is used for the cup shape, which, in its most characteristic form, has a tall lip, shallow bowl, high vertical handles, and stemmed foot. But in pottery there are also many sessile kantharoi (without stems). In some types of kantharoi lip and bowl are not very clearly articulated, while in others the handles are not particularly tall. A few have one handle (*CVA France* 10, Pl. 71).

The kantharos is the shape regularly associated with Dionysus in the ancient authors, and judging from representations in vase paintings we can be certain that the name is correctly applied in modern usage, at least for mainstream varieties. Vase paintings show kantharoi also apparently used for cult purposes, and associated with Dionysus (for example, at the Lenaea), and Herakles (Gericke, p. 24).

Although there are no recorded instances of the name *kantharos* written on a vessel, what is apparently meant to be the same word (*katharos*) is inscribed on the frieze of the Siphnian Treasury at Delphi under one of the figures whose helmet is in the form of a kantharos.

Whether the name was applied to all the ancient variations of the shape we cannot be sure. Evidently *kotylos* could be used as well. There are two cases of the name *kotylos* inscribed on a kantharos (Lazzarini and compare Athenaeus 11. 478 c). But as *kotylos* was apparently a general word for cup, this does not exclude the use of the word *kantharos* as well, as a more specific indication of the shape.

A type of kantharos was made in Protogeometric and Geometric times, the earlier examples with a conical foot, and the shape appears early in Attic and Boeotian black-figure pottery. It was also a favourite in Etruscan Bucchero Ware.

Several variations appear in the sixth and fifth centuries (see Sparkes and Talcott). There is a great spate of kantharos representations painted on pottery from early in the sixth century, but rather surprisingly, not a comparable increase in surviving examples of the shape. One of the new types to appear in the second quarter of the fifth century B.C. was the kantharos decorated with St. Valentin style decoration.

Kantharoi are common in late red-figure and plain black pottery of the fourth century B.C. and later. They become tall and slim, and handles often acquire a strut and spur, and the lip becomes concave in profile.

A common fourth century type is sometimes called a cup-kantharos. Although it is a kantharos shape it has ordinary cup handles. These are attached to the side of the vessel and project horizontally, though the end of the loop of each handle usually curves upwards. Sparkes and Talcott distinguish a number of sub-classes of this type.

It can probably be safely assumed that many of the illustrations of kantharoi on painted pottery are supposed to represent metal vessels. A number of metal kantharoi survive, including examples in gold and silver (see Simon p. 32, and *RE Suppl.* 4, p. 866).

The derivation of the word *kantharos* is not certain. It was also used as the name of a boat, scarab beetle, brooch (possibly in the shape of a scarab), a fish, and part of the Peiraeus. A kantharos device was used on Naxian coins. *Kantharos* might be related to *kanthōn* and *kanthēlios,* "donkey", and *kanthēla,* "panniers" (see Frisk). One common factor linking some of these words might be the typical high handles of the cup (called "ears" in Greek), the donkey's large ears, and perhaps the beetle's wings.

References

Agora, p. 113.

Aristophanes *Peace* 80-82, 143. Play on the words *kantharos* "beetle", and *kanthōn*, "donkey".

Lore Asche, "Der Kantharos" (Diss. Mainz, 1956).

Athenaeus 11. 473d ff. Quotations from various authors. Pamphilus says the kantharos is a kind of cup, the attribute of Dionysus (478c). The kantharos is also described as having a foot attached to a narrow stem (488).

John Boardman, "The karchesion of Herakles", *JHS* 99 (1979): 149-151. The karchesion mentioned by Athenaeus was a footless kantharos. In fifth century Athens the shape was often decorated with scenes dealing with Herakles.

Boston I, pp. 14-18; III, pp. 10-11.
Leo H.M. Brom, *The Stevensweert Kantharos* (The Hague: Nijhoff, 1952). Hellenistic silver kantharos found in the Netherlands in 1942.
Cook, pp. 233, 236-37.
Paul Courbin, "Les Origines du Canthare Attique Archaique", *BCH* 77 (1953): 322-345.
CVA Austria 1 (Pl. 45 f. Several kantharoi, including one with the head of a Negress); *Belgium* 1 (III I c Pl. 5 f. Painted by Duris); *Germany* 3 (Pl. 21 ff.), 6 (Pl. 93 Attic r.f., Pl. 94 St. Valentin), 9 (Pl. 120 Attic Geometric kantharos), 35 (Pl. 47, 9 Attic black-glaze, third quarter of fifth century B.C.), 36 (Pl. 46, 7 Boeotian towards 550 B.C.; Pl. 49-50 Boeotian c. 425 B.C.); *Great Britain* 5 (III I c, Pl. 32-39 esp. St. Valentin, r.f., plastic), 6 (III I, Pl. 371a Christie Painter, 440-430 B.C.).
Frisk, s.v. *kantharos.*
Gericke, p. 22.
Dorothy K. Hill, "The Technique of Greek Metal Vases", *AJA* 51 (1947): 254. Argues against the derivation of the pottery shape from metal ware.
S. Howard and F.P. Johnson, "The Saint-Valentin Vases", *AJA* 58 (1954): 191 ff.
Lazzarini, 357-8.
Macrobius *Saturnalia* 5. 21. 16. The kantharos, Dionysus' vessel).
Euthymios Mastrokostas, "Zu den Namenbeischriften des Siphnierfrieses", *AM* 71 (1956): 77.
RE Suppl. 4, s.v. *kantharos.*
Richter and Milne, p. 25.
Simon, passim.
L. Vance Watrous, "The Sculptural Program of the Siphnian Treasury at Delphi", *AJA* 86 (1982): 165 and Pl. 17.

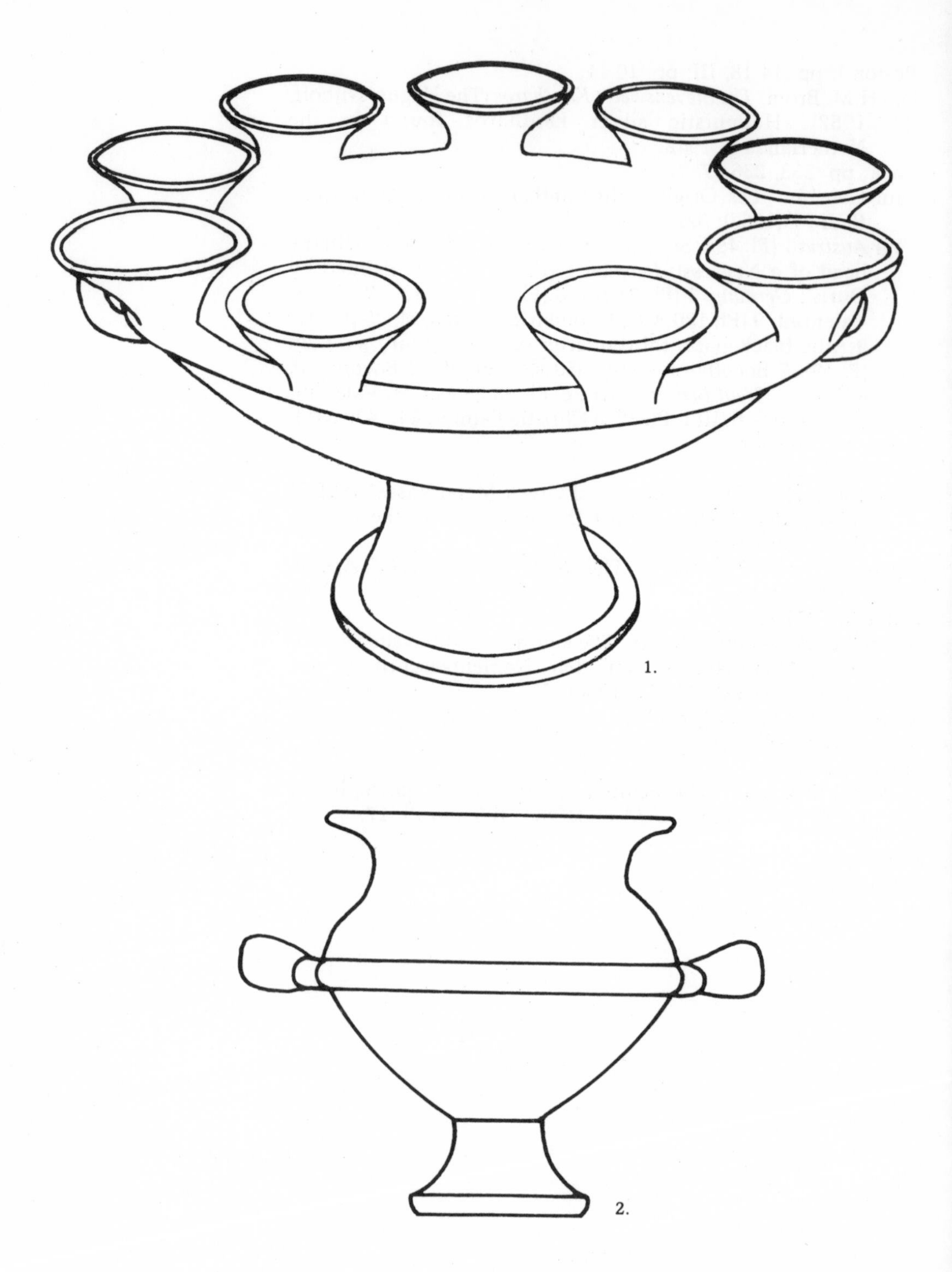

1. Kernos, Attic black-glaze, fourth century B.C. 16 cm.
2. Kernos, Attic, fourth century B.C. 9 cm.

Kernos [κέρνος]
Kerchnos [κέρχνος]

A bowl with a fairly deep concave lip, standing on a conical in-curving base. Horizontal handles may be attached to the belly of the pot, round which there runs a flat, horizontal disk or flange. Small cups (kotyliskoi), varying in number, may be attached to the top of the flange, and to the shoulder of the vessel, though quite often no such additions are made. Sometimes what appears to be a vestigial remnant of these little cups in the form of a series of small clay rings is attached to the vessel, or there may be simply holes perforating the flange or shoulder. In any case there are usually at least two holes in the flange (apart from holes that may appear to be vestigial cups) and occasionally there are two in the base. The holes are possibly used for the attachment of decorative foliage, or for anchoring a string to secure the lid.

The ancient sources refer to kernoi as multiple vessels, but as the plain types without kotyliskoi and those with kotyliskoi were found together for the most part, the same name seems justified for them all. There is a chance, however, that the type without kotyliskoi was the ancient plemochoe.

Kernoi vary a great deal in size, larger ones being about twenty centimetres in diameter or more, the smaller ones very tiny. The proportion between the foot height and total height of the vessel also varies quite a lot. Lids of various types have been found (solid, lattice-work, with triangular holes, or patterned in other ways).

Kernoi are mostly plain vessels, though many of them were probably gilded. At least some Eleusinian kernoi had thin gold-leaf covering. White paint survives on some, and a few have fence-like decorations of red vertical lines. A couple of kernoi of unusual shape are black-glazed.

The main finds are from Eleusis and from the Eleusinion area at Athens, close to the Agora. A few come from Alexandria and one was found at Laurion. Most vessels seem to be fourth century B.C. with perhaps a sprinkling earlier and later.

Many of the surviving kernoi appear to have been votives, and from the circumstances of some of the Athenian finds we can assume that they were buried after a "spring-cleaning" of the Eleusinion.

Literary references and visual representations point to the kernos being used especially in Eleusinian rites, but also in connection with Cybele and Rhea. At Eleusis offerings of fruit and vegetables were placed in these vessels. They were carried on the head in a sort of procession, and illustrations show sprigs of foliage apparently inserted into the holes in the vessels' flanges.

The original meaning of the word *kernos* is unknown, but it seems to have become confused with the similar Greek word *kerchnos*, "rough", possibly because of the appearance of the kernos surface with its kotyliskoi. The two names *kernos* and *kerchnos* were apparently applied to the same shape of vessel.

Gold kerchnoi are mentioned in inscriptions. At Eleusis stone and bronze examples have been found, as well as the pottery types sometimes covered with gold leaf.

In modern literature multiple vessels of different types from elsewhere in the ancient world are also called *kernoi* – as are the large flat stones from Minoan sites (usually circular, with a series of depressions hollowed out near the outer edge of the top surface). There is no substantial evidence of connection between these and the Eleusinian kernoi, though Xanthoudides has claimed possible continuity from the Early Minoan examples to vessels used in the Greek Orthodox Church.

References

Agora, p. 183.

Athenaeus 11. 478d. "Polemon in his treatise on 'The Fleece of Zeus' says: 'After this he (sc. the priest) celebrates the rite, taking the things from the chamber and distributing them to those who went round holding up the kernos. This is a clay vessel which has many small cups attached to it. In them there are sage, white poppies, wheat, barley, peas, pulse, okra, lentils, beans, rice-wheat, oats, preserved fruit, honey, oil, wine, milk, and unwashed sheep's wool. The person who carries it round tastes all of these things in the same way as the mystic fan-bearer.' " (Polemon fl. first half of second century B.C.). Athenaeus elsewhere (11. 476-77.) identifies the source of this passage as Ammonius [first century A.D.?].

CVA Germany 40 (Pl. 38. Kernos-like vessel. Vicinity of Troy (?) L. Helladic III C); *Italy* 10 (Pl. 1, 3. Cretan kernos).

IG I², 313 1. 17; 314 1. 23 (407/406 B.C. Mention of gold kerchnoi in an inventory from the Eleusinion at Athens).

George E. Mylonas, *Eleusis and the Eleusinian Mysteries* (Princeton: Princeton Univ. Press, 1961).

Martin P. Nilsson, *The Minoan-Mycenaean Religion and its Survival in Greek Religion.* 2nd ed. (Lund: Gleerup, 1950), pp. 135-141. A number of Bronze Age kernos-like vessels.

Jerome J. Pollitt, "Kernoi from the Athenian Agora", *Hesp.* 48 (1979): 205-233. A very good recent treatment of the shape.

RE, s.v. *kernos.*

O. Rubensohn. "Kerchnos", *AM* 23 (1898): 271-306.

Homer A. Thompson, "Two Centuries of Hellenistic Pottery", *Hesp.* 3 (1934): 447-450. Includes bibliography of earlier works.

S. Xanthoudides, "Cretan Kernoi", *BSA* 12 (1905–6): 9-23.

Kothon [κώθων]

1. The shape called *kōthōn* in modern literature (See Exaleiptron 1). The name is often applied to the shape dealt with under Exaleiptron 1, but *kōthōn* was certainly not the ancient name for that shape. The ancient kothon was probably some sort of drinking vessel. (See illustration p. 118).

2. The ancient kothon. The drinking vessel called a kothon in ancient times has been variously identified by modern investigators as cup, mug, or a kind of bottle. Part of a cup or mug with the name *kōthōn* incised on it was discovered at Isthmia during the 1957–58 excavation season. This shape is similar to the one illustrated in *CVA Germany* 6, Pl. 96, 1-3: a simple mug with a single handle. The shape is dealt with in detail in *Olympia Forschungen* V.

Although the ancient sources agree in describing the kothon as a drinking vessel of some sort they are not particularly enlightening in other respects and there are problems of interpretation with some of them, for example, we cannot be sure of what is meant by Critias' *ambōnes.* They have been interpreted as indented sides, bumps, or constrictions at the top or neck.

Mingazzini's suggestion for the kothon shape is the Pilgrim's Flask, a round, flat bottle with a narrow neck and one or two handles. This would fit Athenaeus' and Plutarch's description of the Laconian kothon – a vessel suitable for carrying by marching soldiers (it looks rather like some modern water bottles). The main problem with this identification is that the Pilgrim Flask shape, though popular in the Eastern Mediterranean, and later in Egypt and Italy, seems to have been a rarity in Greece.

The use of the word *kōthōnocheilos,* "kothon-lipped", referring to a kylix in Eubulus is evidence that there was something characteristic about the kothon's lip. Then there are the epithets *pachystomos*, "thick-mouthed", used by Heniochus, and *strepsauchēn*, "bent-necked", (or "neck-bending"?) in Theopompus.

According to inscriptions, silver and bronze kothons were made (*RE* s.v. *kōthōn*), which is surprising if the kothon was basically an ordinary sort of drinking vessel (but see *Olympia Forschungen* V).

The Scholiast on Aristophanes *Peace* 1094 implies that by Roman times there was some confusion about the meanings of *kantharos, kōthōn* and *kyathos.* Perhaps the name *kōthōn* was used for a number of different shaped drinking vessels, including the cup shape on which the word was found inscribed. Otherwise it is difficult to reconcile the ancient references.

Compare also *akratokōthōn, kōthōnizein, kōthōnismos* etc.

References

(See also under Exaleiptron)

Aristophanes *Knights* 600 (424 B.C.) "Some bought kothons, others garlic and onions". (The knights on the expedition to Corinth the previous year); *Peace* 1094 "No one gave the soothsayer a shining kothon." (Mock heroic, meant to be reminiscent of Homer. "Shining kothon" replaces the genuinely Homeric "shining krater" of *Il.* 3. 247); Scholiast on *Peace* 1094 "Kothon. Now a poterion (cup), elsewhere a kantharos . . . By kothon he meant what is now called a kyathos."

Athenaeus 11. 66. Athenaeus in a lengthy passage gives the word *kōthōn* as he found it used by several authors. A summary of part of the passage is given here. Athenaeus quotes Critias (last quarter of fifth century B.C.) *Constitution of the Lacedaemonians.* (This is the description on which Plutarch's is based. See below.) In addition to what Plutarch gives, the original version of Critias had "easy to carry in a knapsack". In Archilochus (c. 650 B.C.) the kothon is a vessel into which wine is decanted from a ship's casks. The passages from Aristophanes are quoted, then Heniochus *Gorgons* (c. 350 B.C.): "Let someone take the fire-born, circular, short-eared, thick-mouthed kothon, slave of throat, and pour in wine to drink." Theopompus *Stratiōtides* (just before 400 B.C.): "Shall I then bend back my neck and drink from a crooked-necked kothon?"

Alexis *Spinners* (c. 300 B.C.): "A four-pint kothon, an ancient possession of the house."

CVA Germany 6 (Pl. 96, 1-3. One-handled cup or mug, similar in shape to the one on which the name *kōthōn* was incised).

Daremberg and Saglio, s.v. *cothon.*
Eubulus 56. 3. "Kothon-lipped", an attribute of a kylix.
Hesp. 28 (1959): 335 & Pl. 70i. Fragment of a one-handled cup or mug with the name *kōthōn* inscribed, found at Isthmia. The reference to the shape Berlin 2266 here is incorrect. See Scheibler (1968): 390.
E. Kirsten, *Charites. Studien zur Altertumswissenschaft* (Bonn: Festschrift E. Langlotz, K. Schauenburg, 1957).
Paolino Mingazzini, "Qual'era la Forma del Vaso chiamato dai Greci Kothon?", *AA* (1967): 344 ff. Lists and translates into Italian all the important ancient sources.
Olympia Forschungen V, 169. Discussion of simple mug or cup shape.
Plutarch *Vit. Lyc.* 9. 4. "The Lacedaemonian kothon is very highly thought of on military expeditions, as Kritias maintains. The unpleasant appearance of the water they had to drink was disguised by the kothon's colour, and when mud met the bumps, [?] it was trapped, and what was drunk met the mouth cleaner. The lawgiver was responsible for this too."
RE, s.v. *kōthōn.*
Ingeborg Scheibler, "Exaleiptra", *JdI* (1964): 72-108; "Kothon-Exaleiptra: Addenda", *AA* (1968): 389-397.
Xenophon *Cyrop.* I. 2. 8. "They bring from home bread for food and cardamum for relish, and for drinking, if anyone is thirsty, a kothon to draw water from the river."

1. Krater, Attic Protogeometric, first half of tenth century B.C. 40 cm.
2. Krater, Early Attic, c. 700 B.C. 39 cm.
3. Column Krater, Attic r.f., c. 470–460 B.C. 41 cm.
4. Calyx Krater, Attic r.f., c. 460–450 B.C. 59 cm.
5. Volute Krater, Attic r.f., c. 460 B.C. 63 cm.
6. Bell Krater, Attic r.f., c. 460 B.C. 48 cm.

Krater [κρατήρ]

The krater was a large bowl with a wide mouth. By modern convention the name is applied to types with feet, excluding such shapes as the lebes. Ancient usage was less restrictive and might have included the types that we call *lebēs* and *stamnos*.

One of the commonest functions of the krater, as the name suggests (*kerannymi*, "I mix"), was for mixing wine and water (It was regular ancient practice for wine to be drunk diluted.) The wine and water mixture was dipped or ladled from krater into cups. According to Hesiod it was considered unlucky to put the wine-pourer on a krater at a drinking party.

The krater was a consistently popular shape from the Protogeometric period to late red-figure.

Certain varieties of krater were, like amphorae, used for funerary purposes. Attic Geometric kraters, and Apulian volute kraters of the red-figure period sometimes had this function.

Four of the main krater types are dealt with in separate sections and include their own references.

References

*ARV*2, chaps. 80, 81 etc.

CVA Austria 3 (Attic r.f.); *Belgium* 2 (Attic and South Italian r.f.); *Cyprus* 1 (Mycenaean); *Denmark* 4 (Attic r.f.), 6 (Attic and South Italian r.f.), 8 (Attic r.f.); *France* 1 (Laconian or Cyrenaean black and b.f., Attic r.f.), 4 (Attic r.f.), 5 (Attic r.f.), 6 (Campanian), 8 (Attic r.f.), 16 (South Italian), 18 (geometric); *Germany* 2 (early Attic), 4 (esp. South Italian), 5 (Attic and South Italian), 18 (mainly Attic); *Great Britain* 3

(Attic r.f.); *Italy* 2 (Attic and South Italian r.f.), 14 (Attic r.f.), 15 (Apulian), 17 (Attic and South Italian), 27 (Attic r.f.), 37 (Attic r.f.), 51 (Attic and South Italian), *U.S.A.* 5 (Attic r.f.), 6 (Attic), 7 (Attic and Apulian), 18 (Attic and South Italian r.f.).

Herodotus 1. 51. The silver krater dedicated by Croesus at Delphi, said to have a capacity of 600 amphorae.

Hesiod *Works and Days* 744. "Don't put the wine-pourer on the mixing-bowl when people are drinking, for grievous misfortune goes with that."

S. Papaspyrides-Karouzou, *Aggeia tou Anagyrountos* (Athens: Archaiologike Hetaireia, 1963). Three early b.f. kraters with lids and stands.

RE, s.v. *Mischkrug.*

Richter and Milne, pp. 6 ff.

Simon, passim. Excellent illustrations and bibliography.

A.D. Trendall, *Vasi antichi dipinti del Vaticano* (Vatican City, 1953 and 1955) passim.

A.D. Trendall and Alexander Cambitoglou, *The Red-figured Vases of Apulia* (Oxford: Clarendon, 1978).

Bell Krater

The shape is reminiscent of an upturned bell, which accounts for the modern name.

Although a similar shape had been made at Athens in Protogeometric times it seems to have died out for several centuries, before reappearing in red-figure of the fifth century B.C., at the time when the Berlin Painter was active. It became increasingly common through the fifth and early fourth centuries B.C. when it was the most popular shape of krater. It retained its importance in ordinary red-figure to the end of the red-figure style, especially in southern Italy, and survived in some Hellenistic wares.

Bell kraters are sometimes divided into two classes (*CVA U.S.A.* 10, p. 44). The squatter, more compact type with lugs for handles is possibly derived from Attic vats or tubs. The earliest examples of this type (Berlin Painter) are footless, but later a torus or other type of foot is found. In early examples the foot is often quite distinct from the body which it meets at an angle. The name "lugged krater" is preferred by some for this shape.

The other type of bell krater was possibly called Milesian in ancient times. It appeared in Attic red-figure about the middle of the fifth century B.C. and soon became the fashionable shape. It has proper handles, not lugs. The lower part of its body shows an increasing tendency towards "stemming"; the body contracts in an elegant curve into what is virtually a thick stem above the foot.

As well as the pottery vessels bell kraters in metal are known (e.g. in the Hildesheim silver), and some of the representations on vases look as if they are meant to depict bronze vessels.

References

Agora, p. 55.
Amyx, p. 199. Bibliography, prices, etc.
John D. Beazley, "The Master of the Berlin Amphora", *JHS* 31 (1911). Early bell kraters, p. 283.
Boston II, p. 50.
CVA (See also the general references under *kratēr*.) *Cyprus* 1 (Mycenaean); *France* 2 & 4 (early examples of the shape); *Great Britain* 2; *Poland* 8 (Campanian); *U.S.A.* 18 (Pl. 31-52 passim. A number of later bell kraters).
S. Drougou, "Ein neuer Krater aus Athen", *AA* (1979): 265-82. Beazley's Falaieff Kraters, named after the former Russian owner of some vases of this type. See also *ARV2*, p. 1469.
Anna B. von Follman, *Der Pan-Maler* (Bonn: Bouvier, 1968).
Paolino Mingazzini, "Un nuovo Nome antico per designare una Forma di Vasi", *RM* 46 (1931): 150 ff. Mingazzini suggests that the later class of bell kraters might have been called *kratēres Milēsiourgeis* ("kraters of Milesian workmanship").
RE, s.v. *Mischkrug*.

Calyx Krater

The calyx krater appeared very late in the black-figure period, in the third quarter of the sixth century B.C., but it was a comparative rarity in the black-figure style. The earliest surviving one was made and painted by Exekias, who might have invented the shape. Various suggestions have been made for its source of inspiration, for example, a woman's skirt, a flower (compare the name calyx), and non-Attic pottery shapes (Jacobsthal and Furtwängler suggested Ionia as the source). The Chiot chalice, which was common in the east in the first half of the sixth century, is similar, though a lot smaller.

The calyx krater was a common red-figure shape, and, as usual, the tendency was for it to become slimmer as time went on, its slimness being accentuated by the acquisition of a stem at the base. Late red-figure calyx kraters generally have a more pronounced concave profile than their predecessors, and many of them are rather unsubstantial creations, as are their descendants in the black-glaze pottery of Hellenistic times.

The "calyx" part of a calyx krater is several times taller than the bowl. Sometimes, especially in red-figure, bowl and calyx are made more independent of each other by the calyx being offset. In black-figure the bowl usually has decoration. In red-figure there is usually less, mostly simply a horizontal border at handle level, and palmettes below the handles.

The foot may be of simple rounded profile, but sometimes there is a small flare, and it may be two-tiered. Junction of foot and bowl is sometimes marked by a torus moulding.

Handles, curving upwards, are attached to the bowl.

Their tips are usually rounded, but some later examples are more angular, and occasionally the tips have a pronounced curve back inwards.

References

ARV2, pp. 13-14. Euphronios. This artist favoured the calyx krater shape.

Dietrich von Bothmer, *Bulletin of the Metropolitan Museum* 31 (1972-73): 1 ff. The splendid "Euphronios Krater" acquired in 1972.

C. Boulter, *AK* 6, 71.

Frank Brommer, "Krater Tyrrhenikos", *RM* 87 (1980): 335-339.

Oscar Broneer, "A Calyx-krater by Exekias", *Hesp.* 6 (1937): 469 ff. The earliest surviving calyx krater.

CVA (See also the general references under *kratēr.*) *Austria* 3 (Series of calyx kraters from c. 470 B.C. to late r.f.); *France* 1 (r.f.), 2 (b.f. and r.f.), 19 (Attic b.f.); *Germany* 11 (Late r.f. series); *Italy* 27 (r.f.), 37 (r.f.); *U.S.A.* 17 (b.f., Rycroft Painter).

H. Hinkel, *Der Giessener Kelchkrater* (1967): 38 ff.

Jacobsthal, *Metropolitan Museum Studies* 5, 117-18.

Column Krater

Column kraters appear in Corinthian pottery in the last quarter of the seventh century B.C. Their main distinguishing feature is the double column formed by the handle, which joins shoulder to extended lip (handle-plate). The shape was taken over by Attic potters and became common in Athens in the first half of the fifth century B.C., but declined in popularity in the following century. A slimmer version persisted into late red-figure in southern Italy (especially in Apulia) and in Etruria, to the end of the red-figure period.

On the evidence of words scratched under the foot of Attic red-figure vessels, Athenians in the fifth century B.C. called a vessel of this shape a Corinthian krater (*Korinthios* or *Korinthiourgēs*). As the shape almost certainly originated in Corinth, the name is a natural one for Athenians and was possibly the regular name for this vessel at Athens.

References

Agora, p. 54.

Amyx, pp. 198-99 & n. 79.

ARV2, chaps. 17, 54, 80 etc.

Tomis Bakir, *Der Kolonettenkrater in Korinth und Attika zwischen 625 und 550 v. Chr.* (Würzburg: Konrad Triltsch Verlag, 1974). A very full treatment of the early history of the shape.

John D. Beazley, "Some Inscriptions on Vases IV", *AJA* 45 (1941): 597.

CVA (See also the general references under *kratēr*.) *Austria* 2 (r.f.); *France* 1 & 2 (b.f.), 5 (r.f.), 19 (Attic b.f.), 29 (Pl. 12, Laconian, sixth century B.C.); *Germany* 18 (Pl. 54 ff. r.f.);

Italy 2 (Pl. 15 ff. Attic r.f.), 4 (Pl. 3 ff. Attic r.f.), 5 (Pl. 23 ff. Attic r.f.), 14 (Pl. 40 ff. Attic r.f.), 37 (Pl. 33 ff. Attic r.f.), 41 (III C, Pl. 1 ff. Corinthian); *U.S.A.* 15.

Adolf Greifenhagen, "Frühlukanischer Kolonettenkrater mit Darstellung der Herakliden" (Berlin: Walter de Gruyter, 1969) 123 Winkelmannsprogramm der archäologischen Gesellschaft zu Berlin.

Payne, pp. 300 ff.

RE, s.v. *Mischkrug* (Stangenhenkelkrater).

Andreas Rumpf, *Chalkidische Vasen* (Berlin: De Gruyter, 1927), pp. 45, 123. The name of the column krater.

Volute Krater

The volute krater begins as a variant of the column krater, and establishes itself as a regular species about 600 B.C. The modern name singles out its chief distinguishing feature, the pair of volutes which spring from the shoulder and are attached to the top of the lip.

The most famous early example of the shape is the Francois Vase made by Ergotimos and painted by Kleitias about 570 B.C. or a little later. As the canonical shape had not yet been established when the Francois Vase was made, this early type is sometimes called a proto-volute krater. In the Francois Vase the volutes are still very loose, little more than hooks. In most later examples they tighten up into compact circles like the volutes of Ionic columns.

The canon is firmly established by the beginning of the fifth century B.C. The neck is about a third of the total height of the vessel, and with the exception of the early examples, there is a well-defined lip. The tendency was for the shape to become slimmer, and mouldings of mouth, neck, and foot to become more complex.

In the fourth century a taste for elaboration, especially in Apulia, encouraged the covering of the volute with decoration, for example, rosettes, ivy tendrils or a moulded human face. Apart from the volutes the rest of the vessel also was cluttered with a superabundance of painted decoration. Forerunners of these elaborate and elaborately painted volute kraters can be seen in Athens about 400 B.C.

References

Arias, Hirmer and Shefton. Francois Vase and bibliography to 1962.

Boston II.
CVA France 1 (III Dc, Pl. 6 b.f.), 2 (III Ic, Pl. 17 ff. r.f.); *Germany* 39 (Pl. 41. Attic b.f., c. 500 B.C. Golvol Group); *Italy* 27 (Attic r.f.), 33 (Attic r.f.), 37 (Pl. 3-13 Attic r.f.), 49 (Apulian).
Simon, pp. 69 ff. Francois Vase.

1. Kyathos, Etruscan b.f., c. 500 B.C. 16 cm (incl. handle)
2. Kyathos, Attic r.f., c. 480 B.C. 18 cm. (incl. handle)
3. Kyathos, Hellenistic bronze, second century B.C. 53 cm.

Kyathos [κύαθος]

The kyathos is a ladle or dipper with a single handle. In pottery the handle is usually made in the form of a high, continuous loop, extending well above the lip, and returning to be attached to the side of the vessel. There is sometimes a strut across the handle-loop to strengthen it. Apart from the high handle the shape is similar to a modern tea cup with fairly straight sides and a small foot.

In pottery the tall exposed handle is very vulnerable, and for this reason the shape is more practical in metal. A number of metal kyathoi have survived. In these the handle usually does not form a closed loop but is attached to the top or side of the cup and has a hook at the end, like that of a walking stick, so that the vessel can be hung over the lip of a krater.

Kyathoi are common in the later stages of black-figure pottery, especially from about 530 B.C., and are found amongst Nikosthenes' exports to Etruria. Native Italian wares included a long series of short-handled kyathoi.

There are also some examples in black-glaze and coarser pottery. In these fabrics one type has a shallower bowl and horizontal handle – virtually a large pottery spoon.

The name *kyathos* has also been applied by archaeologists to a different type of vessel – a small, two-handled Protocorinthian kantharos. However, recent usage restricts the name to the ladle type. From the evidence of the name *kyathos* inscribed on this shape, it would seem that modern archaeologists apply the name correctly. So it is odd that the word *kyathea* is written as a graffito on the bottom of a stamnos.

A *kyathos* was also a measure (about one twelfth of a pint). Aristophanes uses the word for what appears to be a small vessel (like a medical cupping glass) which was used for reducing bruises such as black eyes.

The word might be related to *kyar* ("hole"), the ending being similar to that of such words as *gyrgathos* and *lēkythos.* It is not certain whether the names *kyathos* and *kyathis* were applied to different shapes in antiquity. Evidently *kyathis* was Doric.

Compare *antlētēr, kochliarion, kyatheion, kyathion, kyathis, kyathiskos.*

References

ABV, pp. 223 (Nikosthenes), 293 (Psiax), 295, 516, 519, 556, 609, 611, 613. Kyathos in use: p. 186 (Illustration on an oinochoe).

Agora, pp. 143, 229.

ARV², pp. 329, 333. Kyathoi in use: p. 370, n. 13 (The Brygos Painter – illustration on a kylix); p. 1151, n. 2 (The Dinos Painter – illustration on a stamnos).

Aristophanes *Lysistrata* 444, *Peace* 542 (apparently cupping vessels).

Athenaeus 11 480b, "Kyathis, a cup-shaped vessel"; 503b, "pouring in wine with a kyathos".

John D. Beazley & Magi, *Raccolta Guglielmi* i, p. 52-53.

Boardman, *BF*, p. 189.

M. Comstock & C. Vermeule, *Greek, Etruscan and Roman Bronzes in the Museum of Fine Arts, Boston* (Boston: Museum of Fine Arts, 1971)

Cook, pp. 237, 369.

CVA Belgium 1 (III He, Pl. 3-4, b.f.), 2 (III Ic, Pl. 20, 1 Late archaic r.f.); *Great Britain* 15 (Pl. 24-25, b.f.); *Italy* 10 (III Ic, Pl. 6, 3 A kyathos in use), 20 (Attic b.f.).

Margaret Crosby, "A silver ladle and strainer", *AJA* 47 (1943): 209. The name kyathos inscribed on a silver ladle.

Daremberg and Saglio s.v. *cyathus.*

Michael M. Eisman, "New Attributions of Attic Kyathoi", *AJA* 77 (1973): 71; "Attic Kyathos Painters" (Ph.D. diss. 1971, University of Pennsylvania).

Etymologicum Gudianum, s.v. *kyathos.*

Frisk, s.v. *kyathos.*

Hackl, *Münchener archäologische Studien* (Munich, 1909), p. 52. Graffito kyathea on the bottom of a stamnos.

Hesychius. "Kyathos: A drop, a small measure. Or (a vessel) for drawing off liquids. Kyathoi: Skaphiolia [?]. Small iron eggs, like a kyathos. Capacity – two ounces of liquid."

Lazzarini, 373. The name *kyathos* inscribed on ladles.

Reinhard Lullies, *Antike Kleinkunst in Königsberg, Pr.* (Königsberg: Gräfe u. Unzer Verlag, 1935), p. 192. The name *kyathos* inscribed on a bronze ladle.

Olynthus 10, 194-8. Bronze ladles.

RE, s.v. *kyathos*.
Gisela M.A. Richter, "A kyathos by Psiax in the Museo Poldi-Pezzoli", *AJA* 45 (1941): 587 ff.
Richter and Milne, p. 30.
Xenophon *Cyropaedia* 1. 3. 9. "Whenever the king's cup-bearers offer a phiale they draw off wine from it with a kyathos, pour it into their left hand and swallow it."

Kylichnis [κυλιχνίς]

Variants KYLIKIS, KYLIKNIS see also PYXIS

In Attic Greek the word meant "box" and the name was given to a small container for such things as cosmetics, jewellery, and ointment used for medicinal purposes. It is usually assumed to have been round, but the name might have been applied to other shapes as well. That it was used for a physician's ointment box is shown by inventories from the temple of Asclepius at Athens, and this meaning is supported by the ancient lexicographers. Other temple inventories of the female deities support the meaning "jewellery box".

The inscriptions refer to kylichnides made of various types of stone, ivory, bronze, and silver, and the kylichnis itself may be kept in a wooden case. Clay examples must have been common for everyday use, and are mentioned by Hesychius, but as usual the dedicated objects were made of more precious materials.

The word *kylichnis* fell into disuse in the late Classical and Hellenistic period, and is replaced by *libanōtis, phykion,* and *pyxis.*

The origin of the name *kylichnis* is uncertain. It looks as if it might be a variant of *kylix,* "cup", and in fact both Athenaeus (quoting the playwright Achaeus) and Galen give *kylix* as a possible meaning of *kylichnis.* Athenaeus, however, says that Achaeus uses the word *kylichnis* for *kylix* "in an unusual sense". Milne suggests that the word might have meant "cup" in Ionic Greek. However that may be, *kylichna,* a very similar word, did mean "cup". Athenaeus seems to believe in a connection between *kylichnis* and the Greek word meaning "to roll, turn", a notion presumably deriving from the similar sound of the words. Boisacq relates *kylix* and *kalyx* to an Indo-European word meaning "curved".

References

Agora, p. 173.

Athenaeus 11. 480. "The Athenians call the doctor's pyxis a kylichnis because of its being turned on a lathe."

E. Boisacq, *Dictionnaire étymologique de la langue grecque,* p. 533.

Hesychius. "Kylichnides. Pyxides. Others, libanotrides. Others, aggeia made of pottery. Others, kylikes. Others, doctors' pyxides."

IG II-III^2 1394, 11-13; 1533; 1534 B. See also Milne for numerous references

Marjorie J. Milne, "Kylichnis", *AJA* 43 (1939): 247 ff.

Pollux 6. 98. "Kyliskion is a small kylix. Kylichnos [kylichnis ?] is a pyxis." Cf. also 4. 183; 10. 46.

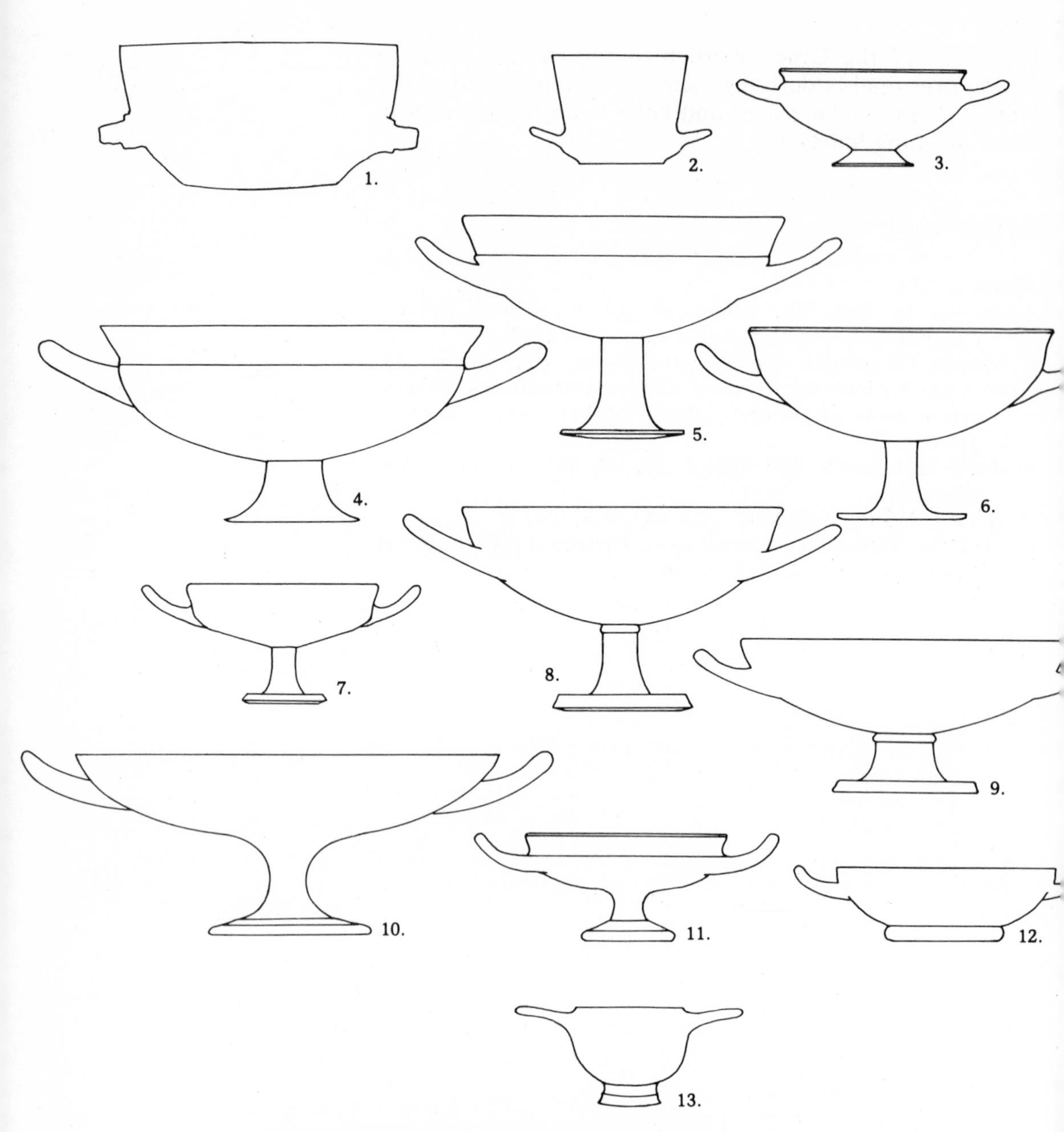

1. Cup, Attic Geometric, second half of eighth century B.C. 10 cm.
2. Cup (lakaina), Laconian, second half of seventh century B.C. 8 cm.
3. Kylix, Ionian, third quarter of sixth century B.C. 7 cm.
4. Kylix (Siana Cup), Attic b.f., c. 550 B.C. 13 cm.
5. Kylix (Lip Cup), Attic b.f., 550–540 B.C. 16 cm.
6. Kylix (Little Master Cup), Attic b.f., c. 540 B.C. 13 cm.
7. Kylix (Kassel Cup), Attic b.f., c. 530 B.C. 8 cm.
8. Kylix (Band Cup), Attic b.f., 550–530 B.C. 13 cm.
9. Kylix (Eye Cup, Type A), c. 520 B.C. 11 cm.
10. Kylix (Type B), Attic r.f., early fifth century B.C. 12 cm.
11. Kylix (Type C), Attic black-glaze, second quarter of fifth century B.C. 8 cm.
12. Kylix (stemless), Attic black-glaze, mid fifth century B.C. 5 cm.
13. Cup, Gnathia Ware, second half of fourth century B.C. 7 cm.

Kylix [κύλιξ] *Cup*

The name *kylix* is used in modern literature for a cup with a comparatively wide, shallow body, The diameter of the base may be about half that of the bowl, but is usually less. The two handles are attached horizontally, but almost invariably their ends curve upwards, often to the level of the top of the lip. For some of the ancient uses of the word, see *Skyphos,* and Cook, p. 233.

A basic division is into stemmed and stemless types. The first category, which includes the more elegant and important cup shapes, has a stem joining bowl and foot. The bowl of a stemless kylix sits directly on the foot. A majority of kylikes have diameters ranging between about twelve and twenty centimetres, but there is quite a significant scattering outside these limits.

The word "cup" as applied to Greek pottery shapes, is used by modern writers in two different senses. Taking the more general definition it means any vessel used for drinking (including the kylix and such shapes as skyphos and kantharos). Others prefer to limit it to the shape dealt with here – the cup par excellence. It is one of the most important shapes of late black-figure and early red-figure.

Stemmed Kylix

There are many different varieties of stemmed kylix, often in vogue for only a few decades before fashions changed and they were superseded. In the neat Protocorinthian cup we see a recognizable prototype of the kylix shape. It is an elegant vessel whose subgeometric decoration is soon abandoned, and at Corinth and even more so at Athens soon flaunts a troupe of padded dancers in the main field. The subsidiary decoration is of Orientalizing type, especially rosettes (on the lip and interspersed with the figures in the field), and more complicated anthemium (intertwined floral decoration) for example, under the handles.

At Athens there follows a succession of handsome shapes, each type with its characteristic style of figured decoration, as well as plainer cups, mainly black, with nothing more than a few reserved bands to relieve the starkness.

Siana Cups (second quarter of the sixth century B.C.)

The name comes from a site in Rhodes. Bowl is beginning to spread wider, lip and foot are taller and more pronounced, and the contours are so designed that the component parts are clearly articulated one from the other. The decoration is often the least successful part of Siana Cups. In one scheme bowl and lip are lumped together and treated as a single field, the decoration ignoring the junction.

Gordion Cups (just before the middle of the sixth century B.C.)

This group is named after the find-place Gordion in Phrygia. They are mostly smallish cups, the prototype of the Lip Cup, with lip fairly straight and offset. Foot is of variable height, but in general not yet as tall and slender as in the following styles. The decoration, which is of a minor type, is in a reserved band at handle-frieze level. The rest of the cup is mainly black.

Lip Cups (made for two or three decades in the middle of the sixth century B.C.)

The stem is now taller, the lip offset and slightly concave (out-curving). There is a groove between lip and bowl on the inside of the vessel, as well as a clearly defined junction between these two members on the outside. The lip is given further emphasis by having the main painted decoration assigned to it.

Band Cups (made mainly in the third quarter of the sixth century B.C.)

Lip, though still concave as in the preceding type, no longer meets the bowl in a distinct join. Otherwise the shape is very similar to that of the Lip Cup. Decoratively however, there is a closer similarity to Gordion Cups. The decoration is on the reserved band at handle level.

Droop Cups (mainly second half of the sixth century B.C.)

Named after the man who first described the class. A more aggressively concave lip and usually rather heavier lines than the preceding types. Often a series of grooves or channels on about one third of the upper part of the stem. The lip is all black, as is the bowl also in some examples, but usually the bowl is decorated lavishly with one major frieze and a series of smaller ones.

Cassel Cups (approximately 530–500 B.C.)

In shape similar to Band Cups, but smaller. The decoration is in a series of bands all over the vessel.

The group of expert miniaturist black-figure painters who concentrated on cup painting around the mid sixth century B.C. are sometimes called Little Masters, and the finest of the shapes painted by them (especially Lip and Band Cups) are also called "Little Master Cups".

Towards the beginning of the red-figure period Eye Cups became popular – an important group whose main feature is the spectacular eyes painted on the outside of the bowl. Other minor types of cups appeared from time to time but were of more limited popularity, such as, the Merrythought Cups with "wishbone" handles towards the middle of the sixth century B.C. (cf. *U.S.A.* 19, Pl. 88).

The kylix appears in vast quantities in black-glaze where it is one of the commonest shapes. Some types appear mainly in black, for example, the Vicup (Bloesch's Vienna Cup, *Agora*, p. 93). The kylix was an important shape in some non-Attic areas too. At Sparta a shape which had a long run of popularity was the lakaina, the "Laconian cup". It had a shallow bowl to which the handles were attached, and a very tall lip. The shape was made from Geometric to Hellenistic times.

The systematizers have divided stemmed kylikes of the mature black-figure and red-figure periods into three main types (A, B and C), so that one can, for example, speak of a Type A Eye Cup. The main characteristics of the three types are as follows.

Type A

Bowl has no lip. Flaring, mostly fairly low stem, at the top of which (at the junction with the bowl) there is a fillet – a small protruding ring moulding. The foot is usually either deep, of concave profile and reserved (in black-figure

Kylix (tondo), Attic red-figure, painted by the Kiss Painter, c. 500 B.C.
Trainer and athlete. Two aryballoi suspended by strings together with sweat-cloths or bags, and a pick used by athletes to prepare the ground for competition.
Courtesy of the Collection of the Baltimore Society of the Archaeological Institute of America.

cups), or of shallower ring type and painted (in red-figure cups). Overall the impression is one of a fairly heavy, solid vessel. Common in late black-figure.

Type B

Bowl curves and merges directly into the stem with no abrupt junction. Foot usually convex in profile. Common in early red-figure.

Type C

Foot convex; there is often a fillet where it joins the stem (in contrast to Type A where the fillet was just below the bowl). Lip often concave. Heavier than Type B, but of mostly smaller dimensions. Commonly all black, or with decoration only on the inside. See Sparkes and Talcott, p. 91, n. 17 for differing definitions of Type C cups.

There are a few hybrids with attributes of more than one type but completely aberrant types are rare.

Stemless Kylix

The stemless cup types were most popular in plain black-glaze, and were common in the fifth and fourth centuries (especially in the middle part of that period). The more important classes of plain black stemless cups are described in *Agora*, pp. 98 ff. Some stemless cups were also painted by Attic red-figure artists during the fifth century B.C.

There are more than a dozen instances of the word *kylix* or one of its variants written on a vessel. The inscriptions range from the Geometric period to about 400 B.C. and are fairly evenly divided between what we call the kylix and kotyle or skyphos shapes, while there is one instance of the name on a Chiot Chalice. It is clear that in ancient usage the definition of kylix was wider than ours, but the shape dealt with here was certainly normally included.

Compare also *kylichnē, kylichnion, kylichnis, kyliskē, kyliskion.*

References

Agora, pp. 5-6, (the word *kylix* inscribed on vases); pp. 88 ff. (detailed discussion of the shape).

*ARV*2, chaps. 3-13, 22-29, 43-48, 65-67, 75-76, 89-90 etc.

Athenaeus 11. passim, esp. 470e, kylix with low sides and short handles; 478e, the kotyle a kind of kylix.

Hansjörg Bloesch, *Formen Attischer Schalen von Exekias bis zum Ende des Strengen Stils* (Bern-Bumpliz: Benteli A.-G., 1940). Attic cup shapes c. 540–450 B.C. Details of differences between Types A, B, and C.

John D. Beazley, *JHS* 49 (1929): 260 (Siana Cups, named after two cups from Siana in the British Museum); 51 (1931): 275 (Siana); 51 (1932): 167 ff.

Boardman, *BF* esp. pp. 31 ff., 58 ff., 106 ff. and 236 ff. (bibliography); *RF* esp. pp. 55 ff., 132 ff., 195 ff. and 237 (bibliography).

Cook, passim, e.g. pp. 81 ff., 233 ff., fig. 30.
CVA Austria 1 (Pl. 1-34 Attic r.f.); *Belgium* 1 (Some Corinthian and Attic b.f. but mainly Attic r.f.); *Canada* 1 (Attic b.f.); *Denmark* 3; *France* 7 (Pl. 20 ff. Sparta and Cyrene, incl. Arcesilaus Cup), 14 (Mainly Attic, Some Ionian b.f.), 17 (Attic b.f. and r.f.), 20 (Attic b.f. and r.f.), 28 (Attic r.f.); *Germany* 1 (Attic r.f.), 4 (Attic r.f.), 17 (Attic b.f.), 18 (Attic r.f.), 21 (Pl. 49-100 Attic r.f.), 22 (Pl. 101-134 Attic r.f. and w.g.), 30 (Attic b.f. and r.f.), 34 (Attic b.f. and r.f.), 41 (Attic b.f.), 47 (Attic b.f.); *Great Britain* 2 (Attic b.f.), 15 (Attic b.f. and r.f.); *Italy* 2 (Attic r.f. and Faliscan), 3 (Attic b.f.), 5 (Attic r.f.), 14 (Attic r.f.), 20 (Attic b.f.), 25 (Attic r.f.), 26 (Attic and Italian r.f.), 28 (Attic r.f.), 30 (Attic r.f.), 33 (Attic r.f.), 36 (Attic b.f.), 38 (Attic r.f.), 41 (Attic b.f.), *New Zealand* 1 (Attic b.f.); *Poland* 4 (Attic b.f.); *Switzerland* 1 (Attic r.f.); *U.S.A.* 4 (Attic b.f.), 5 (Attic b.f. and r.f.), 6 (Attic r.f.), 11 (b.f.), 13 (Attic r.f.), 17 (Attic b.f. and r.f.), 19 (Attic b.f.).
Daremberg and Saglio s.v. *calix*.
Gericke, pp. 13 ff.
Lazzarini, pp. 346 ff.
RE Suppl. Vol. 5 s.v. *kylix*.
Richter and Milne, pp. 24-25.

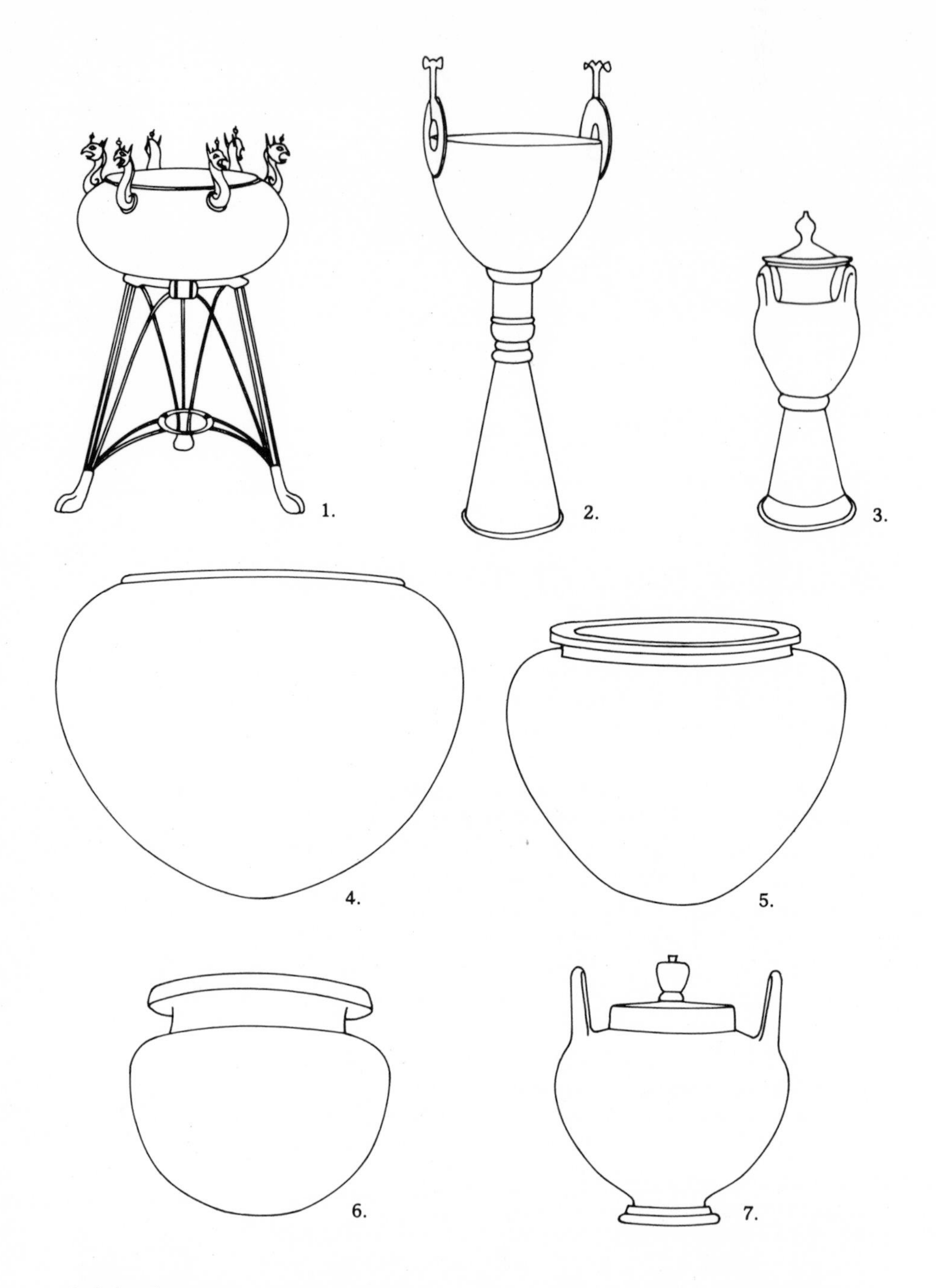

1. Lebes (bronze) on a tripod stand, Samos, c. 650 B.C. 30 cm. (cauldron and griffin heads, but excluding stand).
2. Lebes and stand (clay), Greek (provenance uncertain), early seventh century B.C. 109 cm. (lebes and stand, excluding handles).
3. Lebes gamikos (vessel attached to the stand), Attic r.f., c. 450 B.C. 64 cm. (incl. stand).
4. Lebes, Attic b.f., 575–550 B.C. 33 cm.
5. Lebes, Attic r.f., second half of fifth century B.C. 27 cm.
6. Lebes, Attic r.f., c. 400 B.C. 23 cm.
7. Lebes gamikos, Attic r.f., 430–420 B.C. 27 cm.

Lebes [λέβης]

Also called DINOS [*δῖνος*], LEBES GAMIKOS

The lebes was a deep bowl, usually rounded at the bottom so that it had to be set on a stand. It often had a shallow neck and small lip. Some types had handles, others were handleless.

Lebetes that were awarded as prizes at the games seem to have been made of metal, and the many references to lebetes in inscriptions are mostly to metal vessels. Those used for cooking (often associated with tripod stands) would usually have been of bronze. Stands made of other materials were also used where there was to be no contact with fire.

Vase paintings show that the lebes could be used for the same purposes as a krater – mixing wine and water, and as a general container. The evidence of vase painting also points to a number of religious and cult uses, especially for the type known as the lebes gamikos or lebes nymphikos.

As early as Homer the lebes is the basin used for carrying purifying water and also water for washing the feet.

The lebes gamikos, which was associated with the marriage ritual, traditionally had two tall vertical handles attached to the shoulder and extending to about the lip, or sometimes even above it. Probably it usually had a lid. In many cases the bowls of the tripod lebes and lebes gamikos were permanently attached to their stands.

As far as we can tell the modern usage of the word is much the same as the ancient. It seems highly probable that the ancient term "nuptial lebes" refers to the traditional shape so often portrayed in marriage scenes on vases. The fact that the word *lebēs* is often associated with the word *tripod* (as in the name "tripod lebes") also strengthens the conviction that our attribution is correct, for this shape of vessel regularly appears with tripods in the vase paintings.

The lebes shape is also called *dînos* in much of the modern literature. This usage, though firmly entrenched, has no ancient support.

References

Agora, p. 57.

Amyx, pp. 199-200.

Athenaeus 2. 37-38. "In olden days there were two sorts of tripod, each of which happened to be called a lebes. One was a tripod for standing on the fire, also called the bath-water tripod, as Aeschylus says: 'The tripod-lebes of the house received him – the one that always kept its place over the fire.' The other one was called a krater . . ."

Boardman, *BF*, p. 187; *BSA* 53-54, 160 ff. (lebes gamikos).

Cook, pp. 227, 230.

CVA France 2 (Attic b.f. lebetes and stands), 15 (Attic b.f. lebes gamikos), 19 (Attic b.f.); *Germany* 42 (Pl. 28 lebes gamikos), 43 (Pl. 18 and 36 lebes gamikos); *Great Britain* 8 (Pl. 103-4, r.f.); *Switzerland* 2 (Pl. 34 and 46 lebes gamikos); *U.S.A.* 6 (Pl. 50 ff. lebes gamikos), 19 (Pl. 64 ff. b.f.).

Daremberg and Saglio, s.v. *cratēr* and *lebēs*.

Fouilles de Delphes V, 60 ff.

Gericke, pp. 88 ff.

B. Graef & E. Langlotz, *Die antiken Vasen von der Akropolis zu Athen* 1 (Berlin: De Gruyter, 1925), Ill. 27a (Athens, Acropolis 590). Tripod with a bowl painted on a vase, together with the inscription *lebēs*. A prize at funeral games.

Homer *Il.* 23. 259-60. "From the ships he brought out the prizes – lebetes and tripods, horses and mules . . ."

IG II-III^2, 1544, 63 Lebetes gamikoi (Elis); 1471, 44 Lebes nymphikos (Attic); XI, 161, B, 125 "Round krater" with its stand – presumably a lebes (Delos).

Olympia IV, Pl. 27 ff. Bronze bowl and tripods.

Pausanias 5. 10. 4. The vessel on the roof of the Temple of Zeus at Olympia.

Richter and Milne, pp. 9 ff.

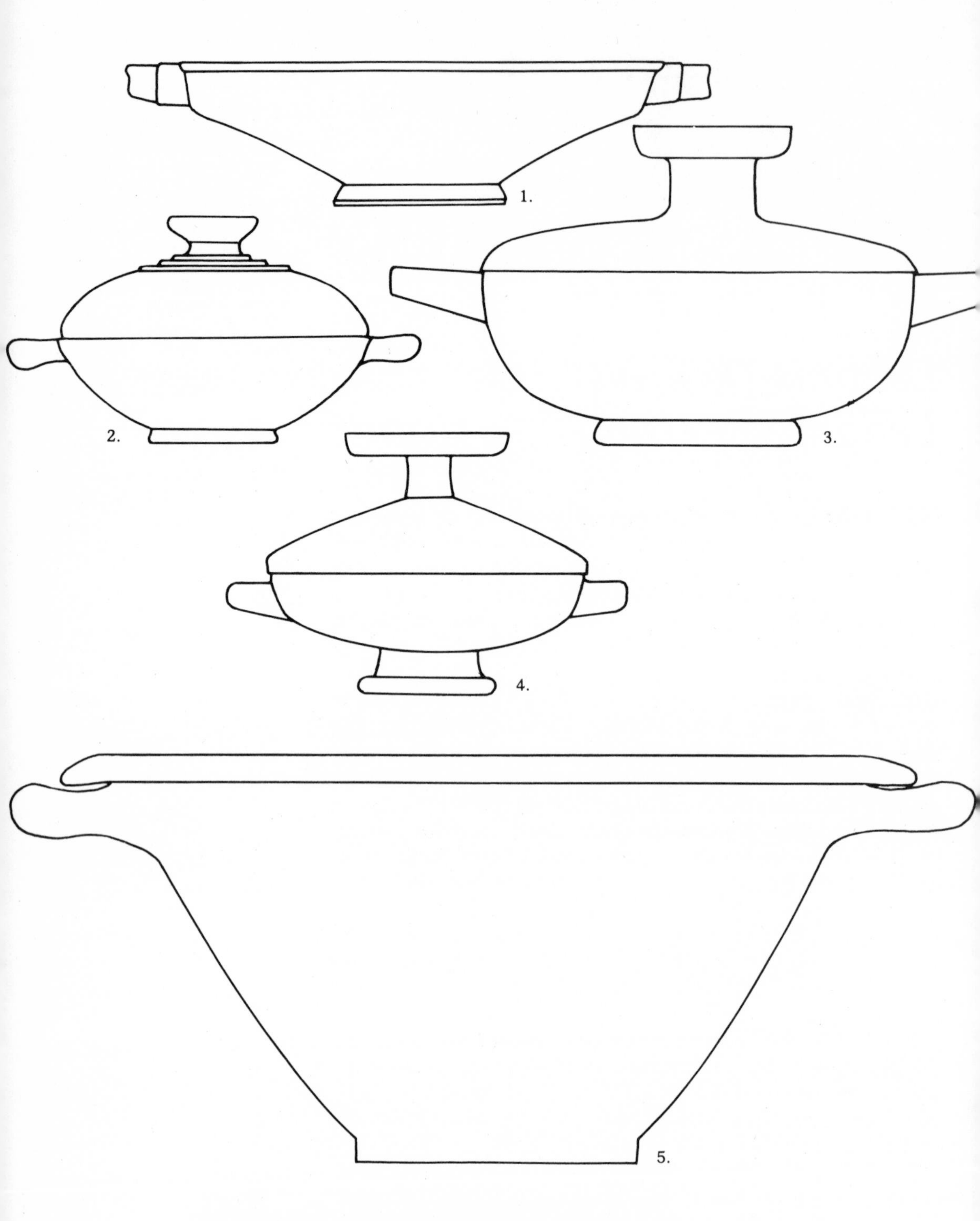

1. Lekane, Boeotian b.f., third quarter of sixth century B.C. 7 cm.
2. Lekane and lid, Corinthian, second quarter of fifth century B.C. 11 cm. (incl. cover)
3. Lekanis, Attic black-glaze, late fifth century B.C. 14 cm.
4. Lekanis and lid, Sicilian r.f., second half of fourth century B.C. 12 cm.
5. Lekane, Attic black-glaze, late fourth century B.C. 19 cm.

Lekanis [λεκανίς] *Lekane* [λεκάνη]

The references to these shapes in literature are not clear enough for us to identify the distinctions between them with absolute certainty. However, we can be sure that they were basins or bowls, even though we cannot be confident that modern usage corresponds with the ancient. Basins, ancient and modern, vary from fairly shallow to very much deeper types. The ancient types usually had two horizontal handles (sometimes one), and a simple spreading or ring foot. Modern writers, assigning the names to shapes according to their own interpretation of the ancient evidence, have not succeeded in establishing complete consistency, and usage has varied.

The word *lekanē* appears to have been used for general purpose dishes or basins. Sparkes and Talcott in *Agora* (p. 211) use the name for the common household basin. Though sometimes shallower, it is often a deep vessel, up to skyphos proportions, and usually lidless. It was solidly constructed, often glazed or partly glazed, and sometimes used for cooking. It was common in a number of types from the sixth to the fourth century B.C., and can be considered to be the coarse pottery counterpart of the lekanis which appears in finer wares. (The word has also been used e.g. *CVA Great Britain* 2, IV Ea, Pl. 11, for a deep, lidded vessel; elsewhere called a *pyxis, CVA Germany* 36, Pl. 50; and a *lebēs gamikos, CVA Italy* 34, IV e, Pl. 2.)

The lekanis is, on the whole, a shallow basin. The ancient evidence, such as it is, would tend to indicate a rather more special vessel than the lekane. It is mentioned as a bridal gift, and a flat, dish-like lidded vessel does appear in what are presumed to be illustrations of weddings. Most types of lekanis were probably regularly fitted with a lid.

The lekanis seems to have been used by women as a toilet article and better quality household container (similar to a pyxis). Like the pyxis it also appears in the grave (*Agora*, p. 164). Its period of popularity at Athens was the same as that of the lekane (sixth to fourth centuries B.C.).

Many lekanis handles, especially in the earlier period are flat and ribbon-like, and at the point of attachment, the ends curve out, away from the wall of the vessel, to form a short spur on either side of the handle-loop. It is widely believed that these ribbon-handles, which are not particularly common in pottery, follow prototypes in other materials: metal (Richter, p. 24), or leather or wicker-work (*Agora*, p. 165). But this tradition in handle-types is as old as the Geometric (compare *CVA Germany* 44, Pl. 21).

An interesting practical development is seen in the design of lekanis lids. It was often possible to use the lid as a dish in its own right, the flattened knob at the top serving as a foot when the lid was turned upside down. The knob was widened and made more stable to cater for this development.

Compare also *lakanē, lekanidion, lekanion, lekaniskē, lekaniskos, lekarion, lekos, lekis, lekiskos, lekiskion,* and Latin *lanx*.

References

Agora, pp. 38, 164 ff, 211 ff.

Amyx, pp. 202 ff. A full treatment of *lekos* and related words.

Aristophanes. *Acharnians* 1110 "Give me a lekanion of hare's meat"; *Clouds* 907 "Give me a lekane" (i.e. "You make me sick").

*ARV*2, chap. 87.

Cook, pp. 141, 237 ff.

Peter E. Corbett, "Attic Pottery of the Later Fifth Century from the Athenian Agora", *Hesp.* 18 (1949): 298 ff. esp. 304 and n. 26.

CVA Germany 29 (Pl. 76 Black-glaze lekanides), 41 (Pl. 33 Attic b.f. lekanis, first quarter of sixth century B.C.), 43 (Pl. 19 South Italian lekanis, c. 200 B.C., handles set at an angle), 47 (Pl. 1 Boeotian b.f., second quarter of sixth century B.C.); *Great Britain* 11 (III c, Pl. 16, 5 Corinthian lekanis, 600–575 B.C.); *Switzerland* 2 (Pls. 49 & 50 Apulian and Sicilian r.f.), 5 passim (Apulian r.f. and black-glaze).

Daremberg and Saglio, s.v. *lêkanê.*

L. Deubner, *JdI* 15 (1900): 152.

Hesychius. *Lekanides* – dishes in which gifts were brought to newly-weds.

Lucian *Erotes* 39. Lekanides of silver.

Photius *Lexicon* s.v. *lekanē*. "The vessel that we call a lekane the ancients called a podanipter (foot-bath). Lakanion and lekanis – vessels with handles to take cooked food etc." Also s.v. *keramion.*

Pollux 6. 10.

RE, s.v. *lekanē* and *lekanomanteia* (a rite claiming to interpret the patterns made by a film of oil on the surface of a basin of water).

Richter and Milne, pp. 23-24.

Annie D. Ure, "Boeotian Orientalizing Lekanai", *Metropolitan Museum Studies* 4 (1932): 18. A good discussion of the shape with a number of references and illustrations. The word *lekanē* is used here for the type that some writers would prefer to call *lekanis.*

1. Lekythos, Cypriot, Sub-Mycenaean. 17 cm.
2. Lekythos (Deianeira shape), Attic, 560–550 B.C. 16 cm.
3. Lekythos, Attic b.f., c. 500 B.C.
4. Lekythos, Attic b.f., second quarter of fifth century B.C. 11 cm.
5. Lekythos, Attic w.g., c. 430 B.C. 31 cm.
6. Lekythos (squat), Attic, c. 430 B.C. 16 cm.

Lekythos [λήκυθος]

The name *lēkythos* is applied nowadays to a number of related shapes, nearly all of them small apart from some of the white-ground funerary types. They were mostly used as containers for oil and perfumed oil, especially by women, and occasionally for table ware as well. A lekythos by the Diosphos Painter has the word *hirinon* (iris perfume ?) painted on it.

The shape has a long history, being represented as early as Protogeometric where it appears to have taken over some of the functions of the Mycenaean stirrup vase. In the full Geometric period it was usually a fairly globular vessel.

In the sixth century B.C., beginning at Corinth and soon followed by Athens, there appears a different variety with longish oval body, short narrow neck which has a ring moulding at its base, splaying mouth with rounded lip, low echinus foot and strap handle (Deianeira Type). This type soon loses the neck-ring, becomes angular at shoulder level, and, in some intermediate types, at the lower part of the body as well. Then the base becomes attenuated, leaving an angular shoulder.

The lekythos is common in black-figure and red-figure. A globular version was especially common in plain black, continuing to the end of the fourth century.

For much of the fifth century B.C., especially from the second quarter on, funerary white lekythoi were made at Athens, presumably for scented oil, and scenes of myth and everyday life gradually give way to those of funerary type. Many white-ground lekythoi are painted with scenes which include grave-stones and show lekythoi in their vicinity. Occasionally the more frugal or more penurious

Lekythos, Attic white-ground, by the Achilles Painter, 450–425 B.C.
36.3 cm.
Woman stands beside a stele. She holds a plemochoe in her right hand and an alabastron in her left.
Courtesy of Ashmolean Museum, 1896.41 (V.545).

made an offering of an almost solid version of the lekythos, of which only the neck was hollow, with space for only a few spoonfuls of oil. This solid type begins with the Beldam Painter.

The earliest white-ground lekythoi are painted in black-figure. Artists soon experimented with washes, dilute glaze and a variety of colours, including matt paint, producing pastel effects which are probably akin to the contemporary work in full-scale free painting on plaster walls. One of the interesting features of a number of white-ground vessels is that artists painted them after firing, and therefore had available a much greater variety of colours (including vegetable dyes) than other pottery painters. But this sort of surface is not practical for a vessel intended for everyday use.

A common shape in red-figure and black-painted wares is the squat lekythos. It was very popular around 400 B.C. but the shape had already been in use for more than a century by then. The usual size for the squat lekythos is ten centimetres or so. The thinner types are mostly late.

On many lekythoi there is a small ridge on the inside of the lip, tending to reduce the size of the mouth. This characteristic seems designed to control the release of small quantities of oil.

The name *lēkythos* (or one of its close variants) has been found written on several ancient vessels – a Protocorinthian pointed aryballos from Cumae (c. 675–650 B.C.), an Attic r.f. aryballos potted and painted by Duris (c. 490–480 B.C.), and an Apulian squat lekythos from Eboli (fourth century B.C.). The word *lechtumuza* (Etruscan diminutive of *lēkythos* ?) was found on an Etruscan bucchero aryballos of the seventh century B.C.

It seems clear that in ancient times the meaning of the word *lēkythos* was not as restricted as in modern usage. It was probably used for a variety of small oil containers including the shape that we call aryballos, as well as the modern lekythos.

The origin of the word is unknown but it is used for an oil-container as early as Homer where Nausicäa's mother is said to give her daughter oil in a lekythos made of gold.

References

AA (1927) p. 74, fig. 2; (1928) col. 571.
Agora, pp. 7, 12, 150 ff.
Annali dell' Inst. di corr. arch. (1831) Pl. D, 1-2. "The lachythos belongs to Dionysus, son of Matalos." (on an Apulian squat lekythos from Eboli, fourth century B.C.).

Aristophanes *Birds* 1589 "There's no oil in the lekythos"; *Ecclesiazusae* 995 "Old Woman: 'Who is he?' Young Man: 'The man who paints lekythoi for the dead' "; *Frogs* 1189 ff. "He's lost his lekythion [little lekythos]"; *Plutus* 810 "The lekythoi are full of perfume"; *Thesmophoriazusae* 139 ff. "What's this? Lekythos and brassiere! How incongruous! What's the meaning of this combination of looking-glass and sword?"

*ARV*2, pp. 447, 274 *Legythos* written before firing on a round aryballos made and painted by Duris (c. 490–480 B.C.). Chaps. 21, 37-40, 64, 72-74 etc.

John D. Beazley, *Attic White Lekythoi* (London: Oxford University Press, 1938). A study of painters and workshops 530–400 B.C. based on shape and pattern.

Boardman, *BF*, esp. pp. 114-15, 147 ff., 189-90.

Ernst Buschor, *Attische Lekythen der Parthenonenzeit. Münchener Jahrbuch.* New Series 2 (Munich: Knorr and Hirth, 1925).

Cook, pp. 230 ff.

CVA Belgium 1 (w.g.), 2 (Attic b.f. and r.f.); *Czechoslovakia* 1 Attic b.f., r.f., w.g.); *Denmark* 3, 4 (Attic r.f., w.g.); *France* 20 (Attic b.f. and r.f.); *Germany* 1 (w.g.), 11 (w.g.), 17 (Attic b.f.), 26 (esp. Attic b.f.), 30 (Attic b.f. and r.f.), 31 (Attic b.f.), 34 (Attic b.f., r.f., w.g.), 41 (Attic b.f.), 47 (Attic b.f.); *Great Britain* 3 (Attic r.f.); *Italy* 14 (Attic r.f.), 40 (Attic b.f. and r.f.), 48 (Attic b.f.), 50 (Attic b.f., r.f., w.g.), 54 (Attic b.f. and r.f.), 56 (Attic esp. b.f.); *Netherlands* 4 (Attic b.f.); *New Zealand* 1 (Attic b.f.); *Norway* 1 (Attic, mainly b.f., some r.f. and w.g.); *Poland* 4 (Attic b.f.); *Rumania* 1 (Attic b.f. and r.f.), 2 (Attic b.f.); *Switzerland* 2 (Attic b.f., South Italian r.f.), 3 (Attic b.f. Some w.g.); *U.S.A* 4 (Attic b.f. and w.g.), 5 (Attic b.f., r.f., w.g.), 10 (Attic various), 15 (w.g.).

Daremberg and Saglio, s.v. *lecythus.*

Lambertus J. Elferink, *Lekythos. Archäologische, sprachliche, und religionsgeschichtliche Untersuchungen* (Amsterdam: Noord-hollandsche Uitgeversmij, 1934). The word *lēkythos* coupled with *lekithos* (egg-yolk). A fanciful discussion of egg symbolism.

Arthur Fairbanks, *Athenian White Lekythoi*, Vols I & II (Ann Arbor: Univ. of Michigan, 1907, 1914).

C.H.E. Haspels, *Attic Black-figured Lekythoi* (Paris: De Boccard, 1936).

Hesychius. "Lekythos . . . a container for myrrh."

Homer *Od.* 6. 79.

JHS 101 (1981): 130 ff.; 102 (1982): 234.

Donna C. Kurtz, *Athenian White Lekythoi. Patterns and Painters* (Oxford: Clarendon, 1975).

______ and John Boardman, *Greek Burial Customs* (London: Thames and Hudson, 1971).

E. Langlotz, *Philologische Wochenschrift* 43 (1923): 1025.

Lazzarini, 360 ff.

I. McPhee, "The Agrinion Group", *BSA* 74 (1979): 159-162. Squat lekythoi from N.W. Greece, fourth century B.C.

Mon. Ant. 22, Pl. 51, 1, Col. 308. "I am the leqythos of Tataie; whoever steals me shall become blind." On a Protocorinthian pointed aryballos from Cumae (British Museum A 1054), c. 675–650 B.C., inscribed after firing.

Payne, pp. 191, 324.

RE, s.v. *lēkythos.*

Richter and Milne, p. 14.

W. Riezler, *Weissgrundige attische Lekythen* (Munich: Bruckmann, 1914).

K. Roth-Rubi "Eine frühe weissgrundige Lekythos", *HASB* 1 (1975): 11-20.

W. Rudolph, *Die Bauchlekythos* (Bloomington, Indiana: 1971).

Bernhard Schmaltz, *Untersuchungen zu den attischen Marmorlekythen* (Berlin: Gebr. Mann Verlag, 1970). Includes evidence relating to the functions of marble and pottery lekythoi.

K. Waldstein, "Der Aryballos im Aryballos", *AA* (1972): 472-73, n. 101. Vessels with false interiors.

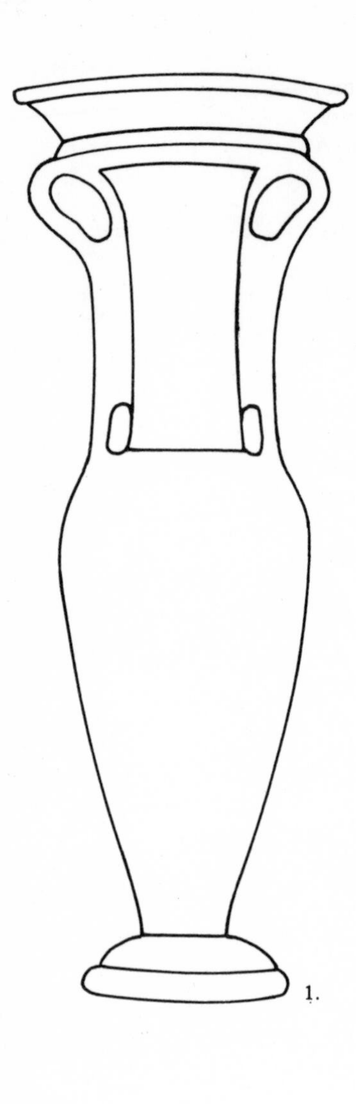

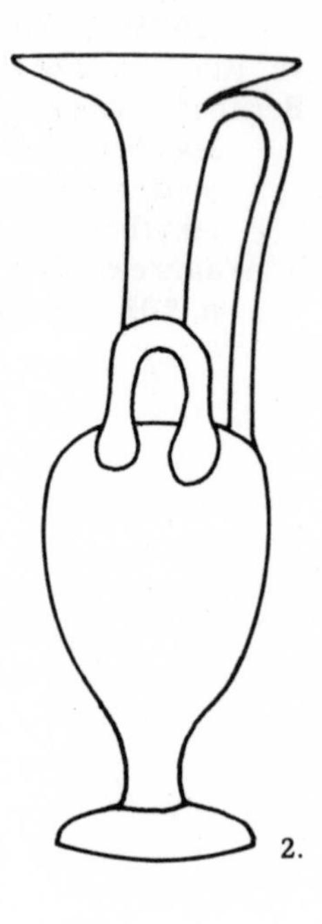

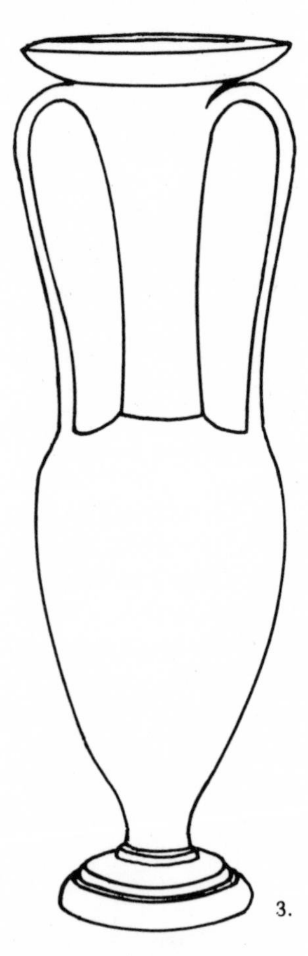

1. Loutrophoros, Attic b.f., c. 520 B.C. 81 cm.
2. Loutrophoros (hydria form), Attic r.f., c. 450 B.C. 28 cm.
3. Loutrophoros, Attic r.f., c. 425 B.C. 48 cm.

Loutrophoros [λουτροφόρος]

The name comes from *loutron*, "bathing water" and *phoros*, "carrying". At Athens it was the name given to the attendant who brought water from the spring Kallirhoe for the ritual washing before a wedding. It has often been claimed that the name was also applied by the ancients to the vessel used in such ceremonies – a very tall, thin neck-amphora. This is not proven. The ancient references to loutrophoroi, apart from adjectival usages, can all be taken to indicate people rather than vessels, and it has been plausibly argued that this should indeed be their interpretation (see Kurtz & Boardman). But in modern usage it is the vessel that is called a loutrophoros.

The name is also applied to a similarly shaped but rarely occurring hydria. The only difference between it and the amphora type is the addition of a third handle, which technically turns the amphora into a hydria.

Loutrophoroi, in the modern sense of the word, were ritual vessels used in the ceremonial of Greek weddings and also at the funerals of young people who died before marriage. Some loutrophoroi used as grave-markers, like other vessels in similar situations (e.g. amphorae) had perforated bases or no bases at all. The reason for this is not known. Suggestions range from a ritual function (to receive drink-offerings) to a simple practical solution for the problem of rain water, which, in a water-tight container, would become stagnant and in freezing conditions might crack the vessel.

Scenes painted on a loutrophoros often fit the vessel's function – illustrations of weddings, funerals and battle scenes, the latter being appropriate for rites in honour of young men killed in battle. Kurtz and Boardman (p. 152) give some evidence to suggest that the loutrophoros might have had a more general function than the widely assumed ritual one. The main argument against this is the impractical shape. Slim and elegant as it was, it was not very suitable for everyday use. The breakage rate can be judged by the fact that a very high proportion of loutrophoroi are restored or exist only in fragments. Whether or not there was a wider function than the ritual one, the funerary (especially prothesis) scenes support the view that the loutrophoros is the ceremonial descendant of the late Geometric amphora painted with similar themes. One of the amphora shapes in late Geometric tends towards loutrophoros proportions. There are some black-figure loutrophoroi but the heyday of the shape was the two centuries or so from the beginning of red-figure. The loutrophoros was never made in large numbers. One or two artists such as Hermonax seemed to specialize in its manufacture, but more typical is the output of artists like the Kleophrades Painter (*ARV* lists one loutrophoros amongst more than twenty amphorae), and the Berlin Painter (*ARV* lists one loutrophoros amongst over a hundred amphorae).

Marble loutrophoroi are fairly common in a funerary context, either free-standing or carved in relief. A lavish fourth century tomb at the Kerameikos (Athens) was crowned with a marble lekythos two metres high.

References

Boardman, *BF*, p. 192; *RF* p. 210.

Cook, pp. 221-22.

CVA Austria 3 (Pl. 148 ff. Attic r.f.); *Denmark* 8 (Pl. 340 ff. Attic r.f.); *Greece* 2 (Pl. 21 ff. Loutrophoroi, amphora and hydria shape); *U.S.A.* 6 (Pl. 49 Attic r.f.), 15 (Pl. 15 f. Attic b.f.).

Daremberg and Saglio, s.v. *loutrophoros.*

Pseudo-Demosthenes 44. 18. Loutrophoros on a tomb a proof that a person died unmarried.

Harpocration, s.v. *loutrophoros.* (Boy bath attendant at a wedding. The loutrophoros on the tomb of those who died unwed was the statue of a boy.)

Hesychius. Loutrophoros vases: hydriai for those who died unmarried; used also for weddings. (Loutrophoros is here used adjectively.)

Donna C. Kurtz & John Boardman, *Greek Burial Customs* (London: Thames and Hudson, 1971), passim. Large tomb monument with marble loutrophoros (p. 111); the evidence for ancient use of the name loutrophoros (pp. 151 ff.).

Pollux 8. 66. Statue of a girl carrying a hydria placed on the tomb of those who died unwed.
RE, s.v. *lutrophoros.*
Richter and Milne, xxii (older bibliography), pp. 5-6.

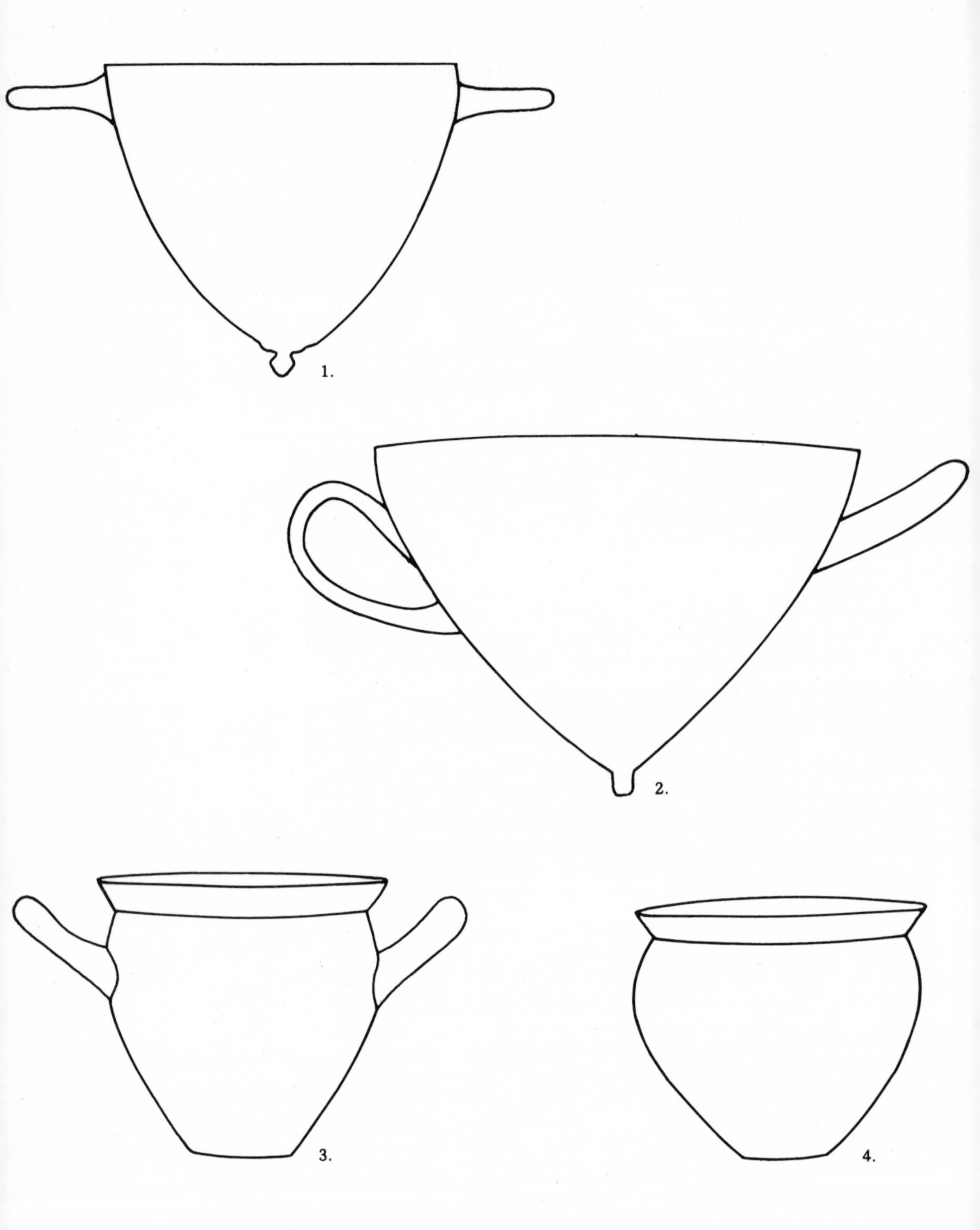

1. Mastos, Corinthian, c. 550 B.C. 9 cm.
2. Mastos, Attic b.f., c. 525 B.C. 10 cm.
3. Mastoid skyphos, Attic b.f., first quarter of fifth century B.C. 9 cm.
4. Mastoid skyphos, Attic b.f., first quarter of fifth century B.C. 8 cm.

Mastos [μαστός]

MASTOID CUPS

The mastos was a breast-shaped vessel, (*mastos*, "breast") mentioned occasionally by such writers as Athenaeus, and found in inscriptions. There is little doubt that the breast-shaped cups which survive are the vessels called *mastos* by ancient writers. Modern archaeologists have kept the same name.

The mastos is really a variation on the skyphos shape, usually having two handles, one vertical and the other horizontal or nearly so. The shape does not appear to have been very common. It had a brief period of popularity in the black-figure period. Nearly all the surviving Attic black-figure mastoi come from the third quarter of the sixth century B.C. Mastoid vases continued longer. Of the few older Corinthian mastoi that have been found one was dated by Payne to Middle Corinthian (before 585 B.C.).

The true mastos has no base and therefore cannot stand upright. But there are quite a few mastoid cups, similar to the mastos except that they have a flat base.

References

ABV, p. 257 Mastos Group; p. 262 Mastos Painter; pp. 557-560 mastoids.

*ARV*2, p. 773. Mastoi of the potter Sotades.

Athenaeus 487b. "Mastos. According to Pamphilus, Apollodorus of Cyrene says that the Paphians call the drinking-cup (poterion) by this name."

John D. Beazley, *Greek Vases in Poland* (Oxford: Clarendon Press, 1928), p. 4.
Frank Brommer Festschrift (1977) Pls. 37, 38.
CVA Belgium 2 (III H e, Pl. 15 Attic b.f.); *France* 10 (Pl. 67-68 Attic b.f. Mastos and mastoid cups); *Germany* 30 (Pl. 51, b.f. Mastoid cups), 35 (Pl. 10 Corinthian mastos); *Great Britain* 15 (Pl. 23 Mastoid cups); *Italy* 57 (Pl. 35, 36 Mastoid cups).
Edith H. Dohan, "Unpublished Vases in the University Museum, Philadelphia", *AJA* 38 (1934): 530 ff. Something of the history of the shape and references to other mastoi.
IG 7, 1307; 1408, 2; 3498.
Payne, p. 312.
Richter and Milne, p. 30.

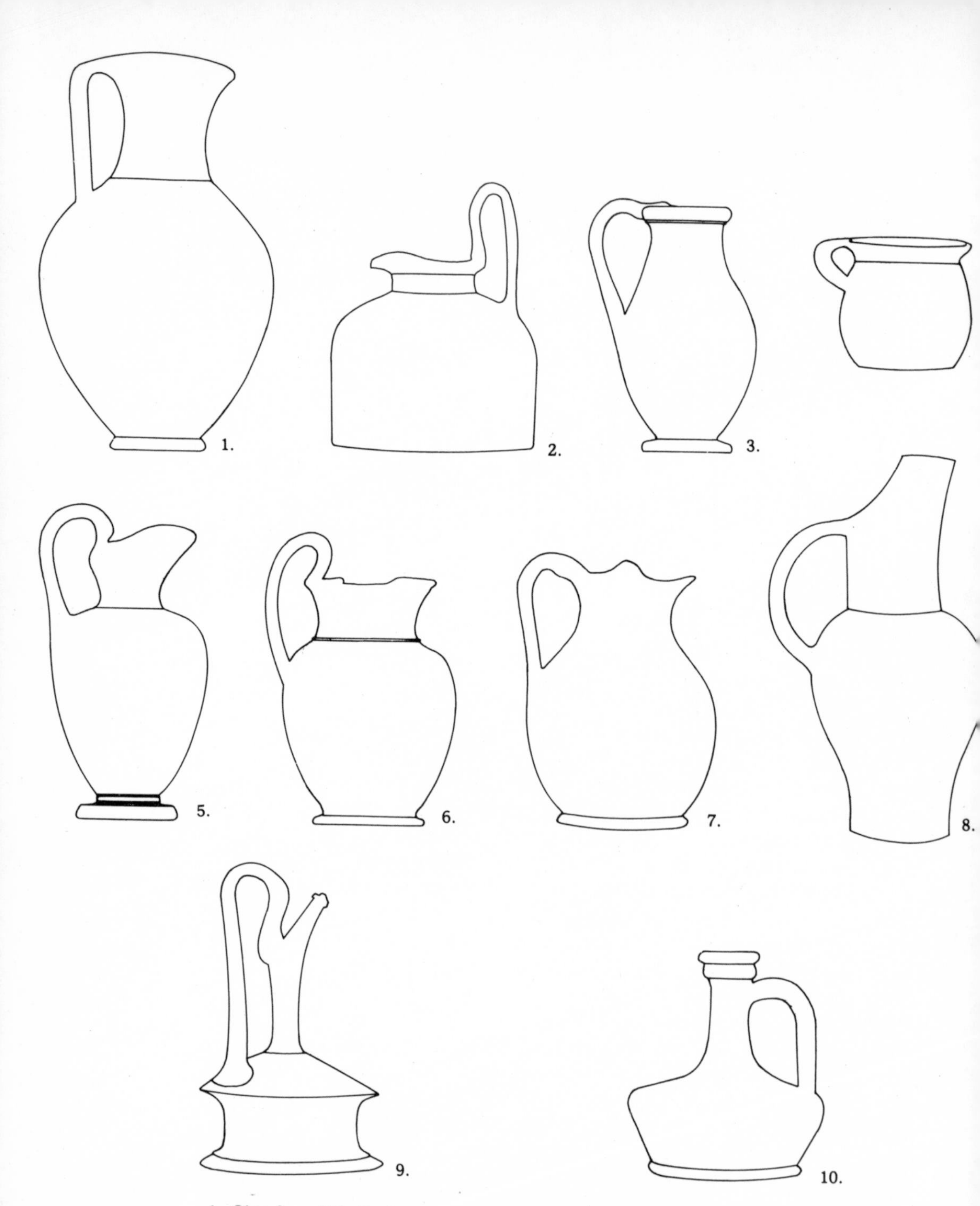

1. Oinochoe, Attic Protogeometric, late tenth century B.C. 30 cm.
2. Oinochoe (squat), Corinthian, c. 600 B.C. 22 cm. (incl. handle)
3. Olpe, Attic b.f., c. 500 B.C. 18 cm. (incl. handle)
4. Mug, Attic r.f., c. 500 B.C. 9 cm.
5. Oinochoe, Attic b.f., beginning of fifth century B.C. 24 cm (incl. handle)
6. Oinochoe, Attic b.f., beginning of fifth century B.C. 21 cm. (incl. handle)
7. Chous, Attic r.f., second quarter of fourth century B.C. 20 cm.
8. Oinochoe, Etruscan r.f., c. 400 B.C. 28 cm.
9. Epichysis, Apulian, second half of fourth century B.C. 21 cm.
10. Lagynos, Cypriot, second century B.C. 15 cm.

Oinochoe [οἰνοχόη]

Also CHOUS (Pl. CHOES), OLPE, EPICHYSIS, LAGYNOS
Jug, Pitcher

Oinochoe is a compound word formed from *oinos*, "wine" and *cheō*, "I pour". The word is applied nowadays to most ancient Greek jug shapes, including the small type that goes by the name of *chous,* but usually excluding the types called *olpē* and *lagynos*.

Jugs of one type or another are common in Greek pottery at all periods. Beazley distinguishes ten different basic shapes, and this system has been regularly adopted, at least for red-figure (see also Boardman). Richter and Milne distinguished five main types with additional variants, and Gericke also follows this system.

As with amphorae there is a basic division between oinochoai with off-set neck and those of one-piece type (for example, Beazley's Shape 1 has neck off-set; Shape 3 is a plump one-piece variety). Fashions in handle-types changed with the shapes, and also independently of them. Beazley's Shape 2 is similar to Shape 1 but has a lower handle. Other distinctive features are type of mouth (Shape 3 has a trefoil mouth, Shape 5a round, Shape 6 beaked), and proportion of height to width (Shape 5 is a narrow one, Shapes 3 and 8 usually short and plump).

Some oinochoai with hardly any spout look like mugs (such as Beazley Shape 8) and are sometimes called mugs in the modern literature. A plump, globular jug with low foot and trefoil mouth is commonly called a chous (Beazley Shape 3).

Although Beazley's classification provides a basic framework it is not completely comprehensive for all types of oinochoe. Sparkes and Talcott, in describing plain pottery, were obliged to add more types. Others prefer to use Beazley's classification in part, and to supplement it with information of a more geographical or chronological type (e.g. Cook, p. 225 – "the canonical oinochoe of the Middle Wild Goat style").

Archaeologists use the name *olpē* for a jug common in Corinth (and areas under Corinthian influence) and Attica from the seventh century B.C. This is a jug of practical shape, with a low centre of gravity, whose sides are formed in a continuous curve. The Corinthian shape is mostly taller, the Attic shorter and squatter. In the second half of the sixth century B.C. it developed into the chous.

Ancient writers use the name *epichysis* for some sort of pourer. Modern usuage restricts this name to a slender oinochoe made mainly in Apulia in the late red-figure period.

The word *lagynos* is used for a special late shape with flattish body, tall, chimney-like neck, small lip, and wide, flat handle, extending from near the lip to the top of the body.

Evidence from ancient writers and illustrations on vessels showing oinochoai in use agree that one of the main functions of the shape was the decanting of wine, and its ladling from a larger container to a cup, as in Euripides *Troades.* This is the function that gives the shape its name, and it was no doubt a very important use in social life (symposia etc.) and ritual. But of course oinochoai were used for pouring many other liquids besides wine.

It is likely that the ancients applied the name *oinochoē* to similar sorts of shapes to the ones included in current usage. It is not likely, however, that the ancient and modern boundaries were exactly the same. The exclusion of *olpē* from the oinochoe shape is very arbitrary. To complicate matters, the current definition of the word *olpē*, though well established in modern literature, has little to justify it. All that we really know about the word *olpē* is that it was applied at Corinth to the aryballos shape. (One or two vessels of that shape have the name *olpē* inscribed on them.)

The ritual use of oinochoai is well established, especially for the plump type of chous which was used at the Festival of Choes (part of the Anthesteria) at Athens. At that festival larger choes (about three and a half litres capacity) were used in the ceremony, and smaller ones (often appropriately decorated with children's scenes) were given to the young.

Oinochoai are also depicted standing with lekythoi at the grave, and libation scenes show that the libation was carried in an oinochoe and transferred from the oinochoe to a phiale before being poured on to the ground.

References

Agora, p. 7, olpe, pp. 58 ff., oinochoe, mug, olpe.
*ARV*2, xlix-l, chaps. 67, 72 etc.
Boardman, *BF* 187; *RF* 208 f.
Boston I & II, p. 40.
Cook, pp. 224 ff.
CVA Belgium 2 (III I d, Pl. 5 Oinochoai with vertical spouts); *Denmark* 2 (Corinthian, Italo-Corinthian and East Greek); *France* 1 (II d c, Pl. 4 ff. Rhodian oinochoai), 7 (Corinthian and imitation, Ionian); *Germany* 6 (Attic r.f.), 17 (Attic b.f. trefoil mouthed oinochoai and olpai), 22 (Attic r.f.), 26 (Various, esp. late), 34 (Attic geometric), 44 (Attic geometric), 47 (Pl. 18 ff. Attic b.f. olpai, oinochoai); *Great Britain* 11 (III C N Pl. 30, 23 & 25 Lagynoi second century B.C.), 15 (Pl. 56 Apulian epichysis, second half of fourth century B.C.); *Italy* 1 (Chigi Vase), 36 (esp. Attic olpai), 48 (Pl. 6-41 Attic b.f. oinochoai and olpai), 55 (Pl. 1-27 Protocorinthian and Italo-geometric); *New Zealand* 1 (Attic geometric and b.f., Corinthian oinochoai, pitchers); *U.S.A.* 2 (Pl. 6 Corinthian olpe; Pl. 8 Geometric olpe), 5 (Pl. 2 Geometric jug, Pl. 30 Etruscan b.f. olpe).
Daremberg and Saglio, s.v. *oinochoē*.
L. Deubner, *Attische Feste* (Berlin: 1932), pp. 96 ff.
Etymologicum Gudianum. Oinochoē "ladle".
Euripides *Troades* 820 ff. Ganymede fills Zeus' kylikes, using golden oinochoai.
John R. Green, *Bulletin of the Institute of Classical Studies* 19: 1 ff.
Herbert W. Parke, *Festivals of the Athenians* (London: Thames & Hudson, 1977), pp. 107 ff.
K. Peters, *JdI* 86: 103 ff.
E. Pfuhl, *Malerei und Zeichnung der Griechen* Vol. 2 (Munich: Bruckmann, 1923), p. 518. Toy oinochoai (bibliography).
Phrynichus *Praeparatio Sophistica*. Oinochoe, a vessel like a pitcher for pouring wine into cups.
Richter and Milne, pp. 18 ff.
G. Van Hoorn, *Choes and Anthesteria* (Leiden: Brill, 1951).

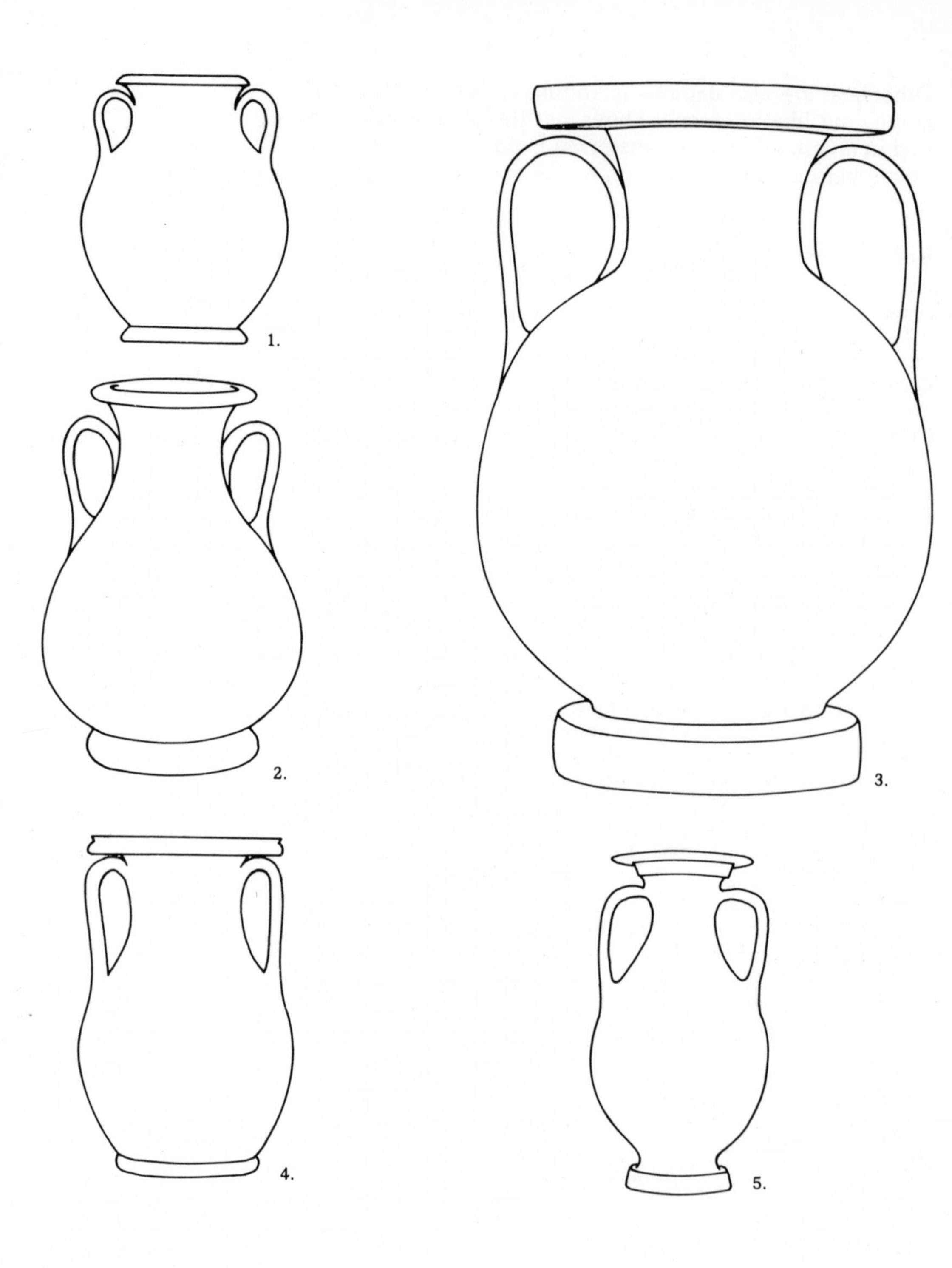

1. Pelike, Attic black-painted, 500–480 B.C. 23 cm.
2. Pelike, Attic r.f., second quarter of fifth century B.C. 32 cm.
3. Pelike, Apulian r.f., first quarter of fourth century B.C. 58 cm.
4. Pelike, Attic r.f., first half of fourth century B.C. 25 cm.
5. Pelike, Apulian (Gnathian), c. 340 B.C. 25 cm.

Pelike [πελίκη]

The word *pelikē* in the modern archaeological literature refers to a shape like an amphora, usually one-piece (Type C), with sagging belly. Instead of the widest part of the pot being at shoulder level or at most, half way down as is normal in the amphora, the pelike is widest in its lower half. In other respects the amphora and pelike shapes are so similar that some prefer to include the pelike in the amphora category.

The mouth of the pelike, proportionally to the size of the pot, is often wider in the later examples. The simple foot is usually echinoid or disk-like. With few exceptions pelikai were not meant to take a lid.

The shape first appears in Athens in the last quarter of the sixth century B.C. (in the early red-figure period), and for a generation or so pelikai appear in both black-figure and red-figure, whereupon the former disappear. Red-figure pelikai are very popular in the late Classical period, especially the small variety which is particularly common in South Italian pottery.

Changes in shape are gradual during the period of its use. The changes in decoration are rather more noteworthy.

Scenes painted on pelikai show some of the uses to which this shape was put. From this evidence we can infer that it was popular as an oil-container. But other illustrations show that it was used for holding wine, and for drawing water. For this latter purpose a rope was passed through the handles.

The lexicographers give several different meanings for the ancient *pelikē* – *chous* or other similar vessel, *kylix* (or at least a cup of some sort), *lekanē*, and a vessel resembling the Panathenaic amphora. Whatever the shape or shapes

Pelike, Apulian red-figure. The Suckling Group, 375–350 B.C. 36.5 cm. Bride bathing and preparing to dress. She holds an alabastron with perfume over a basin. The woman on the right is seated on the hydria which has been used to fetch the bath water. Courtesy of Ashmolean Museum, G269 (V.550).

of the ancient pelikai they were apparently different from the amphora type to which the name is applied nowadays. The modern usage goes back to Eduard Gerhard (mid nineteenth century).

The derivation of the word is uncertain. That given by Photius and Pollux (from the Greek word *pelekaō*, "I shape with an axe") can hardly be taken seriously.

Compare also *pelichnē, pelix, pella, pellis, pelika, pelyx.*

References

Agora, p. 49.

ARV2, chaps. 83, 84 etc.

Athenaeus 495a. "Callistratus in his commentary on the *Thracian Women* of Cratinus gives cup as the meaning of pelike. Crates in the second book of his *Attic Dialect* writes as follows: 'As we said, choes used to be called pelikai. The shape of the vessel when it was called pelike was formerly like that of Panathenaic jars. Afterwards it had the shape of an oinochoe, the sort that are set before guests at a festival, in fact the sort they used to call olpai and used for pouring out wine . . .' "

Regina-Maria Becker, "Formen Attischer Peliken von der Pioneergruppe bis zu Beginn der Frühklassik" (Böblingen: Codex Verlag, 1977). Doctoral dissertation containing text, catalogue and many line-drawings.

Dietrich von Bothmer, "Attic Black-figured Pelikai", *JHS* 71 (1951): 40-47; review of R. Becker, *Formen Attischer Peliken AJA* 83 (1979): 361-2.

CVA Austria 2 (Pl. 68-84 Attic); *France* 9 (Pl. 42-50 Attic), 12 (III I d Pl. 40-49 Attic); *Germany* 4 (Pl. 9 Attic b.f. neck-pelike; Pl. 39 Apulian r.f.), 6 (Pl. 70-83 Attic), 18 (Attic and South Italian); *Italy* 17 (Attic); *Switzerland* 2 (Apulian), 5 (Apulian); *U.S.A.* 18 (Attic and South Italian), 19 (Pl. 59 Attic b.f.).

Gericke, pp. 68 ff.

Hesychius. "Pelika. A type of wooden cup. (Named) from being made with an axe. Others (use the word for) a wooden lekane."

Photius. "Pelika. The Boeotians call the wooden lekane a pelika, because of its being made with an axe. Apollodorus (gives the name to) a kind of cup."

Pollux 10. 67. "One should add the pelikai in the *Thracian Women* of Cratinus, which were apparently cups or small jugs . . ."

______ 10. 78. ". . . The tragedians call the lekane a pella, the Aeolians a pelika, especially the wooden one, from the word 'to be made with a pelekys (axe)' ".

RE s.v. *amphora; Suppl. 7* s.v. *pelike*.

Richter and Milne, pp. 4-5.

Schiering, p. 19.

A.D. Trendall, *Vasi antichi dipinti del Vaticano* (Vatican City, 1953 and 1955).

______ and Alexander Cambitoglou, *The Red-figured Vases of Apulia* (Oxford: Clarendon, 1978).

Phiale [φιάλη]

The word has been connected by ancient and modern writers (not very convincingly in either case) with *pinō*, "I drink".

The phiale is a plate-like vessel, whose depth is rather greater than the average plate or saucer, but not so deep as to be described as a bowl. It has no foot or handle. The trend was for the phiale to become shallower, the average ratio of diameter to height changing from 5:1 in the early sixth century to 7:1 in the classical period (Luschey, p. 31).

A special variety of phiale, which has a raised knob in the centre of the inside is called a *phialē mesomphalos*. The modern name was probably also the ancient name for these vessels.

The phiale, especially the mesomphalos variety, appears from illustrations to be the favoured shape for pouring libations. In vase painting it is often shown in departure scenes, human and divine, as in the sending of Triptolemos, and is pictured in the hands of Nike and other deities. It is also seen illustrated on vases as a drinking vessel of both men and immortals, and perhaps occasionally as an unguent container. Hellenistic inscriptions from Delos point to its being used as a collection plate for money.

Phialai were made at Corinth from about 600 B.C. on, and at Athens soon afterwards. The shape was at no time very popular in pottery, but judging from the evidence of inscriptions and tell-tale indications like flutings when a phiale is drawn on other vases, it was commonly made of metal. The concentric horizontal ribs that are a feature especially of some archaic phialai are also copied from metal work.

The word phiale was also used for a shield.
Compare also *phialidion, phialion, phialis, phialiskē, phielē.*

References

Agora, pp. 12, 105.
Aristotle *Poetics* 21. 12; *Rhetoric* 3. 4. 4.
Athenaeus 11. 500f-502a. This reads in part: "(Homer) does not mean the cup called phiale but a flat, lebes-like vessel made of bronze, probably with two handles one on each side. . . . Asclepiades of Myrlea says the word phiale derives from piale by substitution of a letter – the vessel that gives plenty to drink (*piein, halis*), for it was bigger than the cup."
ARV[2], p. 772 . Fluted phiale by the potter Sotades.
Cook, p. 237.
CVA Germany 22 (Pl. 135, 1-3 Attic r.f. c. 460 B.C.); *Great Britain* (9 III C Pl. 2 & 3 Corinthian and Italo-Corinthian); *Italy* 37 ("Patera" and "Patera ombelicata"); *U.S.A.* 2 (III C Pl. 5, 11 Corinthian), 13 (Pl. 23 Illustration of Nike carrying phiale and oinochoe to altar), 15 (Pl. 48, 2 Calenian mesomphalic phiale, 3rd century B.C.).
Daremberg and Saglio, vol. 4.
Gericke, pp. 27 ff.
Herodotus 2. 151. "Psammetichus who stood last did not have a phiale. So he took off his helmet, held it out, and poured the libation with it."
Homer *Iliad* 23. 243 & 270.
H. Luschey, *Die Phiale* (Bleicherode, Nieft: Diss. München, 1939).
Monuments et mémoires publies par l'Academie des inscriptions et belles-lettres. Commission de la Foundation Piot V (Paris, The Academy, 1894–) p. 42. *Phi* (for phiala?) inscribed on a Boscoreale bowl.
J.L. Myres, *Handbook of the Cesnola Coll. of Antiquities from Cyprus* (New York: Metr. Mus. of Art, 1914) n. 4552.
Pindar *Nemean* 9. 120 f. "Let someone mix and dispense in silver phialai the sweet inspirer of revelry, the potent child of the vine."
Pollux 6. 95. "Let them carry the phialai on finger-tips, and carefully bring them to the banquet guests."
RE Suppl. 7.
Gisela M.A. Richter, *AJA* 45 (1941): 363 ff.; 54 (1950): 357 ff.; 63 (1959): 241 ff. Metal prototypes.
Richter and Milne, p. 29.
Simon, pp. 13, 17 etc.
D.E. Strong, *Greek and Roman Gold and Silver Plate* (London: Methuen, 1966): 55 ff., 75 ff., 80 ff. Metal phialai, including ribbed "Achaemenid" type.
F. Wolters, *AM* 38 (1913): 195-96. Cypriot silver bowl with the inscription "I am the phiale of Epiorus".
Xenophanes *Eleg.* 1. 3. Phiale for unguent.

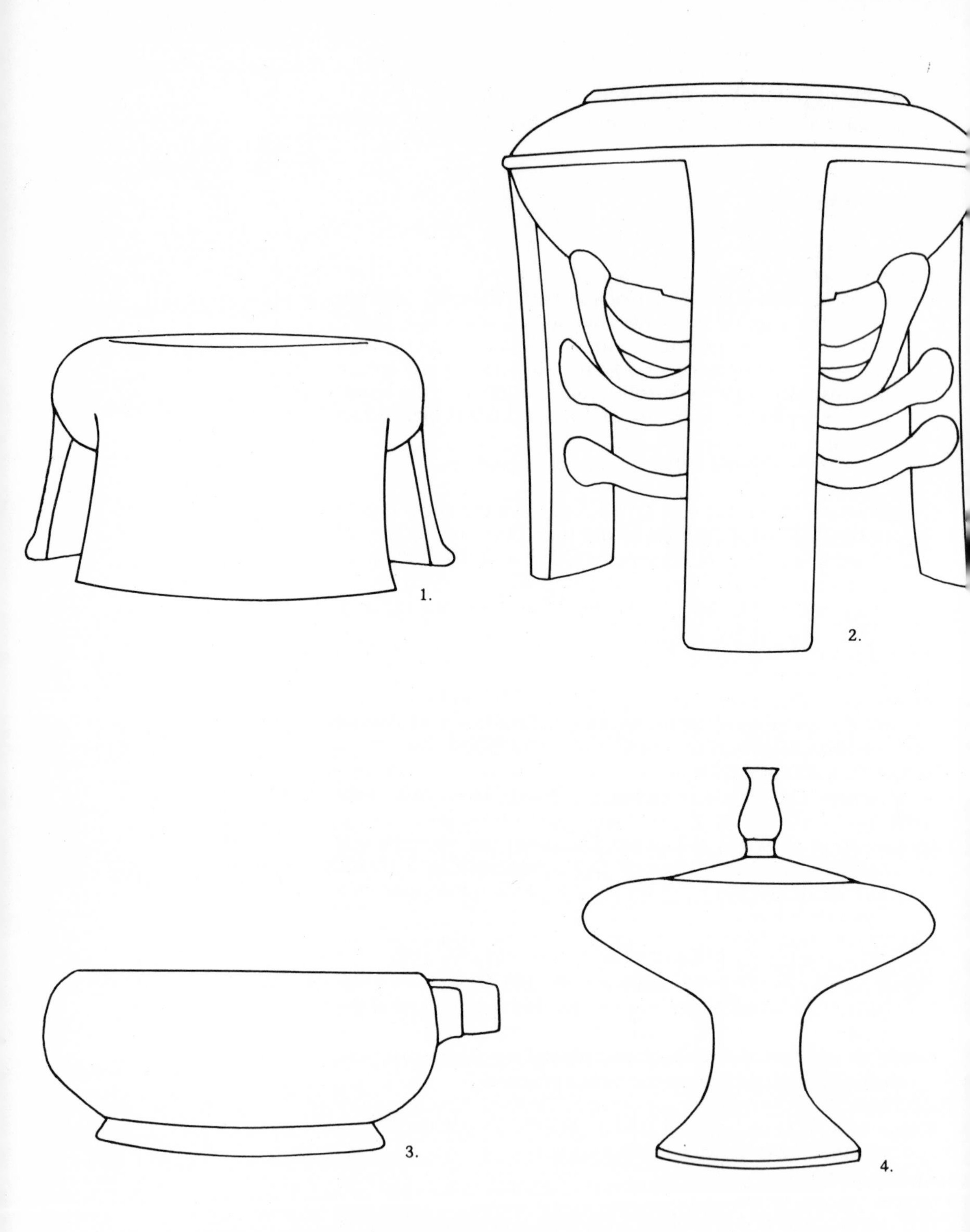

1. Tripod Exaleiptron or Kothon, Boeotian, c. 570 B.C. 12 cm.
2. Tripod Exaleiptron or Kothon, Boeotian, 570–560 B.C. 22 cm.
3. Kothon (with ribbon handle), Corinthian, second half of sixth century B.C. 7 cm.
4. Kothon or Plemochoe (Type B), Attic, c. 500 B.C. 15 cm.

Plemochoe [πλημοχόη]

See also EXALEIPTRON

The name is presumably from *pimplēmi*, "I fill" and *choē*, "jug, pourer". In the modern literature the word is used for the shape described under Exaleiptron 3. Richter and Milne take the name to refer to this shape's efficiency as a pourer when full or nearly so, as the incurving lip would make pouring difficult when the level dropped. But this is not a convincing explanation. The shape does not make a very satisfactory pourer, no matter how much liquid is in it. If the derivation is from *plēmē*, "flood-tide" perhaps the name refers to a ritual that has something to do with the tides or moon.

However, it is by no means certain that this vessel was the ancient plemochoe, as the descriptions that survive are open to various interpretations. Athenaeus' adjective "toplike" fits the shape, but the name *kotyliskos* (little cup), which was said to be an alternative, is not very suitable.

To find a vessel whose shape combines features that could be described both as toplike and like a little cup has been the aim of diligent searching, which has turned up a number of ingenious but not altogether convincing suggestions. The fact that no front runner has emerged from amongst the candidates might cause us to question the trustworthiness of the ancient descriptions. It seems likely that the sources, which are late, are not reliable, unless, as is possible but not very probable, it was a very rare shape which has not survived.

Scheibler mentions several possible shapes – most of which depend on the accuracy of the ancient statement that the plemochoe and kotyliskos were one and the same. Except perhaps for the kernos shape the vessels suggested are fairly nondescript – not the sort one would expect to find used in an important ritual.

The overturning of the two vessels at the Plemochoae (the last day of the Eleusinia), described by Athenaeus, has been described as a ceremony designed to ensure rain – an interesting guess, but there is little evidence to back it up.

References

P. Amandry, "Collection Helene Stathatos", *Les Bijoux antiques* No. 230, Pl. 34. A scene which it has been claimed (see Möbius and Scheibler) shows a representation of the Plemochoae. The two vessels are shaped like small calyx kraters.

Athenaeus 11. 496 a-b. "A plemochoe is a pottery vessel, shaped like a top, on a steady foot. According to Pamphilus (first century A.D.) some call it a kotyliskos. They use it at Eleusis on the last day of the mysteries, which they call after it Plemochoae. On that day they fill two plemochoai and after setting them up, one towards the East and one towards the West, they overturn them, reciting a mystic formula as they do so. The author of the *Pirithoos* . . . mentions them in the following passage: '. . . so that we may pour these plemochoai with words of good omen into the chasm of the earth.' "

Daremberg and Saglio s.v. *kernos, plemochoe.*

Delt. 15 (1933/4) Paratema 33, Ill. 34 (left); 16 (1960): 2, Ill. 43. These references are to cups found at Eleusis, cited by Scheibler as possibilities for the plemochoe shape.

Gericke, p. 82.

H. Möbius, "Amandry, Collection Helene Stathatos", *Gnomon* 27 (1955): 39.

H.W. Parke, *Festivals of the Athenians* (London: Thames and Hudson, 1977), pp. 71-72. Plemochoai possibly used in a rain rite.

Pollux 10. 74 "It is a pottery vessel, with a base which is not pointed but firm and stable. They use it on the last day of the mysteries which they call after it Plemochoe."

RE s.v. *plemochoe.*

Richter and Milne, pp. 21-22.

Ingeborg Scheibler, "Exaleiptra", *JdI* (1964): 72-108; "Kothon-Exaleiptra: Addenda," *AA* (1968): 389-397.

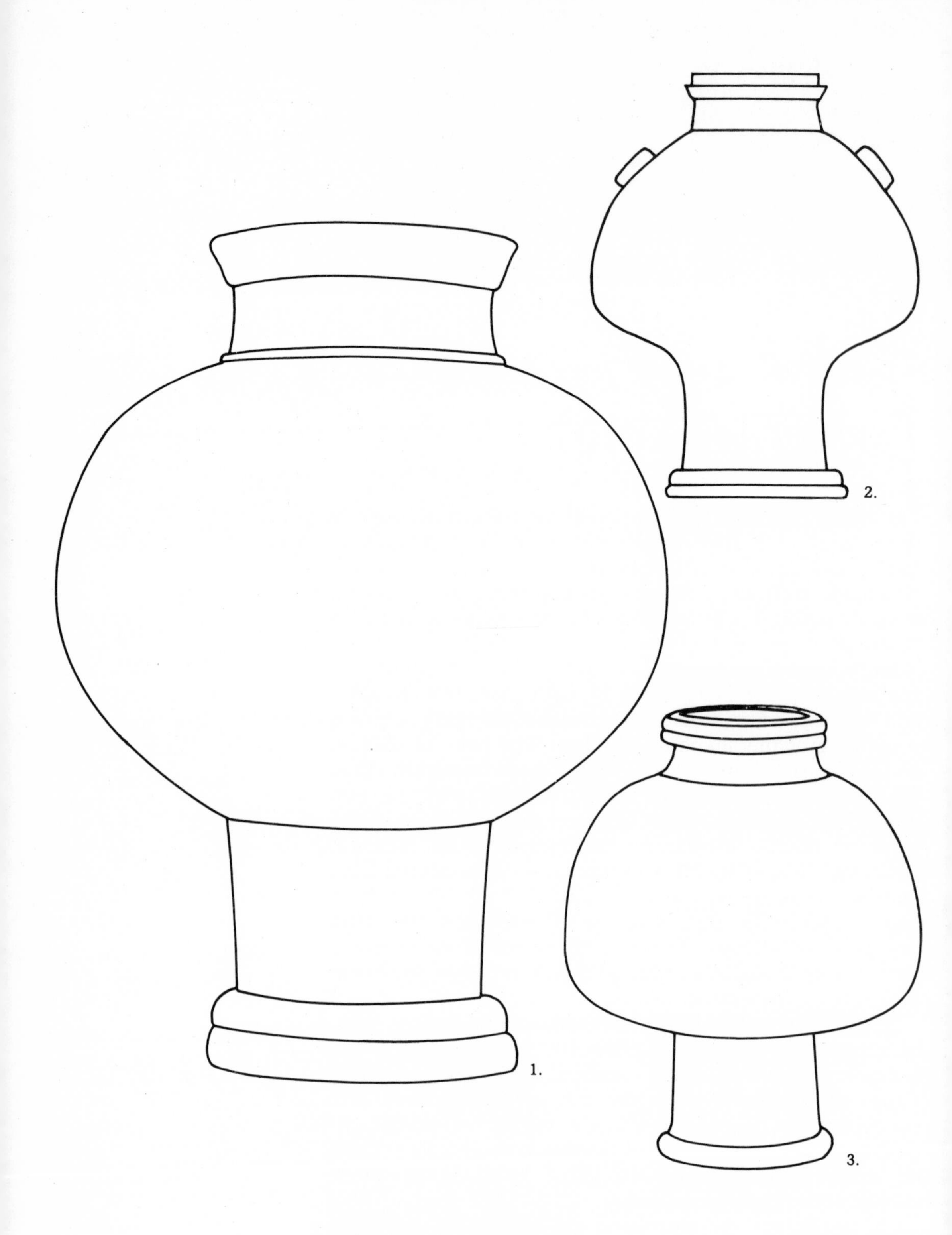

1. Psykter, Attic b.f., c. 520 B.C. 39 cm.
2. Psykter, Attic black-glaze, c. 500 B.C. 26 cm.
3. Psykter, Attic r.f., c. 480 B.C. 28 cm.

Psykter [ψυκτήϱ]

The name *psyktēr* is conventionally applied nowadays to the mushroom-shaped vessel (wide body on a tall, usually flaring foot), used in antiquity as a wine cooler. The shape was made at Athens from the last quarter of the sixth century, or perhaps a little earlier, to the middle of the fifth century B.C.

The bodies of earlier psykters usually have drop-like profile and no shoulder to speak of; later there is a shoulder and the belly tends to sag more, so that the body acquires something of the shape of an apple. The neck is shallow and there are one or two torus (ring) mouldings at the base of the vessel and at the mouth. There may be two diminutive handles, rather like eyelets, sometimes in the form of double tubuli, for stringing a cord through. Psykters without handles usually have a one-torus base and two torus lip; psykters with handles, a two-torus base and one-torus lip. But there are exceptions to the main line of development. The normal height is about thirty centimetres though there are some rather larger and some miniatures. They often had lids.

Stella Drougou lists only about a hundred psykters, some of them fragmentary, all made of pottery except one example in bronze. A little over half are painted in black-figure or red-figure, while the rest are black-glazed. She concluded that the psykter was a luxury vessel for the symposium, where the custom of floating it in a krater provided amusement (compare the game *kottabos* sometimes played with floating *oxybapha*).

The oldest surviving psykters are miniatures and were painted by the Swan Group of vase painters. Towards 500 B.C. the best painted psykters in red-figure were made in workshops with which the Pioneer Painters were associated. Groups around Euthymides and Kleophrades are identified in the earlier period. They were followed in the fifth century by individual works of some of the better painters of the time such as Myson, Duris and the Pan Painter, but the general quality soon declined.

Representations of psykters painted on other vessels often show them floating in kraters, mainly calyx kraters. Which vessel held the wine and which the cold water has been a subject of debate. There are protagonists for both points of view: von Bothmer, Diehl, Greifenhagen, and Pfuhl for psykter as wine-container; Bakalakis, Boardman, Drougou, Hampe, and Lane for psykter as water-container. (See Drougou, pp. 31-2 and notes.)

Psyktēr means cooler (*psychō*, "I cool") and it seems likely that the name was applied in antiquity to various shapes of cooling vessel, perhaps including the amphora with double walls (*CVA U.S.A* 14, Pl. 33-5). There is no early literary evidence for the name being used specifically for the mushroom-shaped vessel. The one reference where a psykter is actually mentioned, and where the description might fit the shape now called by that name, is very late – a Scholiast on Clement of Alexandria (who himself belongs to the second century A.D.). Earlier references to psykters give inconclusive and contradictory evidence. Pollux's psykter is a footless vessel with astragaliskoi. An inscription, on the other hand, mentions a vessel which has a foot.

Compare also *oinopsyktēr, psygeus, psyktērias, psyktērion, psyktēridion, psyktēriskos*.

References

Agora, p. 52.

Alexis (Kock Fragment 9).

Athenaeus 4. 131c; 4. 142d; 5. 199; 6. 230d; 10. 431f; 11. 503b.

Beazley, *V. Amer.* Discussions of psykters with special reference to Oltos and Phintias; "Groups of early Attic Black-figure", *Hesp.* 13 (1944): 56-7. The Swan Group.

Dietrich von Bothmer, *Bulletin of the Metropolitan Museum* 19 (1960–1): 138 ff.; review of Howard Comfort, "Attic and South Italian painted Vases at Haverford College", *AJA* 61 (1957): 309-10.

Boston II, pp. 1 ff.; III, p. 92.

CVA France 12, passim; *Germany* 31 (Pl. 184, 3 Special form. Short psykter without the high foot); *Great Britain* 8 (III Ic Pl. 104 f.); *Italy* 40 (III I, Pl. 1, 2 Attic r.f.).

Daremberg and Saglio s.v. *psykter.*

Stella Drougou, *Der Attische Psykter* (Würzburg: Konrad Triltsch

Verlag, 1975). The best and fullest discussion of the shape. Many illustrations.
Anna B. von Follman, *Der Pan-Maler* (Bonn: Bouvier, 1968) pp. 27-28. Summary of the shape's development.
Gericke, p. 46.
Adolf Greifenhagen, *Jb. Berliner Museen* 2 (1961): 117 ff.; 3 123 ff.
Hesychius s.v. *kalathos.*
IG II-III^2, 1542, 22-23.
Otto Jahn, *Beschreibung der Vasensammlung König Ludwigs in der Pinakothek zu München* (München: Jos. Lindauer'sche Buchhandlung, 1854), p. 96. Jahn uses psykter as the general term for a cooling vessel, of no particular shape.
George Karo, "Notes on Amasis and Ionic Black figured Pottery", *JHS* 19 (1899): 141.
Wilhelm Klein, *Euphronios* (Vienna: Carl Gerold's Sohn, 1886), pp. 106 ff., 267. One of the earliest modern writers to use the name *psykter* for the mushroom-shaped vessel.
Menander (Kock Fragment 510).
Moeris. "A psykter full of cold water, or what we call a prochyma."
Plato *Symposium* 213e. (Apparently influenced by this passage, some ancient writers assumed that the psykter was a kind of cup, e.g. Hesychius and Scholiast on Plato 50.)
Pollux 6. 99.
Richter and Milne, pp. 12-13.
Konrad Schauenburg, "Ein Psykter aus dem Umkreis des Andokidesmalers", *JdI* 80 (1965): 76 ff.
Scholiast, Clemens of Alexandria. Paedagogus II. III 35. 1-2 (Ed. Stählin 177). "A psykter consists of a cylinder, with column-like stem resting on a cylindrical prism base, which gives stability and allows ease of handling."
Scholiast Plato 50.
Strattis (Kock Fragment 57).
Johann L. Ussing, "De nominibus vasorum graecorum disputatio" (Inaugural Dissertation, Hauniae, 1844), pp. 76 ff.
Eugene Vanderpool, "The Rectangular Rock-cut Shaft", *Hesp.* 15 (1946): 322-23. Black-glazed psykters from the Kolonos Agoraios.

1. Pyxis, Attic Geometric, third quarter of eighth century B.C. 20 cm.
2. Pyxis, Boeotian, second half of eighth century B.C. 12 cm. (incl. lid).
3. Pyxis, Boeotian, first half of eighth century B.C. 23 cm. (incl. lid).
4. Pyxis ("Powder Pyxis"), Corinthian, 625–600 B.C. 6 cm. (incl. slip-over lid).
5. Pyxis ("Stamnos Pyxis"), Corinthian, c. 550 B.C. 31 cm.
6. Pyxis (Type A), Attic black-glaze, c. 525 B.C. 14 cm. (incl. lid).
7. Pyxis (Type B), Attic black-glaze, c. 400 B.C. 5 cm.
8. Pyxis (Type A), Attic w.g., 470–460 B.C. 12 cm. (incl. lid).
9. Pyxis (Type C), Attic r.f., c. 400 B.C. 8 cm. (incl. lid).
10. Pyxis (Type D), Attic r.f., c. 400 B.C. 5 cm. (incl. lid).
11. Pyxis, Apulian r.f., second half of fourth century B.C. 12 cm. (incl. lid).
12. Pyxis, West Slope Ware, Late Hellenistic, 9 cm.

Pyxis [πυξίς]

See also EXALEIPTRON

The name is used for a number of small, box-like containers with lids. They are mostly round or nearly so, but they have many different vertical profiles.

A type of pyxis was made in Mycenaean times and is represented in finds in the Athenian agora. (See Exaleiptron 2 for a special variety of pyxis.)

In Attic Protogeometric pottery Desborough distinguishes two main varieties. The more common globular pyxis stands on a low foot, slightly spreading or of torus profile. The top of the vessel turns outwards forming a small lip. A low, conical lid sits on top and there are no handles. The other variety of Protogeometric pyxis is widest towards the top. Two vertical handles are attached to the shoulder, and often rise above the top of the vessel.

In early Geometric pottery the descendant of the globular Protogeometric type appears (without lip) and there is also an ovoid type, broad at the top and narrowing to a blunt point at the bottom. The main later Geometric type is the flat pyxis. In this shape the tendency is for the walls to straighten out. Some Geometric examples are unusually large compared with other pyxides earlier and later, and the knob of the lid is replaced by figures of horses.

Beazley's four-type division has usually been followed for the classification of Attic post-Geometric pyxides (see *Agora*, pp. 173 ff.). This is, however, different from the Richter and Milne classification and excludes Corinthian. Following Beazley the types are as follows.

Type A (mainly sixth-fifth century B.C.).

Concave sides (in appearance something like a large cotton-reel). The container stands on three or four feet (early), or a continuous ring foot (late).

Type B (popular for about a century, from the second quarter of the fifth to the second quarter of the fourth century B.C.)

This is a flat cylindrical vessel with vertical sides and protruding rim at the bottom. The large slip-over lid which covers the whole vessel and rests on the protruding rim is almost as tall as the vessel itself. This pyxis looks like a powder box and one of the sub-types is called a powder-pyxis. Though a box-like vessel and generally called a pyxis, Type B is not really closely related to Type A.

Type C (mid fifth to mid fourth century B.C.)

This is the later version of Type A, also with concave sides but much lower than is normal with Type A. The low ring foot has a smaller diameter than the rest of the vessel, and the lid rests on an off-set lip without overlapping the sides.

Type D (late fifth to fourth century B.C.)

Small cylindrical vessel with fairly straight sides. The lid is flattish and disk-like, fitting just inside, or sometimes slightly overlapping the top of the box. Type D is closest to Type B except that the lid does not slip down over the outside of the vessel. In appearance it is a simple round box.

Red-figured pyxides are very common at Athens, but were rarely exported.

Concave pyxis types begin to be made at Athens at about 600 B.C., becoming more popular towards the middle of the sixth century. They seem initially to be the result of Corinthian influence. In Corinth the succession of fashions is from the straight-sided (late eighth century), concave (seventh century), to convex (sixth century).

The evidence from illustrations on pottery and grave reliefs points to the pyxis being used mainly by women. Some of the boxes found in graves contained cosmetics including cinnabar and psimythion (white lead). They were also trinket boxes and jewellery containers, and were possibly used for medicines and incense.

Illustrations on red-figure pyxides are often of weddings or funerals. Pyxides were dedicated in large numbers to Artemis Brauronia.

As well as the ordinary pottery types, materials out of

which pyxides were manufactured include metal (especially bronze and silver), marble, and alabaster.

The name *pyxis* was used by late Greek and Roman writers in the sense of "box" (corresponding to the modern archaeological usage), but the word also had other meanings in ancient times, for example, a small tablet. In classical Athens the regular word for small containers of this sort appears to have been *kylichnis*, later *libanōtis*. *Pyxis* is related to the Greek word *pyxos*, "box-wood", Latin *buxus*, English box. The name would seem to imply the early existence of similar wooden containers as prototypes for later pyxides in other materials. Some small wooden boxes have been found, and pyxides made of other materials are sometimes reminiscent of wooden types (Roberts, pp. 3, 7, and n. 16).

Compare *exaleiptron, kylichnis, libanōtis, libanōtris, pyxidion, pyxos.*

References

Agora, pp. 173 ff.
______ 8, pp. 14-15, 62.
ARV2, p. 890, no. 173 (Type A); p. 1328 (Types B, C, D).
Athenaeus 11. 480c. "The Athenians also call the doctor's pyxis a kylichnis because of its being turned (*kekylisthai*) on a lathe."
Boardman, *BF*, p. 191.
Barbara Bohen, "Attic Geometric Pyxis" (N.Y. Uni., Ph.D., 1979).
Coldstream, pp. 11, 47.
Cook, p. 232.
CVA Austria 1 (Pl. 48 ff. Attic r.f.); *Germany* 6 (Pl. 96 ff. Attic pyxides and lids), 22 (Pl. 136 ff. Attic Types A and C), 36 (Boeotian, Corinthian), 39 (Pl. 6 geometric), 44 (Attic geometric), 46 (Pl. 32 ff. Attic r.f.).
Daremberg and Saglio, s.v. *pyxis.*
Vincent R. d'A. Desborough, *Protogeometric Pottery* (Oxford: Clarendon, 1952), pp. 106, 112.
Etymologicum Magnum, s.v. *pyxis.*
Gericke, p. 86.
Hesp. 2 (1933): 367; 8 (1939): 383 (Mycenaean pyxides in the Athenian agora).
Josephus *Bellum Judaicum* 1. 30, 7.
Lucian *Erotes* 39.
Martial *Epigrams* 9. 37.
Quintilian 8. 6. 35.
Richter and Milne, p. 20.
Sally R. Roberts, *The Attic Pyxis* (Chicago: Ares Publishers, 1978). Very full discussion, catalogue and illustrations of the concave Attic pyxis, Types A and C and variants, from 600 B.C.
Suidas. *Pyxis* "small tablet".
Simon, passim esp. pp. 35-36. (Geometric "horse pyxis").

1. Rhyton (mule's head), Attic r.f., c. 500 B.C. 23 cm.
2. Head Vase, Attic r.f., second half of fifth century B.C. 21 cm.
3. Plastic Vase (almond), Attic, early fourth century B.C. 12 cm.
4. Rhyton (goat's head), Apulian, second half of fourth century B.C. 20 cm.
5. Rhyton (griffin's head), Apulian, second half of fourth century B.C. 19 cm.
6. Drinking Horn, Apulian, second half of fourth century B.C. 27 cm.

Rhyton, Figure Vases and other Plastic Vases [ῥυτόν, κέρας, κριός]

See also ASKOS

The name *rhyton* is applied to several classes of drinking and ritual vessels, including the drinking-horn (*keras*) and vases moulded into animal shapes.

The animal vases usually have an animal's head or protome attached to the bottom of a cup or mug, and are fitted with a handle. The heads are often made in two-piece moulds, some of which continue in use for a very long time. Animal-head vases may have a stem or a foot (these are normal in the early period); later ones usually do not have a foot, and were able to stand only if they were turned upside down.

Sometimes a human, rather than animal head is attached to a vessel of otherwise regular shape (aryballos, kantharos, mug etc; compare *ARV*, App. 1). These vases are usually called figure-vases or head-vases (head-kantharos etc.). Often there are two heads to a vessel in a Janiform arrangement.

The Greeks also made vessels in other odd shapes, for example, boots, horses' feet, fruits such as pomegranate, knuckle-bones, and lobsters' claws (*ARV*, p. 970 The Class of the Seven Lobster Claws). Duck vases are especially common in Etruria and Southern Italy in the fourth century B.C., and the usual Eastern "sway-backed" type of askos looks rather like a duck.

German archaeologists commonly use the term *Salbgefäss* (unguent-vessel) to cover many of these types.

The drinking-horn is often represented in black-figure vase-painting painted on other vases. This becomes

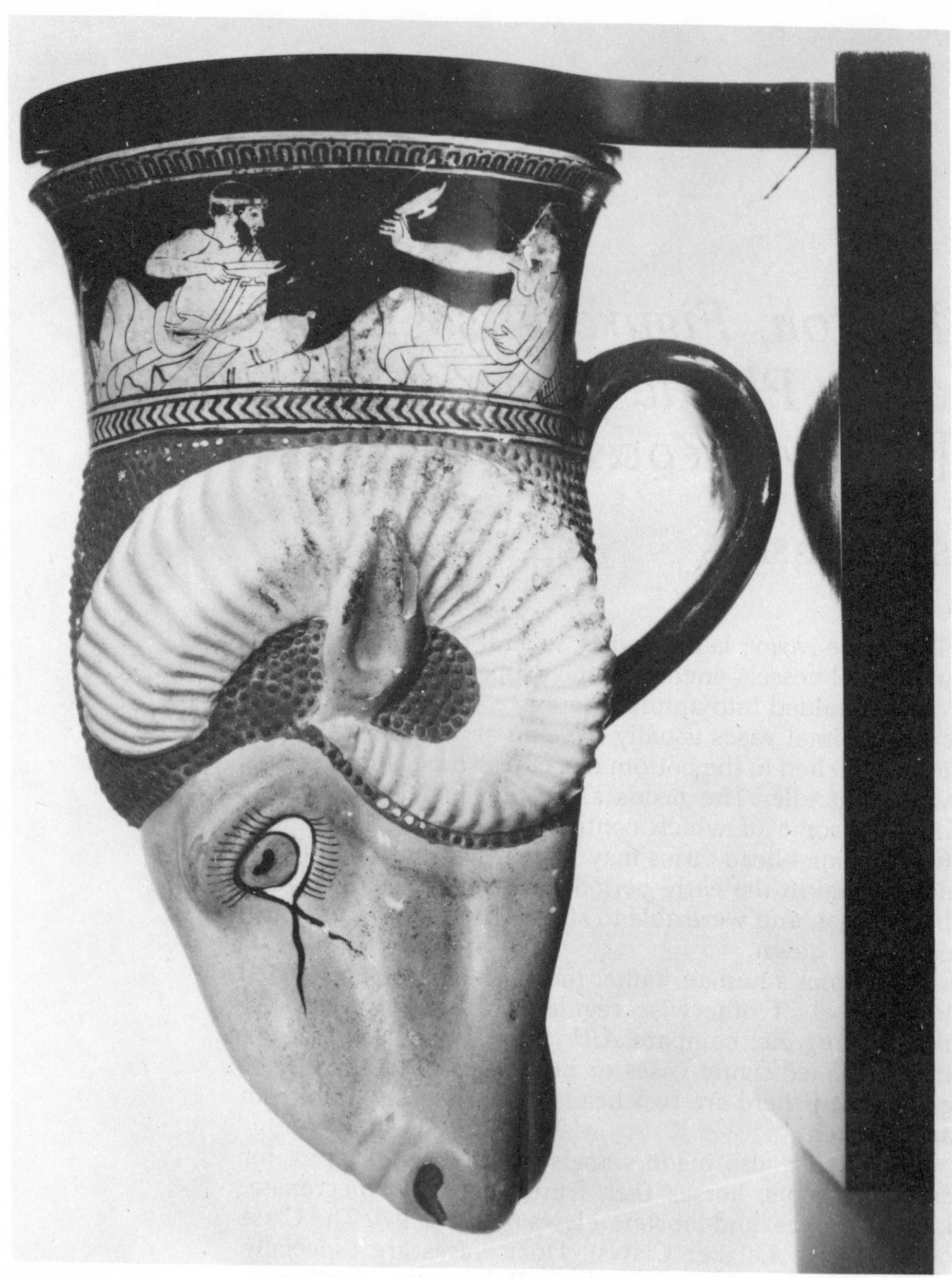

Rhyton (ram's head), Attic red-figure, by the Syriskos Painter, 480–470 B.C. 21 cm. Banqueters with kylikes, one of which is being used in a game of kottabos (throwing wine at a target).
Courtesy of British Museum E795.

increasingly common in the late archaic period. It is often, but not exclusively, associated with Dionysus.

Animal head vases were made at Athens from the end of the sixth century. Their shape seems to have been inspired by similar vessels from the east. The Sotades Painter was a prolific painter (and possibly maker) of rhyta. Other well-known artists like the Phiale and Villa Giulia Painters made some rhyta but were less productive than the Sotades Painter. The Penthesilea workshop produced a fairly large number of pieces, but the standard was often mediocre. In the later red-figure period most interest in the shape is in Apulia.

Athenaeus connected the word *rhyton* with *rhysis* (flowing), referring to an attribute of many of the horn-shaped vessels – a small hole at the bottom through which drink (or libation) might flow. Some modern authorities following the definition of Athenaeus restrict the word *rhyton* to vessels with such a hole ("Ausgussloch" – Tuchelt, p. 115). Others, accepting a wider definition, assign the name according to the shape of the vessel alone, whether or not it has such a hole. Most Attic animal-head vases made of clay do not have this drain-hole yet they are conventionally called rhyta, at least in English-speaking countries. However, this was almost certainly not the ancient practice. In Athens such vessels were perhaps given the name of the animal whose head is portrayed. (See Hoffmann *AK*, 1961 25 & n. 46.) As one might expect the name *keras*, "horn" was also used for the horn-shaped vessel in antiquity.

A number of rhyta are known from the Minoan and Mycenaean period in materials such as metal and stone, especially in the shape of a bull's head. They were possibly used for sacrificial and libation purposes. Temple inventories from the later Greek period also mention vessels of metal (Tuchelt, p. 117).

References

Agora, p. 3, n. 4. Names of containers from fish, animals etc.

ARV2, App. 1.

Athenaeus 496e-497e. This reads in part: "The rhyton was formerly called a horn. It appears to have been first made by King Ptolemy Philadelphus, as an attribute for statues of Arsinoe. She carries such an object in her left hand, full of all sorts of fruit . . . Dorotheus of Sidon says that rhyta are like horns, but have a hole bored in them . . . People drink from the stream that trickles gently down from them. They get their name from this flow (rhysis)."

John D. Beazley, "Charinos", *JHS* 49 (1929): 28 ff. Head vases.

W.R. Biers, "A Group of Leg Vases", *AJA* 84 (1980): 522-4.

CIG 2852, 39 ff.

CVA France 12 (This fascicule contains many Corinthian plastic vases, esp. III Cc, Pl. 1-7), 15 (Pl. 26-47 passim. A very extensive collection of Attic and South Italian); *Germany* 33

(Many East Greek and Boeotian head, animal, and other odd-shaped vases); 35 (Pl. 45, 9 & 10), 36 (Pl. 47, 48 Boeotian); *Great Britain* 5 (Pl. 36 ff. Many plastic kantharoi and rhyta), 9 (III c 7 ff., II D 7 ff.); *Italy* 57 (Pl. 5 & 6 Various animals); *U.S.A.* 1 (Vases in shape of woman's head, dog's head, booted foot, shell etc.), 2 (Pl. 26 Attic r.f. oinochoe in form of woman's head. "Providence Group", 490–480 B.C.), 5 (Pl. 58, 2-3 Attic plastic oinochoai).

Gericke, p. 19.

R.A. Higgins, *Catalogue of the Terracottas in the Department of Greek and Roman Antiquities, British Museum* (London: British Museum, 1954).

Herbert Hoffmann, *Attic Red-figured Rhyta* (Mainz: Philipp von Zabern, 1962); *Tarentine Rhyta* (Mainz: Philipp von Zabern, 1964); "The Persian Origin of Attic Rhyta", *AK* 4 (1961): 21-26.

T. Homolle, *BCH* 6 (1882): 115-16.

IG II-III^2, 1443, 132; 1444, 1-4, 22; 1640, 21. The word *rhein*, "to flow" used of leaky vessels.

M.I. Maximowa, *Les vases plastiques dans l'antiquité* (Paris: Geuthner, 1927).

Theodor Panofka, *Die Griechischen Trinkhörner und ihre Verzierungen* (Berlin: Trautwein'sche Buchhandlung, 1851).

I. Richter, *Das Kopfgefäss. Zur Typologie einer Gefässform* (Köln: Magister-Arbeit, 1967).

Richter and Milne, p. 28.

Helmut Sichterman, *Die Griechische Vase. Gestalt, Sinn und Kunstwerk* (Berlin: Hessling, 1963).

Maria Trumpf-Lyritzaki, *Griechische Figurenvasen des reichen Stils und der späten Klassik* (Bonn: H. Bouvier & Co., 1969).

Klaus Tuchelt, *Tiergefässe in Kopf – und Protomengestalt*, Bd. 22 (Berlin: Istanbuler Forschungen, 1962).

Webb, passim. Many small figured vases in faience.

E.R. Williams, "Figurine Vases from the Athenian Agora", *Hesp.* 47 (1978): 356 ff. Fourth century B.C. and later.

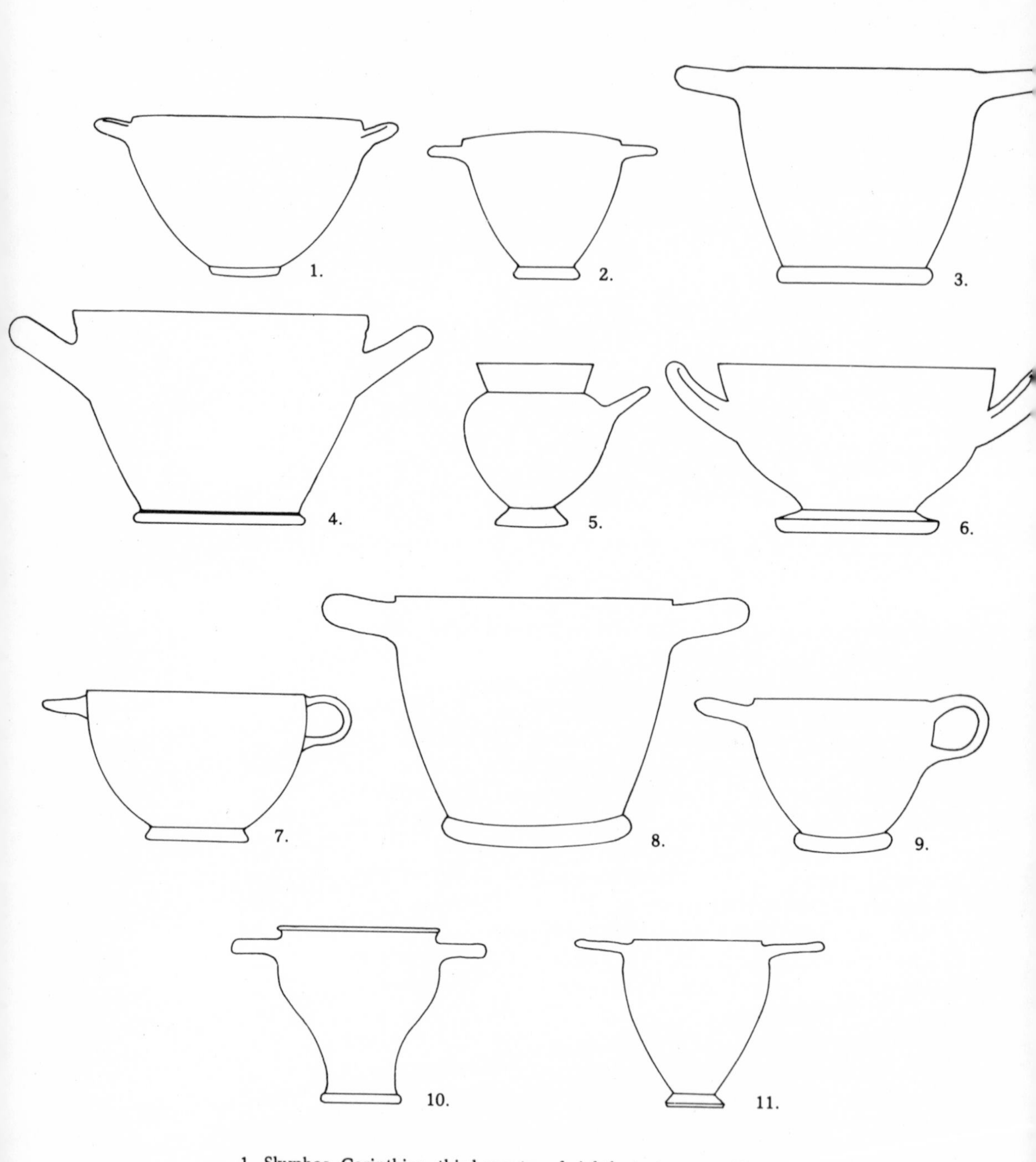

1. Skyphos, Corinthian, third quarter of eighth century B.C. 10 cm.
2. Skyphos, Corinthian, last quarter of seventh century B.C. 9 cm.
3. Skyphos, Attic black-glaze, second quarter of fifth century B.C. 13 cm.
4. Skyphos, Attic b.f., third quarter of sixth century B.C. 10 cm.
5. Deep cup, Chalcidian, third quarter of sixth century B.C. 13 cm.
6. Cup-Skyphos, Attic b.f., c. 510 B.C. 11 cm.
7. Skyphos, Attic r.f., c. 500 B.C. 8 cm.
8. Skyphos, Attic r.f., c. 450 B.C. 15 cm.
9. Glaux, Attic r.f., c. 450 B.C. 8 cm.
10. Skyphos, Attic black-glaze, c. 330 B.C. 11 cm.
11. Skyphos, Attic black-glaze, late fourth century B.C. 10 cm.

Skyphos [σκύφος]

KOTYLE, GLAUX and other deep cup shapes

The word *skyphos* might be cognate with the Greek word *kypellon*, Latin cupa, English cup.

The name *skyphos* is used nowadays of a deep cup with two handles, and usually standing on a low foot. The handles are attached horizontally at the level of the rim, or lower down the sides. If there is a lip it is small and unobtrusive. The maximum diameter of the vessel is not much different from its height, in contrast to the kylix where the diameter far exceeds the bowl height.

Cups with some of the characteristics of skyphoi are found in Mycenaean, Protogeometric and Geometric pottery. After the canonical shape emerges in Late Geometric and Protocorinthian the subsequent changes are relatively minor until the late Classical period. Skyphoi remain popular at Corinth to the end of red-figure. At Athens the shape becomes very common in the red-figure period, and a narrow form with concave sides persists in Hellenistic pottery. In the west, especially in Etruria, the late red-figure skyphos acquires a very exaggerated concave profile in its lower part.

Though some fine artists made and painted skyphoi, the shape never attained the glamour of the kylix. The side of a skyphos presents an approximately triangular field which was difficult to decorate effectively with figures. Some of the most successful were the Hermogenean skyphoi of the Little Master period. These, and the group called Band Skyphoi, carried the same sort of decoration as Band Cups (see Kylix).

Cup-skyphos or kotyle (Type C), Attic red-figure, painted by Epictetus, 520–500 B.C. 9.5 cm.
Two boys kneeling beside a column krater. One holds a skyphos and kylix; the other dips an oinochoe into the krater.
Courtesy of British Museum G276.

The skyphos becomes very popular in plain pottery – the poor man's cup, the most common shape for ordinary drinking cups at Athens for most of the Classical period. In the Corinthian type, which was copied at Athens and regularly made there also, the lip curves slightly inwards, the walls are thin, and there is a low ring foot. The usual decoration of the Corinthian-type plain skyphos is simply a reserved band at the bottom on the outside of the wall. This band is sometimes enlivened with vertical lines, cross-hatched, or painted with added red. Under the foot there are a couple of concentric circles with a dot in the centre. Handles develop from bell-shaped to horse-shoe and finally triangular types. The Attic-type skyphos, which is emerging by the mid sixth century B.C., has an overall heavier appearance; wall, foot and handles are more massive. In the earlier examples there are a few red lines below the handles, and handle-panels are reserved. Eventually the whole surface is black-glazed, though the underside of the foot may be unglazed and decorated in various ways. In the late fifth and earlier fourth centuries the lower part of the skyphos wall becomes concave, and there is often a concave lip as well.

A popular red-figure variant of the fifth century B.C. has one handle horizontal and one vertical. Its usual decoration is an owl set between two sprigs of olive, hence the name *glaux* (the Greek word for owl). A cup with this sort of decoration, but with two horizontal handles, is called "owl-skyphos" or "owl-kotyle".

One-handled cups of skyphos shape are given the obvious name "one-handled skyphoi". They look more like our regular cups, but are mostly thicker, heavier, and larger. Amongst the more unusual types one might mention the "Chalcidian Deep Cup" (*CVA Great Britain* 15, pl. 30).

A fairly shallow skyphos with a small concave lip is called a "cup-skyphos". It appears in both black-figure and red-figure.

In ancient times the word *skyphos* was certainly used for some sort of cup. Apparently it could be used for the kylix shape, for on an Attic black-glaze kylix in the Achaeological Museum, Syracuse, the words, "this skyphos Porkos dedicates", are found in the dedicatory inscription. The Euripides reference (*Cyclops* 390-91), though describing a vessel of mammoth dimensions, at least gives approximately the correct proportions for the shape that we call skyphos. But the name *kylix* was also written on skyphoi. It seems likely from the evidence available that in ancient usage *skyphos* was a general name for "cup", not reserved specifically for the shape to which we apply the name.

Some modern authorities use the name *kotylē* for the shape here called *skyphos*.

References

Agora, pp. 8, 81 ff. A very detailed discussion of the important plain types.

ARV² chaps. 49, 68 etc.

Athenaeus 11 (passim). A good deal of information of variable value. He mentions the fact that the skyphos was popular amongst the Boeotians. Handles sometimes made in the shape of Herculean knots.

Cook, pp. 233 ff., 235-6 (kotyle).

CVA Austria 1 (Pl. 35-44 Attic r.f.); *Germany* 41 (Attic b.f.), 42 (Corinthian), 43 (Pl. 22 Gnathian), 47 (Pl. 34 ff. Attic); *Great Britain* 5 (Pl. 28-32 Attic r.f.); *Italy* 11 (Pl. 39-45 Campanian); *Switzerland* 3 (Attic b.f.); *U.S.A.* 4 (Pl. 22 ff. Attic b.f.), 5 (Corinthian, Attic b.f. and r.f.).

Daremberg and Saglio s.v. *scyphus.*

Euripides *Cyclops* 390-91. A skyphos four cubits deep and three cubits wide.

Hesychius. "Skyphos. A type of *potērion* or *ekpōma*." (Both words mean "cup".)

Lazzarini, 355-56.

Macrobius *Saturnalia* 5. 21. 16. "The skyphos is Hercules' cup."

Richter and Milne, pp. 26 ff. The Richter and Milne classification is based on the position of the handles. Type I: Handles set below the lip and curving upwards. Type II: Horizontal handles set at rim level.

H.R.W. Smith, *Der Lewismaler* (Leipzig: Keller, 1939), p. 7, n. 3. The problem of skyphos/kotyle.

1. Stamnos, Attic r.f., early fifth century B.C. 30 cm.
2. Stamnos, Attic r.f., c. 450 B.C. 33 cm.
3. Stamnos, Attic r.f., third quarter of fifth century B.C. 43 cm.

Stamnos [σταμνος]

Apparently related to *histēmi,* "I set, stand". The name is conventionally applied to a vessel, wider in the upper part of the body, with a short, wide, usually off-set neck, and a wide mouth. There is regularly a low foot which may be ring-shaped, disk-like or more complicated. Footless stamnoi are rare. The handles are horizontal loops and attached to the widest part of the vessel at belly to shoulder level. But while the handle-grip is attached horizontally, its ends almost invariably curve upwards. In a very early black-figure example, the Hirsch Stamnos (Niarchos Collection, Paris), the handles are vertical loops.

Some types of amphora with horizontal handles come fairly close to the stamnos, the main difference being in the proportions of the neck, which in the amphora is taller and narrower. Such vessels are often called stamnoid amphoras.

A vessel like the later stamnos was made in Attica in the Bronze Age and Geometric period. It is the sort of practical shape that must have been used in most periods in the coarser kitchen wares. But the great period of Attic high-quality, painted-pottery stamnoi is the century or so ending about 420 B.C. There are some good examples in black-figure, but it is more characteristically a red-figure shape.

Attic stamnoi seem to have been exported in quantity to Magna Graecia where they were very popular, and it is in this area that most Attic stamnoi have been found. They were also made locally in Etruria and elsewhere in the west, and production continued there after the flow of imports from Attica had ceased.

One of the stamnos' main functions was to contain wine.

Stamnos (detail), Attic red-figure, painted by Smikros, c. 500 B.C. 38.5 cm.
Two men with pointed amphorae beside a lebes on a stand flanked by oinochoai. The stand is perhaps fancifully constructed of a kylix and kothon.
Courtesy of Bruxelles, Musées royaux d'art et d'histoire, inv. A717 (photo ACL, Bruxelles)

From illustrations which show wine being ladled from stamnoi it is clear that it was also used as a krater for mixing wine and water. Its wide neck and mouth are convenient for such a function. Some modern archaeologists have accordingly classified the stamnos as a type of krater (e.g. Buschor p. 141). When used as a storage vessel it had a lid. Besides wine the stamnos was also used for other liquids including oil, and like the amphora and hydria it was sometimes used as a burial urn. This applies especially to the Etruscan bronze stamnoi. There is reason to believe that it was equivalent to half an amphora as a measure of capacity, though this equation might not have been a general one.

The stamnos was certainly associated with some Dionysiac festivals such as the Lenaea (see Gericke) but it is not clear whether its function there was any more important than that of other types of wine container.

The evidence for the ancient usage of the word *stamnos* points to its being a general word for shapes of approximately this type, and probably including amphorae. Where a measure of capacity is implied, it might have been smaller than the amphora. The letters *sta* which might be short for stamnos are found as a graffito on a Nolan amphora (*ARV*, p. 488, 73). Other possibilities for this shape's specific ancient name are *bikos, kados, krōssos* and *hyrchē*. The modern definition of *stamnos* dates to the nineteenth century.

References

Amyx, pp. 190 ff.

Aristophanes *Frogs* 22, "Dionysus son of Stamnios"; *Lysistrata* 196, "a Thasian stamnion of wine".

Boardman, *BF*, p. 187; *RF*, p. 209.

BCH 6 (1882): 6 ff. & 60 ff. (money kept in stamnoi); 14 (1890): 413 (oil container); 62 (1938): 149 (bronze stamnoi); 79 (1955): 365-66 (measure of capacity).

CVA France 1 (r.f.), 2 (r.f.), 4 (Attic r.f.), 5 (r.f.); *Germany* 18 (Pl. 86 ff. One-piece stamnos or pelike?), 20 (Pl. 238 ff. r.f.), 30 (Pl. 72 ff. Attic r.f.); *Great Britain* 3 (Pl. 26 ff. r.f.), 4 (Pl. 19-25 Attic r.f.); *Italy* 1 (Attic and South Italian r.f.); *U.S.A.* 18 (Pl. 14 ff. b.f.).

Daremberg and Saglio, s.v. *stamnos*.

Etymologicum Magnum, "Amphora: a stamnion with two handles, one on either side".

Gericke, p. 43.

Stamnoi (California: J. Paul Getty Museum, 1980).

Hesychius, "Bikos: a stamnos with handles".

Cornelia Isler-Kerenyi, *Stamnoi* (Lugano: Cornèr Banca, 1976, 77).

Moeris, s.v. *amphorea*. Amphoreus given as the Attic name for the two-handled vessel that was elsewhere known as a stamnos.

Barbara Philippaki, *The Attic Stamnos* (Oxford: Oxford University Press, 1967). An excellent discussion and history of the shape from the second quarter of the sixth century to the last quarter of the fifth century B.C.

RE, s.v. *stamnos*.

Richter and Milne, pp. 8-9.

1. Thymiaterion (as illustrated on an Attic r.f. vase), c. 470 B.C. 70 cm. (estimate)
2. Thymiaterion, Apulian fourth century B.C. 26 cm.

Thymiaterion [θυμιατήριον]

THURIBLE

Thymiaō, "I burn incense". The name *thymiatērion* is applied to what is believed to have been the ancient *thymiatērion*, "incense-burner". It is a shallow bowl attached to a tall stand which often flares out at the bottom to a base that is wide enough to prevent tipping. Also popular were short legs in the form of lions' paws.

The profile is usually enlivened with a couple of mouldings about half way up, or towards the top of the stand. In some cases the decoration is rather more complex. The Meidias Painter depicts an elaborate thymiaterion on one of his hydriai (Florence, Museo Archaeologico, Inv. 81947).

Wigand divides thymiateria of the Greek and Roman period into half a dozen categories on the basis of differences in size, elaboration of shape, and type of foot.

In pottery, thymiateria are fairly common in the red-figure period and later. Some varieties are probably taken over from Egypt and the Near East. As might be expected of essentially religious objects, many thymiateria were also made of metal, including precious metals. Most thymiateria probably had perforated lids.

The word is used by many ancient authors (see *RE*, and compare also *thymiatēr, thymiatris, kichētos*, and in Latin *turibulum*).

References

Agora, p. 182.

CVA Denmark 6 (Pl. 263, 4); *Germany* 4 (Pl. 40, 13).
Daremberg and Saglio, s.v. *turibulum.*
Demosthenes 24. 183.
Diodorus 13. 3. 2 (before the departure of the Athenian expedition to Sicily, "the whole circuit of the harbour was filled with incense-burners").
Herodotus 4. 162.
RE s.v. *thymiatērion.*
K. Wigand, "Thymiateria", *Bonn. Jahrb.* 122 (1912): 1-97.

1. Bird bowl, Boeotian, first half of sixth century B.C. 23 cm.
2. Pilgrim flask, Etruscan, sixth century B.C. 28 cm.
3. Situla, East Greek, second quarter of sixth century B.C. 46 cm.
4. Fusiform unguentarium, Apulian, c. 300 B.C. 19 cm.
5. Chalice, Etruscan Bucchero, first half of sixth century B.C. 16 cm.
6. Lydion, East Greek, c. 530 B.C. 9 cm.
7. Unguentarium, Tarantine, Hellenistic. 12 cm.
8. Fish plate, Campanian r.f., mid-fourth century B.C. 5 cm.

Miscellaneous Shapes

Basket

Apart from the kalathos many other types of basket are mentioned occasionally by ancient writers, for example, *kibōtos, kistē, phormos.*

"Bottle"

CVA U.S.A. 18 (Pl. 50)

Epinētron (Onos)

The shape nowadays conventionally called *epinētron* or (wrongly) *onos* is often included in pottery books. It is not a container but a protective covering for the thigh and knee used in working wool, no doubt especially in the carding process. Made of pottery, it was decorated in the same way as other pottery pieces. There may be a moulded female head at the end.

ABV, pp. 480, 481; *ARV*2, pp. 1225, 1250, 1503.
G. Bakalakis, *Jahreshefte des Österreichischen Archäologischen Instituts* 45 (1960): *Beiblatt* 199 ff.
Boardman, *BF*, p. 191.
CVA Poland 1 (III He Pl. 15).
E. Gullberg and P. Aström, *The Thread of Ariadne.* A Study of Ancient Greek Dress, *Studies in Mediterranean Archaeology 21* (Göteborg: Paul Aströms Förlag, 1970) pp. 13-14, Ill. 3.
D.M. Robinson, *AJA* 49 480 ff.
Simon, pp. 146-47.

Ēthmos

A vessel which might have been the ancient *ēthmos* is known in a few surviving examples. It is spherical, with a sieve base, and a hollow bail handle connects with the inside of the vessel which is sealed at the top. There is a small hole in the top of the handle. The vessel can be dipped into a liquid and allowed to fill through the sieve. If the hole in the handle is then sealed with the thumb the liquid can be transferred to another container (kept in place during the process of transfer by atmospheric pressure), and then released by opening the hole in the handle.

CVA U.S.A. 7 (Pl. 3).

Holmos

A mortar used for pounding grain etc. Probably often made of wood or stone.

Amyx, p. 235.

Lēnos

Large vessel or trough used for treading grapes. Usually made of stone.

Amyx, p. 241.

Loutērion

Washing basin, often on a stand and commonly made of stone. Used for washing and as a sponge bath.

Amyx, p. 221.
Sparkes and Talcott, p. 218.

Lydion

The name *lydion* is used nowadays for a small, handleless, high-footed vessel mainly from the East Greek area and belonging to the sixth century B.C. Its ancient name is unknown, as is its function, though it might have been used for unguents.

CVA Germany 28 (Pl. 303), 42 (Pl. 26); *Italy* 53 (Pl. 37, 38); *Switzerland* 3 (Pl. 87).
A. Rumpf, "Lydische Salbgefässe", *AM* 45 (1920).

Perirrhantērion

A special basin or louterion, kept in a sacred place, and holding water for purification.

Amyx, pp. 24, 221.
Sparkes and Talcott, p. 218.

Phidaknē and Phidaknis

These are the Attic forms. Elsewhere also *pithaknē*, *pithaknis*. Diminutive of *pithos*?

A fairly large storage vessel, possibly of variable shape. Capacity was probably generally between that of amphora and pithos, though there might have been a considerable size range, with consequent overlap and blurring in terminology at both ends of the scale. Phidaknai listed in the Attic Stelai have a capacity of up to a dozen or so amphorae.

There is also quite a lot of other information about these vessels on the Stelai. Five empty second-hand phidaknai are sold for prices ranging from four to eleven drachmas each. A reference to a *phidaknē dedemenē* is to a vessel bound in some way, perhaps with a rope, or with hoops like a cask, or in a wicker basket, or repaired with lead. (A large pithos excavated at Olynthus was repaired with approximately forty lead cleats.) One entry on the Stelai refers to the *stoma* of a phidakne. This word should mean "mouth", and possibly refers to the top part of a broken phidakne with intact mouth. It would seem to imply that such a piece of pottery was saleable for use as a well-head or something similar.

Compare *phidaknion*, *pithaknion*, *pitharion*, *pitheōn*, *pithiskos*, *pithos*, *pithōn*.

Amyx, pp. 170 ff. Very good discussion of shape, size, price, and many references.
Aristophanes *Knights* 792. Phidakne used as living quarters.
Athenaeus 11. 483d. A reasonably large vessel.
Hesychius, s.v. *phidaknē*, *pithaknai*.
Olynthus 8, 316. Repair of vessel with lead.
Pollux 10. 74 & 131.
Suidas, s.v. *pithaknē*.

Phormiskos

The *phormiskos*, according to the current usage of the word, is not really a container. The name is conventionally applied to a shape hung in Attic graves of the sixth century B.C. It is bulbous, and the thin end is pierced to take a string. The name is wrongly used nowadays. The ancient phormiskos was a basket.

Boardman, *BF*, p. 192.
CVA Germany 47 (Pl. 22, 6 & 7) *Italy* 7 (Pl. 24) "ardanion simbolico".
R. Hampe, *AA* (1976) 192 ff.

Plate

Plates are fairly common (some archaeologists used to use the word *pinax* for plate, but it is now obsolete in this sense). Fish plates, an important group of plates mostly from southern Italy, were decorated with fish and other sea creatures.

Cook, pp. 237-8.
CVA Great Britain 2 (IV Ea Pl. 12); *Italy 11* (Pl. 1-6); *U.S.A.* 5 (Pl. 31), 6 (Pl. 22), 7 (Pl. 26).

Situla

The name *situla* is applied to deep, cylindrical pots.

CVA Great Britain 13 (II Dm Pl. 1-10 and p. 29).

"Teapot"

An east Greek "kettle" or "teapot" shape with bail handle and spout occurs occasionally.

CVA Germany 33 (Pl. 179, 1 & 2).

Tub

Agora, p. 216.

Shapes mainly found in coarser wares

The following shapes occur especially in the ordinary coarser wares, and were used for kitchen and other household purposes. For more affluent householders they were made of bronze and other more precious materials: –

chytra: a two-handled cooking pot (Amyx, p. 211; Sparkes and Talcott, p. 224)

deutēr: cauldron or kneading basin (Amyx, p. 232)

kados: The kados was used as a wine and water container, including the dipping of water from wells. The name is sometimes used nowadays for the sturdy, household-ware amphorae that have often been recovered from wells. *Kados* might have been a synonym for *amphora* (Amyx, p. 186; Sparkes and Talcott, p. 201).

kardopos: a mixing or kneading basin or trough (Amyx, p. 239)

lopas: shallow cooking pot with a lid (Sparkes and Talcott, p. 227)

mykē: a mushroom-shaped vessel ? (Amyx, p. 208)

oxybaphon: small dish

pithos: an important househould storage vessel, sometimes set in the ground. It was wide-mouthed and deep-bodied, usually handleless. Sizes range from the tall, capacious ones found in Crete (dating to Minoan times) to the small ones found in the Agora at Athens which look like dumpy handleless amphorae. If, as has been suggested, the word *pithos* is related to the Greek words for "pine" and "pitch", the name comes from the material (pitch or resin) used to make it water-tight. See Amyx, pp. 168 ff., Sparkes and Talcott, pp. 193 ff. and *CVA France* 1 (II Ad Pl. 1 & II Bb Pl. 1 Cretan and Theran).

pyelos: bathtub (Amyx, p. 252)
sipyē: possibly a large crock with a lid (Amyx, p. 195)
skaphē: broad, shallow basin (Amyx, p. 231)
thermantērion: kettle or cauldron (Amyx, p. 218)
thermaustis: kettle or cauldron (Amyx, p. 219)
triptēr: possibly a run-off basin for grape juice or olive oil (Amyx, p. 247)

CORPUS VASORUM ANTIQUORUM INDICES

Index of CVA fascicules according to Countries

The index contains three possible entries: Significant representation, Contents, and Shapes. If the contents consist mainly of one type or a few types of pottery the entry "Significant representation" is omitted, as the listing under "Contents" will usually be similar. When the shapes dealt with in a fascicule are mixed, with no high representation of any particular one, the entry "Shapes" is omitted.

AUSTRIA 1 (Plates 1-50), *VIENNA 1* (Kunsthistorisches Museum)
CONTENTS: ATTIC (R.F.).
Shapes: Many kylikes, but also skyphoi, kantharoi, one-handled beaker, rhyta, pyxides, lids.

AUSTRIA 2 (Plates 51-100), *VIENNA 2* (Kunsthistorisches Museum)
CONTENTS: ATTIC (R.F.).
Shapes: Amphorae, stamnoi, pelikai, column krateres, fragmentary kylix, one-handled pot (or oinochoe), rhyton.

AUSTRIA 3 (Plates 101-150), *VIENNA 3* (Kunsthistorisches Museum)
CONTENTS: ATTIC (R.F.).
Shapes: Many krateres, but also hydriai, loutrophoroi, kylix.

BELGIUM 1 (Plates 1-48), *BRUSSELS 1* (Musées Royaux du Cinquantenaire)
SIGNIFICANT REPRESENTATION: ATTIC (B.F. AND R.F.).
Contents: Cypriot, Mycenaean, Corinthian, Boeotian, Attic (b.f., r.f., w.g.), Apulian (incl. Gnathian).

Shapes: Mixed. A number of Attic b.f. amphorae, r.f. kylikes, w.g. lekythoi.

BELGIUM 2 (Plates 49-96), *BRUSSELS 2* (Musées Royaux du Cinquantenaire)
SIGNIFICANT REPRESENTATION: ATTIC (B.F. AND R.F.), SOUTH ITALIAN.
Contents: Mysian, Cypriot, Mycenaean, Corinthian imitation, Chalcidian, Boeotian, Attic (geometric, Attic-Corinthian, b.f., r.f., w.g.), Etruscan, Apulian, Lucanian (incl. Paestan).
Shapes: Mixed. A number of b.f. lekythoi, r.f. krateres.

BELGIUM 3 (Plates 97-151), *BRUSSELS 3* (Musées Royaux du Cinquantenaire)
SIGNIFICANT REPRESENTATION: ATTIC (B.F. AND R.F.).
Contents: Mysian, Trojan, Cretan, Cypriot, Mycenaean, Melian, Ionian (incl. plastic), Laconian, Corinthian (and imitation), Italogeometric, Boeotian, Attic (geometric, b.f., w.g., r.f., plastic, black-glaze, stamped, relief), Hellenistic (incl. Hadra hydriai), Apulian and various Italian (incl. plastic), Campanian, Etruscan, Etrusco-Ionian, Imitation Etruscan and Attic.

CANADA 1 (Plates 1-42), *TORONTO 1* (Royal Ontario Museum, Toronto)
SIGNIFICANT REPRESENTATION: ATTIC (B.F.).
Contents: Attic (b.f., patterned vases), Boeotian (b.f.), "Tanagran" by Teisias.
Shapes: Mixed (incl. a number of b.f. amphorae, kylikes).

CYPRUS 1 (Plates 1-40), (Cyprus Museum 1)
SIGNIFICANT REPRESENTATION: MYCENAEAN.
Contents: Aegean (Mycenaean, a little Minoan and late Cypriot III) mainly from sites in Cyprus.
Shapes: Mixed. A number of krateres, three-handle jars, stirrup-jars.

CYPRUS 2 (Plates 41-81), (Private Collections 1)
CONTENTS: MYCENAEAN.

CZECHOSLOVAKIA 1 (Plates 1-50), *PRAGUE 1* (University Charles)
SIGNIFICANT REPRESENTATION: ATTIC (GEOMETRIC, B.F., R.F.).
Contents: Thessaly Neolithic (Dimini Style), Early Helladic, Middle Helladic, Minoan, Mycenaean, Protogeometric, At-

tic (geometric, b.f., r.f., w.g.), Non-Attic geometric (Corinthian style, Smyrna, Boeotian), Corinthian and Imitation (Italo-Corinthian, Attic or Boeotian). Many sherds and incomplete vessels.
Shapes: Mixed, (incl. lekythoi).

DENMARK 1 (Plates 1-49), *COPENHAGEN 1* (Musée National)
SIGNIFICANT REPRESENTATION: CYPRIOT, CRETAN, MYCENAEAN.
Contents: Egyptian, Proto-Elamite, Syrian, Trojan, Cypriot, Cretan, Melian, Rhodian, Amorgian, Cycladic, Thessalian, Mycenaean.

DENMARK 2 (Plates 50-98), *COPENHAGEN 2* (Musée National)
SIGNIFICANT REPRESENTATION: MYCENAEAN, ATTIC (GEOMETRIC), EAST GREEK, CORINTHIAN.
Contents: Mycenaean, Greek (geometric, incl. Santorini, Rhodes, Thessaly, Boeotia), Attic (geometric), East Greek Orientalizing (incl. Milesian, Rhodian, Samian), Greek figure vases, Argive, Sicyonian (Protocorinthian), Corinthian, Italo-Sicyonian, Italo-Corinthian.
Shapes: Mixed, (incl. oinochoai).

DENMARK 3 (Plates 99-146), *COPENHAGEN 3* (Musée National)
SIGNIFICANT REPRESENTATION: ATTIC (B.F. AND R.F.).
Contents: Chalcidian, Ionian, Boeotian, Small Greek archaic vases, Laconian, Attic (b.f., r.f., plastic).
Shapes: Mixed, (incl. a number of b.f. amphorae, lekythoi, kylikes).

DENMARK 4 (Plates 147-195), *COPENHAGEN 4* (Musée National)
SIGNIFICANT REPRESENTATION: ATTIC (R.F. AND W.G.).
Contents: Attic (r.f. and w.g.), Greek various (incl. Boeotian), Greek black-glazed, Hellenistic various, Sicilian and Sardinian, Italian (Bronze Age to early Iron Age).
Shapes: Mixed, (incl. krateres, lekythoi).

DENMARK 5 (Plates 196-235), *COPENHAGEN 5* (Musée National)
SIGNIFICANT REPRESENTATION: FALISCAN AND ETRUSCAN.
Contents: Faliscan, Proto-Etruscan, Etruscan, Apulian (geometric and derivative), Campanian.

DENMARK 6 (Plates 236-275), *COPENHAGEN 6* (Musée National)
SIGNIFICANT REPRESENTATION: R.F. VASES OF SOUTH ITALY (ESP. CAMPANIAN AND APULIAN).
Contents: Proto-Italiote, Lucanian, Paestan, Campanian, Apulian, South Italian (various).
Shapes: Mixed, (incl. a number of krateres).

DENMARK 7 (Plates 276-315), *COPENHAGEN 7* (Musée National)
SIGNIFICANT REPRESENTATION: BLACK-PAINTED WARES.
Contents: Vases from South Italy and North Africa: Apulian (incl. Gnathian), Campanian, Black-painted (various), Calene Ware, Canosa Ware, Plastic, Hellenistic (various), Sicilian, Punic, North African (Pre-Roman and Roman).

DENMARK 8 (Plates 316-362), *COPENHAGEN 8* (Musée National)
CONTENTS: ATTIC (B.F., R.F., AND A LITTLE W.G.).
Shapes: Mixed, (incl. a number of r.f. krateres).

FRANCE 1 (Plates 1-49), *PARIS, LOUVRE 1* (Musée du Louvre)
SIGNIFICANT REPRESENTATION: PROTO-ELAMITE, RHODIAN, ATTICO-CORINTHIAN, ATTIC (R.F.).
Contents: Proto-Elamite (1st Period), Rhodian, Cretan, Theran, Cyrenaean, Laconian, Attico-Corinthian, Attic (r.f.).
Shapes: Mixed, (incl. a number of Attic b.f. amphorae, r.f. krateres, stamnoi).

FRANCE 2 (Plates 50-98), *PARIS, LOUVRE 2* (Musée du Louvre)
CONTENTS: PROTO-ELAMITE (2ND PERIOD(, ATTICO-CORINTHIAN, ATTIC (B.F. AND R.F.).
Shapes: Mixed. (Large vessels – amphorae, dinoi, krateres etc.).

FRANCE 3 (Plates 99-131), *COMPIEGNE 1* (Musée de Compiègne)
SIGNIFICANT REPRESENTATION: ATTIC (B.F. AND R.F.), SOUTH ITALIAN (VARIOUS).
Contents: Egyptian, Graeco-Phoenician, Cypriot, Rhodian, Melian, Greek uncertain, Corinthian and imitation, Attic (geometric, Attico-Corinthian, b.f., r.f., mixed, w.g., plastic), Italian uncertain, Etruscan, Imitations of Greek b.f. and r.f., Italian relief, Graeco-Italian, Apulian (incl.

Gnathian, plastic), Campanian, Lucanian, Gallo-Roman, German (Danube Valley).

FRANCE 4 (Plates 132-180), *PARIS, LOUVRE 3* (Musée du Louvre)
SIGNIFICANT REPRESENTATION: ELAMITE, ATTIC (B.F. AND R.F.).
Contents: Proto-Elamite (1st and 2nd period), Elamite uncertain, Cypriot, Persian, Attic (b.f. and r.f.).
Shapes: Amphorae (b.f.), krateres and stamnoi (r.f.).

FRANCE 5 (Plates 181-229), *PARIS, LOUVRE 4* (Musée du Louvre)
SIGNIFICANT REPRESENTATION: CYPRIOT, ATTIC (B.F. AND R.F.).
Contents: Vases of the Troad, Mysian, Cypriot, Attic (b.f. and r.f.).
Shapes: Amphorae (b.f.), krateres and stamnoi (r.f.).

FRANCE 6 (Plates 230-284), (Collection Mouret)
SIGNIFICANT REPRESENTATION: ATTIC (R.F.), SOUTH ITALIAN, IBERIAN (GALLIC).
Contents: Attic (b.f. and r.f.), Corinthian, Graeco-Italian, Apulian, Campanian, Attico-Italian, Greek Hellenistic, Iberian and indigenous (Gallic), Rhodian, Pottery of the Roman period (Gaul).
Shapes: Mixed, (incl. r.f. krateres).

FRANCE 7 (Plates 285-332), *PARIS* (Bibliothèque Nationale 1)
SIGNIFICANT REPRESENTATION: CORINTHIAN, ATTIC (B.F.).
Contents: Egyptian, Theran, Cypriot, Rhodian, Melian, Uncertain, Cycladic, Argive, Mycenaean, Corinthian and imitation, Spartan and Cyrenaean, Chalcidian, Ionian, Boeotian, Attic (geometric, Protoattic, b.f.).
Shapes: Mixed, (incl. a number of Attic b.f. amphorae, Corinthian oinochoai).

FRANCE 8 (Plates 333-384), *PARIS, LOUVRE 5* (Musée du Louvre)
CONTENTS: CYPRIOT, ATTIC (B.F. AND R.F.).
Shapes: Attic amphorae (esp. b.f.), krateres (r.f.).

FRANCE 9 (Plates 385-434), *PARIS, LOUVRE 6* (Musée du Louvre)
CONTENTS: CORINTHIAN, ATTIC (B.F. AND R.F.).
Shapes: Mainly b.f. and r.f. hydriai; r.f. amphorae and pelikai.

FRANCE 10 (Plates 435-482), *PARIS* (Bibliothèque Nationale)
SIGNIFICANT REPRESENTATION: ATTIC (B.F.).
Contents: Nearly all Attic b.f. except for a small quantity of Rhodo-Ionian (plastic), relief, Attic mixed styles and r.f.

FRANCE 11, PARIS, LOUVRE 7 (Musée du Louvre)
LOUVRE – INDEX OF THE FIRST SIX FASCICULES.

FRANCE 12 (Plates 483-529), *PARIS, LOUVRE 8* (Musée du Louvre)
CONTENTS: CORINTHIAN (INCLUDING PLASTIC), ATTIC (B.F. AND R.F.).
Shapes: Many Corinthian aryballoi, alabastra and plastic vases; Attic mainly psykteres, amphorae, pelikai.

FRANCE 13 (Plates 530-592), *SEVRES 1* (Musée national de Sèvres)
SIGNIFICANT REPRESENTATION: ATTIC (VARIOUS), SOUTH ITALIAN.
Contents: Egyptian, Proto-Elamite, Syrian and Palestinian, Mysian, Theran, Cypriot, Melian, Cycladic, Mycenaean, Graeco-Phoenician, Rhodian, Corinthian (incl. imitation and plastic), Attic (b.f., r.f., w.g., black-glaze), Hellenistic (incl. plastic), Uncertain, Relief, Glazed, Cyrenaean, Etruscan, Italian (geometric and without decoration), Italo-Ionian b.f., Apulian (various), Attico-Italian, Campanian (various), Lucanian, Gnathian, Vases of the Roman period, Pottery of Gaul (Bronze and Iron Age), Gallo-Roman, Roman pottery from the Rhine area, European (various), Pottery stamps.

FRANCE 14 (Plates 593-640), *PARIS, LOUVRE 9* (Musée du Louvre)
CONTENTS: IONIAN, CORINTHIAN, ITALO-CORINTHIAN, CAERETAN , ATTIC B.F.
Shapes: Mixed, (incl. Corinthian alabastra, Caeretan hydriai, Attic b.f. kylikes and r.f. hydriai).

FRANCE 15 (Plates 641-688), *PARIS, PETIT PALAIS 1* (Palais des Beaux-Arts de la Ville de Paris)
SIGNIFICANT REPRESENTATION: ATTIC (B.F. AND R.F.), SOUTH ITALIAN (ESP. FIGURE VASES).
Contents: Lucanian geometric, Italo-Corinthian, Italian (various), Chalcidian, Etruscan, Attic (b.f., r.f., plastic, w.g.), Greek plastic, Rhodian, Italian, Apulian, South Italian and Etruscan (various esp. rhyta and plastic vases).

FRANCE 16 (Plates 689-728), *PARIS, MUSÉE NATIONAL RODIN*
SIGNIFICANT REPRESENTATION: CORINTHIAN, ATTIC (B.F. AND R.F.), SOUTH ITALIAN
Contents: Cypriot, Protocorinthian and Corinthian, Naucratite, Attic (geometric, b.f., r.f., w.g.), Etruscan, Etrusco-Corinthian, Apulian (various), Lucanian, Gnathian, Campanian, Italian (various), Calenian, Hellenistic, Gallo-Roman, Hadra hydria.
Shapes: Mixed, (incl. r.f. krateres).

FRANCE 17 (Plates 729-776), *PARIS, LOUVRE 10* (Musée du Louvre)
CONTENTS: ATTIC (B.F. AND R.F.).
Shapes: Kylikes.

FRANCE 18 (Plates 777-824), *PARIS, LOUVRE 11* (Musée du Louvre)
CONTENTS: ATTIC (GEOMETRIC AND B.F.).
Shapes: Krateres (geometric), amphorae and hydriai (b.f.).

FRANCE 19 (Plates 825-872), *PARIS, LOUVRE 12* (Musée du Louvre)
CONTENTS: ATTIC (B.F.).
Shapes: Krateres (esp. Column krateres). A few dinoi.

FRANCE 20 (Plates 873-924), *LAON 1* (Musée de Laon)
CONTENTS: ATTIC (GEOMETRIC, B.F., R.F.).
Shapes: Mixed (incl. a number of kylikes, lekythoi).

FRANCE 21 (Plates 925-972), *PARIS, LOUVRE 13* (Musée du Louvre)
CONTENTS: CORINTHIAN (FROM PROTOCORINTHIAN GEOMETRIC TO CORINTHIAN B.F.).

FRANCE 22, PARIS, LOUVRE 14 (Musée du Louvre)
LOUVRE – INDEX.

FRANCE 23 (Plates 973-1019), *PARIS, LOUVRE 15* (Musée du Louvre)
SIGNIFICANT REPRESENTATION: BOWLS AND VASES WITH RELIEF DECORATION.
Contents: Attic, Boeotian, Asia Minor, Antioch, Calenian Ware.

FRANCE 24 (Plates 1020-1067), *LIMOGES* and *VANNES* (Musée Adrien Dubouché, Société Polymathique)

(a) *LIMOGES*
SIGNIFICANT REPRESENTATION: ATTIC (B.F. AND R.F.), ETRUSCAN AND SOUTH ITALIAN.
Contents: Chalcidian, Ionian, Laconian, Uncertain, Corinthian, Attic (incl. b.f., r.f., black-glaze), Etruscan (various), Apulian, Sicilian, Paestan, Gnathian and other South Italian (incl. black-painted, terra sigillata, Gallo-Roman, uncertain).
(b) *VANNES*
CONTENTS: ATTIC (B.F.), VARIOUS.

FRANCE 25 (Plates 1068-1123), *PARIS, LOUVRE 16* (Musée du Louvre)
CONTENTS: GREEK GEOMETRIC (MOSTLY ATTIC AND BOEOTIAN).

FRANCE 26 (Plates 1124-1178), *PARIS, LOUVRE 17* (Musée du Louvre)
SIGNIFICANT REPRESENTATION: BOEOTIAN
Contents: Boeotian various (from geometric to black-glaze), Euboean various, Uncertain.

FRANCE 27 (Plates 1179-1227), *PARIS, LOUVRE 18* (Musée du Louvre)
SIGNIFICANT REPRESENTATION: ATTIC (GEOMETRIC, SUB-GEOMETRIC, PROTOATTIC), NON-ATTIC GEOMETRIC.
Contents: Attic (geometric, sub-geometric, Protoattic), Argive (geometric), Rhodian (geometric), Cycladic (geometric), Cretan (geometric), Uncertain.

FRANCE 28 (Plates 1228-1283), *PARIS, LOUVRE 19* (Musée du Louvre)
CONTENTS: ATTIC (R.F. AND A FEW HYBRIDS).
Shapes: Kylikes.

FRANCE 29 (Plates 1284-1335), *RENNES 1* (Musée des Beaux-Arts et d'Archéologie de Rennes)
SIGNIFICANT REPRESENTATION: ATTIC (B.F. AND R.F.).
Contents: Asia Minor, Mycenaean, Cypriot, Geometric (various), Rhodian, East Greek, Corinthian, Laconian, Attic (b.f., r.f., black-glaze), Etruscan (various), Etrusco-Corinthian, Italian r.f. (Apulian, Campanian, Paestan), Italian (black-glaze and painted decoration), Coarse pottery of Greece and Italy.

FRANCE 30 (Plates 1337-1338), *BOURGES* and *TOURS*

(Musée des Beaux-Arts à Tours, Musée du Berry à Bourges)
(a) *TOURS*
CONTENTS: ATTIC (GEOMETRIC, B.F., R.F., W.G., BLACK-GLAZE), IONIAN, CORINTHIAN, LACONIAN, ETRUSCAN, GNATHIAN, CAMPANIAN, ITALIAN.
(b) *BOURGES*
CONTENTS: CORINTHIAN, IONIAN, ATTIC (B.F., R.F., BLACK-GLAZE), ITALO-GEOMETRIC, ETRUSCAN, APULIAN, PAESTAN, GNATHIAN, ITALIAN BLACK-GLAZE.

GERMANY 1 (Plates 1-46), *BONN 1* (Akademisches Kunstmuseum)
SIGNIFICANT REPRESENTATION: ATTIC (R.F. AND W.G.).
Contents: Attic (r.f., black-decorated and black-glaze, w.g.), Greek various.
Shapes: Mixed, (incl. a number of kylikes, lekythoi).

GERMANY 2 (Plates 47-94), *BERLIN* 1 (Antiquarium)
CONTENTS: EARLY ATTIC (ESP. FROM AEGINA, AND INCL. MANY FRAGMENTS).
Shapes: Mixed, (incl. a number of amphorae, krateres).

GERMANY 3 (Plates 95-146), *MUNICH 1* (Museum Antiker Kleinkunst)
CONTENTS: ATTIC (B.F.).
Shapes: Belly amphorae.

GERMANY 4 (Plates 147-194), *BRAUNSCHWEIG 1* (Herzog Anton Ulrich Museum)
SIGNIFICANT REPRESENTATION: ATTIC (B.F. AND R.F.), SOUTH ITALIAN.
Contents: Greek incl. Attic (geometric, b.f., r.f., black-decorated, black-glaze, w.g.), East Greek, Chalcidian, Protocorinthian, Corinthian, Hellenistic, Italian incl. Faliscan, Etruscan (Red ware, unpainted, Bucchero, b.f., r.f.), South Italian, Campanian (b.f. and r.f.), Apulian (r.f.), Lucanian (r.f.), Italian glazed ware, Italian various, Gnathian, Italian black decorated, Italian polychrome (Canosa, Sicily), Cypriot, Egyptian, Various.
Shapes: Mixed, (incl. alabastra, Attic kylikes, South Italian krateres).

GERMANY 5 (Plates 195-248), *VIENNA 1* (Universität, and Collection Matsch)
(a) *UNIVERSITY*
SIGNIFICANT REPRESENTATION: ATTIC (R.F.).

Contents: Erythraean, Hittite, Cypriot, East Greek, Corinthian and imitation, Attic (geometric, early Attic, b.f., r.f., w.g.), Greek black-glaze, Hellenistic, Villanovan, Etruscan (Bucchero, b.f. with added red), Italian r.f., Faliscan? (r.f.), Daunian, Italian black-glaze, Various.
(b) *COLLECTION MATSCH*
CONTENTS: BOEOTIAN (GEOMETRIC, B.F., CORINTHIAN IMITATION), PROTOCORINTHIAN, CORINTHIAN, ATTIC (B.F., R.F., W.G., PLASTIC), GREEK BLACK-GLAZE, LUCANIAN R.F., APULIAN (R.F., PLASTIC, INDIGENOUS WARES), GNATHIAN, ITALIAN BLACK-GLAZE, ETRUSCAN (BUCCHERO, B.F., WITH ADDED RED).

GERMANY 6 (Plates 249-298), *MUNICH* 2 (Museum Antiker Kleinkunst)
CONTENTS: ATTIC R.F. (MOSTLY FIFTH CENTURY B.C.).
Shapes: Mixed, (incl. a number of Nolan amphorae, oinochoai, pelikai).

GERMANY (WEST) 7 (Plates 299-340), *KARLSRUHE 1* (Badisches Landesmuseum)
SIGNIFICANT REPRESENTATION: ATTIC (B.F. AND R.F.), CORINTHIAN.
Contents: Late Mycenaean, Attic (geometric, Protoattic, b.f., r.f., w.g.), Head vases, b.f. (C4 B.C.), "Megarian" bowls, Small decorated vases, Black-glaze, Boeotian, Corinthian various.

GERMANY (WEST) 8 (Plates 341-384), *KARLSRUHE 2* (Badisches Landesmuseum)
SIGNIFICANT REPRESENTATION: SOUTH ITALIAN.
Contents: Cypriot, East Greek, Italian incl. native wares (Villanovan, Impasto, Bucchero, Daunian, Messapian, Peucetian), geometric, Etruscan, Early Italian, Apulian, Campanian, Paestan, Lucanian, Campanian b.f., Gnathian, Glazed with clay-coloured painted decoration, Unglazed relief-decorated, Glazed with relief decoration, Figure vases.

GERMANY (WEST) 9 (Plates 385-434), *MUNICH 3* (Museum Antiker Kleinkunst)
SIGNIFICANT REPRESENTATION: ATTIC (GEOMETRIC AND PROTOATTIC), FIGURE VASES (VARIOUS).
Contents: Attic (geometric, Protoattic, b.f.), Cycladic, Corinthian, Italo-Corinthian, Boeotian, Figured unguent vessels (Corinthian, Ionian, Magna Graecia, Rhodian).

GERMANY (WEST) 10 (Plates 435-478), *HEIDELBERG 1* (Universität)
SIGNIFICANT REPRESENTATION: EAST GREEK, CORINTHIAN, BOEOTIAN, ATTIC (B.F.)
Contents: Rhodian, Clazomenian, Chiot, Melian and other East Greek fabrics, Figured unguent vessels, Laconian, Corinthian, Corinthian imitations (local, Boeotian, Italian), Boeotian, Attic (Protoattic and b.f.).

GERMANY (WEST) 11 (Plates 479-530), *SCHLOSS FASANERIE 1* (Adolphseck)
CONTENTS: ATTIC (B.F., R.F. AND A LITTLE W.G.).
Shapes: mixed, (incl. a number of krateres, lekythoi).

GERMANY (WEST) 12 (Plates 531-586), *MUNICH 4* (Museum Antiker Kleinkunst)
CONTENTS: EARLY ATTIC (R.F.).
Shapes: Amphorae (belly amphorae Types A & C, Panathenaic, pointed), including a few bilinguals.

GERMANY (WEST) 13 (Plates 587-640) *MANNHEIM 1* (Reiss-Museum)
SIGNIFICANT REPRESENTATION: ATTIC (B.F. AND R.F.).
Contents: Greek incl. Mycenaean, Attic (geometric, b.f., r.f., w.g.), Protocorinthian and Corinthian, East Greek, Boeotian, Hellenistic; Italian incl. Impasto/Etruscan face urns and "Red Ware", Etruscan Bucchero, Corinthianizing (Etruscan), South Italian black-painted, Early Italian r.f., Apulian r.f., Paestan r.f., Campanian r.f., Lucanian r.f.; Greek and Italian black-glaze, Gnathian and Calenian, Various.

GERMANY 14 (Plates 641-693), *LEIPZIG 1* (Archäologisches Institut der Karl-Marx-Universität)
SIGNIFICANT REPRESENTATION: CYPRIOT, ATTIC (GEOMETRIC), CORINTHIAN (INCL. IMITATIONS).
Contents: Egyptian, Cypriot (Bronze and Iron Age), Attic (geometric, imitations of Corinthian), Corinthian (incl. Italian and Etruscan imitations), East Greek.
Shapes: Mixed, (incl. a number of Corinthian alabastra and aryballoi).

GERMANY (WEST) 15 (Plates 694-741), *MAINZ* (Universität)
SIGNIFICANT REPRESENTATION: PROTOATTIC, ATTIC (B.F.).
Contents: Attic (protogeometric, geometric, Protoattic, b.f.),

Non-Attic protogeometric, Cypriot early Iron Age, Argive Monochrome, Protocorinthian, Italian Protocorinthian, Corinthian, Boeotian (geometric and archaic).

GERMANY (WEST) 16 (Plates 742-785), *SCHLOSS FASANERIE 2* (Adolphseck)
SIGNIFICANT REPRESENTATION: SOUTH ITALIAN (ESP. APULIAN).
Contents: Egyptian, Cypriot, Greek incl. Mycenaean, East Greek, Magna Graecia, Protocorinthian, Corinthian, Uncertain b.f., Pseudo-Chalcidian, Boeotian, Attic (b.f., r.f. figure vases), Greek various; Italian incl. Faliscan, Etruscan, Campanian, Campanian r.f., Italian r.f. (early Italian, Paestan, Lucanian etc.), Apulian r.f., Gnathian, Italian black-painted, Various, Italian w.g. and polychrome (Canosan and related); Attic and Italian glazed wares, Hellenistic and Roman relief pottery and glazed wares.

GERMANY 17 (Plates 786-827), *ALTENBERG 1* (Staatliches Lindenau-Museum)
SIGNIFICANT REPRESENTATION: CORINTHIAN, ATTIC (B.F.).
Contents: Corinthian, Ionian, Fikellura, Attic (b.f.).
Shapes: Mixed, (incl. a number of Attic b.f. amphorae, kylikes, lekythoi, oinochoai).

GERMANY 18 (Plates 828-876), *ALTENBERG 2* (Staatliches Lindenau-Museum)
SIGNIFICANT REPRESENTATION: ATTIC (R.F.).
Contents: Attic (r.f., w.g., black-glaze, stamped, relief), Uncertain, Boeotian, Italian late geometric, Italo-Corinthian, Italian (r.f.).
Shapes: Mixed, (incl. krateres, kylikes, pelikai).

GERMANY 19 (Plates 877-923), *ALTENBERG 3* (Staatliches Lindenau-Museum)
SIGNIFICANT REPRESENTATION: APULIAN.
Contents: Apulian, Lucanian, Campanian, Gnathian, Paestan, South Italian uncertain, South Italian black-glaze, Tarentine, Greek/South Italian relief ware, Calenian, Messapian and similar, Canosan, Etruscan (Bucchero, Impasto Nero, b.f., painted, r.f.), Hellenistic, Relief wares, Uncertain.

GERMANY (WEST) 20 (Plates 924-977), *MUNICH 5* (Museum Antiker Kleinkunst)
CONTENTS: ATTIC (R.F.).
Shapes: Neck amphorae, hydriai, stamnoi.

GERMANY (WEST) 21 (Plates 978-1029), *BERLIN 2* (Antiquarium)
CONTENTS: ATTIC (R.F.).
Shapes: Kylikes.

GERMANY (WEST) 22 (Plates 1030-1079), *BERLIN 3* (Antiquarium)
CONTENTS: ATTIC R.F. (ALSO A LITTLE W.G.).
Shapes: Kylikes, phiale, plate, lid, pyxides, skyphoi, oinochoai.

GERMANY (WEST) 23 (Plates 1080-1124), *HEIDELBERG 2* (Universität)
SIGNIFICANT REPRESENTATION: ITALIAN.
Contents: Italian prehistoric (various), Impasto, Etruscan incl. undecorated, Bucchero, Red ware, b.f., glazed with clay colour or white paint (also non-Etruscan), Etruscan and Faliscan r.f., Etruscan various, Upper Adriatic ? Proto-Apulian and Early Apulian, Apulian, Paestan, Proto-Lucanian, Lucanian, Sicilian and Campanian, Hellenistic, Gnathian, Various.

GERMANY 24 (Plates 1125-1172), *GOTHA 1* (Schlossmuseum)
SIGNIFICANT REPRESENTATION: CORINTHIAN, ETRUSCAN, ATTIC (B.F. AND R.F.).
Contents: Trojan, Cypriot, Mycenaean, Corinthian, Italian Corinthianizing (Etruscan), Etruscan (Impasto, Bucchero, b.f.), Laconian, Attic (geometric, b.f., and r.f.).

GERMANY (WEST) 25 (Plates 1173-1212), *FRANKFURT AM MAIN 1*
SIGNIFICANT REPRESENTATION: ATTIC (GEOMETRIC, B.F.), CORINTHIAN.
Contents: Cypriot, Mycenaean, Attic (geometric, Protoattic, b.f.), East Greek and Rhodian, Corinthian, Italo-Corinthian.
Shapes: Mixed, (incl. a number of Attic b.f. amphorae).

GERMANY (WEST) 26 (Plates 1213-1283), *STUTTGART 1* (Württembergisches Landesmuseum).
SIGNIFICANT REPRESENTATION: CORINTHIAN, ATTIC (GEOMETRIC, B.F., R.F.), APULIAN, ITALIAN BLACK-GLAZE.
Contents: Aegean, Cypriot, Greek incl. Late Mycenaean, Attic (geometric, b.f., r.f., w.g., figure vase, with painted ornament, black-glaze), Boeotian (geometric, b.f., black-glaze), Corinthian, East Greek, Relief ware, Hellenistic;

Italian incl. Daunian, Messapian, South Italian uncertain, Corinthianizing, Etruscan (Red ware, Red decoration on w.g., Bucchero, with black ornament, r.f.), Impasto, Early Italian r.f., Apulian (r.f. and black-decorated), Campanian (r.f. and black-decorated), Gnathian, Italian clay coloured and white-painted, Italian black-glaze, Calenian relief pottery, Apulian relief pottery, Italian head vases, Various. *Shapes:* Mixed, (incl. a number of Attic b.f. lekythoi, oinochoai).

GERMANY (WEST) 27 (Plates 1284-1331), *HEIDELBERG 3* (Universität)
SIGNIFICANT REPRESENTATION: MYCENAEAN, ATTIC (GEOMETRIC AND EARLY ATTIC).
Contents: Cycladic (Syros Group), Minoan, Mycenaean, Protogeometric, Attic (geometric, Protoattic), Boeotian, Rhodian, Cycladic, Cretan, "Argive Monochrome", Argive, Protocorinthian, Corinthian, Lydian, Italian; Finds from Aegina (incl. Mycenaean, protogeometric, Attic geometric, and Protoattic, Argive, Protocorinthian, Corinthian, East Greek); Finds from the Argive Heraion (incl. Mycenaean, Argive, Protocorinthian, Corinthian, East Greek, Cycladic); Finds from Asine (Argive); Finds from Mycenae (Argive); Finds from Amyclae (Laconian); Finds from Thera (incl. Cretan, East Greek, Protocorinthian, Corinthian, Theran, Cycladic); Finds from Paros (Cycladic); Finds from Naucratis (incl. Aeolic grey ware, "Rhodian"); "Rhodian" of unknown origin.

GERMANY (WEST) 28 (Plates 1332-1379), *MUNICH 6* (Museum Antiker Kleinkunst)
SIGNIFICANT REPRESENTATION: BOEOTIAN, EAST GREEK, ISLANDS IONIAN.
Contents: Cycladic, Boeotian, East Greek, East Greek figured unguent vessels, Islands Ionian ("Chalcidian"), Laconian cups, Ionian cups, Samian face kantharos, North Ionian, Various East Greek, Clazomenian sarcophagi.

GERMANY 29 (Plates 1380-1431), *GOTHA 2* (Schlossmuseum)
SIGNIFICANT REPRESENTATION: ATTIC (R.F.), SOUTH ITALIAN (ESP. APULIAN).
Contents: Attic (r.f., w.g. and later fabrics), Apulian, Gnathia Ware, Apulian figure vases, Canosan, Lucanian, Campanian, Italian black-glaze, Teano fabric, Calenian, Alexandrian, Uncertain, Roman relief ware, Terra sigillata, Roman various.

GERMANY (WEST) 30 (Plates 1432-1476), *FRANKFURT AM MAIN 2*
CONTENTS: ATTIC (B.F. AND R.F.).
Shapes: Mixed, (incl. a number of kylikes, lekythoi).

GERMANY (WEST) 31 (Plates 1477-1524), *HEIDELBERG 4* (Universität)
SIGNIFICANT REPRESENTATION: ATTIC (B.F.).
Contents: "Chalcidian", Boeotian b.f., Early Attic, Attic b.f., Fragments of Panathenaic amphorae, Fragments from Naucratis, Attic black-glazed with ornament, Attic black-glazed with black-red, "Attic-Boeotian" black-glazed, black-glazed various.
Shapes: Mixed, (incl. a number of lekythoi).

GERMANY (WEST) 32 (Plates 1525-1576), *MUNICH 7* (Museum Antiker Kleinkunst)
SIGNIFICANT REPRESENTATION: ATTIC (B.F.).
Contents: Tyrrhenian amphorae and related, Panel neck-amphorae, Neck-amphorae of the Affecter, Special forms, Standard neck-amphorae of the second half of the sixth century B.C.
Shapes: Neck-amphorae.

GERMANY (WEST) 33 (Plates 1577-1632), *BERLIN 4* (Antiquarium)
SIGNIFICANT REPRESENTATION: EAST GREEK, BOEOTIAN.
Contents: Cycladic, Cretan, East Greek (incl. geometric, Orientalizing, Figured unguent vessels, "Vroulia", "Fikellura", North Ionian b.f., "Chalcidian", East Greek various; Laconian, Boeotian (incl. geometric, Bird-bowl Group, b.f.).

GERMANY (WEST) 34 (Plates 1633-1680), *HANOVER 1* (Kestner-Museum)
SIGNIFICANT REPRESENTATION: ATTIC (B.F. AND R.F.).
Contents: Attic (geometric, b.f., r.f., w.g.).
Shapes: Mixed, (incl. a number of kylikes, lekythoi).

GERMANY (WEST) 35 (Plates 1681-1728), *KASSEL 1* (Antikenabteilung der Staatlichen Kunstsammlungen).
SIGNIFICANT REPRESENTATION: ATTIC (B.F. AND R.F.).
Contents: Mycenaean, Thessalian (geometric), Attic (geometric), East Greek (sub-geometric), Corinthian, East Greek, Boeotian, Laconian, Workshop of Andros,

Clazomenian, Fikellura, Attic (b.f., r.f., various, w.g.), Boeotian r.f., Kabirion fabric.
Shapes: Mixed, (incl. a number of amphorae).

GERMANY (WEST) 36 (Plates 1729-1780), TÜBINGEN 1 (Antikensammlung des archäologischen Instituts der Universität).
SIGNIFICANT REPRESENTATION: CORINTHIAN, RHODIAN, BOEOTIAN.
Contents: Corinthian (geometric), Theran (geometric), Cycladic (geometric), Boeotian (geometric), Rhodian (Fikellura and figured unguent vessels), Clazomenian (incl. sarcophagi), Protocorinthian, Corinthian, Argive, Laconian, Euboean, Boeotian (Bird-bowl Group, b.f., r.f., various, Kabirion fabric), non-Attic r.f.

GERMANY (WEST) 37 (Plates 1781-1848), *MUNICH 8* (Antikensammlungen)
CONTENTS: ATTIC (B.F.).
Shapes: Neck-amphorae (late archaic).

GERMANY (WEST) 38 (Plates 1849-1892), *KASSEL 2* (Antikenabteilung der staatlichen Kunstsammlungen).
SIGNIFICANT REPRESENTATION: VASES FROM THE EXCAVATIONS ON SAMOS (1894), SOUTH ITALIAN (ESP. APULIAN).
Contents: Vases from the excavations on Samos (1894), Cypriot (Bronze Age, Iron Age), Villanovan, Impasto, Faliscan, Etruscan Bucchero, Daunian, Italian, Corinthianizing, Messapian, Sicilian, Unguentaria, "Pontic Vases", Italian black-glaze, Volcani Group, Calenian Relief Pottery, Gnathian, Apulian, Sicilian, Paestan, Caeretan (Genucilia Group), Boeotian (various), West Slope Ware, Hadra Vases, w.g. lagynos, Hellenistic (relief, various), Hellenistic/Roman.

GERMANY (WEST) 39 (Plates 1883-1926), *WÜRZBURG 1* (Martin von Wagner Museum)
SIGNIFICANT REPRESENTATION: ATTIC (GEOMETRIC AND B.F.).
Contents: Mycenaean, Protogeometric, Attic (geometric, Protoattic, b.f.), Sherds from Orchomenos; East Greek (incl. Rhodian, Chiot, Clazomenian sarcophagus, East Greek various, Aeolian, North Ionian), Corinthian (incl. geometric, Protocorinthian), Boeotian (b.f.), Attic (b.f.).

GERMANY (WEST) 40 (Plates 1927-1966), *BONN 2* (Akademisches Kunstmuseum)

SIGNIFICANT REPRESENTATION: MINOAN AND MYCENAEAN.
Contents: Minoan (Neolithic from Crete, Prepalatial, Protopalatial, Neopalatial, Postpalatial, sherds from Prinias), Cycladic, Mainland (incl. Neolithic, Early and Middle Helladic, Mycenaean/Late Helladic), Finds from Tell el-Amarna, Grave find from Keos.

GERMANY (WEST) 41 (Plates 1967-2014), *HAMBURG 1* (Museum für Kunst und Gewerbe)
SIGNIFICANT REPRESENTATION: BOEOTIAN (GEOMETRIC AND B.F.), ATTIC (GEOMETRIC AND B.F.).
Shapes: Mixed, (incl. a number of amphorae, kylikes, lekythoi).

GERMANY (WEST) 42 (Plates 2015-2057), *MAINZ* (Römisch-germanisches Zentralmuseum 1)
SIGNIFICANT REPRESENTATION: GEOMETRIC, CORINTHIAN, ATTIC (B.F., R.F., BLACK GLAZE).
Contents: Mycenaean, Geometric, Corinthian, Corinthian and Etrusco-Corinthian plastic vases, Etrusco-Corinthian, Etruscan b.f., East Greek and Magna Graecia, Boeotian (incl. black and b.f.), Attic b.f., Attic and Greek various, Attic (r.f., w.g., black), Corinthian black, Black various.

GERMANY (WEST) 43 (Plates 2058-2101), *MAINZ* (Römisch-germanisches Zentralmuseum 2)
SIGNIFICANT REPRESENTATION: ETRUSCAN AND SOUTH ITALIAN.
Contents: Etruscan r.f., South Italian r.f. (incl. Lucanian, Apulian, Campanian, Paestan, Sicilian), South Italian with white slip or polychrome, Gnathia Ware, Etruscan polychrome, Teano Ware, Campanian (b.f., polychrome, without painting), Etruscan (black), Apulian (black), Campanian (black), South Italian (black), Calenian relief pottery, Latin, Hellenistic (black), West Slope Ware and predecessors, Late Hellenistic polychrome, Lagynoi, Megarian relief pottery, Pergamene relief pottery, Eastern terra sigillata, A Tunisian grave-find.

GERMANY (WEST) 44 (Plates 2102-2145), *TÜBINGEN 2* (Antikensammlung des archäologischen Instituts der Universität)
SIGNIFICANT REPRESENTATION: ATTIC (GEOMETRIC AND B.F.).
Contents: Cypriot, Mycenaean, Attic (geometric and b.f.).
Shapes: Mixed, (incl. many b.f. amphorae).

GERMANY (WEST) 45 (Plates 2146-2201), *BERLIN 5* (Antikenmuseum)
CONTENTS: ATTIC (B.F.).
Shapes: Amphorae. Also a few lids.

GERMANY (WEST) 46 (Plates 2202-2245), *WÜRZBURG 2* (Martin von Wagner Museum)
CONTENTS: ATTIC (R.F., W.G., BLACK-GLAZE).

GERMANY (WEST) 47 (Plates 2246-2297), *TÜBINGEN 3* (Antikensammlung des archäologischen Instituts der Universität)
CONTENTS: BOEOTIAN LEKANIS, ATTIC B.F.
Shapes: Mixed, (incl. a number of kylikes, lekythoi, oinochoai). Many fragmentary pieces.

GERMANY (EAST) 1 (Plates 1-61), *SCHWERIN* (Staatliches Museum)
SIGNIFICANT REPRESENTATION: ATTIC (B.F. AND R.F.), SOUTH ITALIAN.
Contents: Corinthian, Pseudo-Chalcidian, Attic (b.f., r.f., black-glaze), Etruscan, Lucanian, Campanian, Apulian, Italian black-glaze, Calenian, Various.

GERMANY (EAST) 2 (Plates 62-111), *LEIPZIG 2* (Antikenmuseum der Karl-Marx-Universität)
SIGNIFICANT REPRESENTATION: ATTIC (B.F.).
Contents: Attic (b.f., w.g.), Chalcidian (b.f.), Laconian (b.f.), Boeotian (b.f.), Italian (b.f.), Etruscan (b.f. and with black and white ornamentation).

GREAT BRITAIN 1 (Plates 1-44), *LONDON* (British Museum 1)
CONTENTS: CYPRIOT (BRONZE AGE AND MYCENAEAN), ATTIC B.F. (PANATHENAIC AMPHORAE), APULIAN (GNATHIA VASES).

GREAT BRITAIN 2 (Plates 45-92), *LONDON* (British Museum 2)
CONTENTS: CYPRIOT (SUB MYCENAEAN AND IRON AGE), ATTIC (B.F.), CAMPANIAN (R.F.).
Shapes: Mixed, (incl. krateres, kylikes).

GREAT BRITAIN 3 (Plates 93-142), *OXFORD 1* (Ashmolean Museum)
CONTENTS: ATTIC, MAINLY R.F., TOGETHER WITH SOME PLASTIC, BLACK, BLACK PATTERN AND W.G.
Shapes: Mixed, (incl. krateres, lekythoi).

GREAT BRITAIN 4 (Plates 143-190), *LONDON* (British Museum 3)
CONTENTS: ATTIC (B.F. AND R.F.).
Shapes: Amphorae and stamnoi.

GREAT BRITAIN 5 (Plates 191-238), *LONDON* (British Museum 4)
CONTENTS: ATTIC (B.F. AND R.F.).
Shapes: B.f. amphorae; r.f. mainly kotylai, kantharoi, rhyta and figure vases in the form of human or animal heads.

GREAT BRITAIN 6 (Plates 239-284), *CAMBRIDGE 1* (Fitzwilliam Museum)
SIGNIFICANT REPRESENTATION: ATTIC (B.F. AND R.F.).
Contents: Attic (geometric, archaic, Protoattic, b.f., r.f., w.g., plastic, black-glaze), Boeotian (geometric, archaic, black-glaze), Laconian and imports, Protocorinthian, Corinthian and imitations, Italo-Corinthian, Figure vases, Rhodian, Chalcidian, Etruscan (Imitation of Ionic, b.f., Bucchero), Italiote (incl. early and black-glaze), Early pottery of north and central Italy, Native Apulian, Gnathian, Vases of Paestum, Campania, Apulia, Canosa.

GREAT BRITAIN 7 (Plates 285-332), *LONDON* (British Museum 5)
CONTENTS: MYCENAEAN (EXCLUDING CYPRUS AND CRETE), ATTIC R.F.
Shapes: Nolan amphorae, early r.f. hydriai.

GREAT BRITAIN 8 (Plates 333-380), *LONDON* (British Museum 6)
CONTENTS: ATTIC (B.F. AND R.F.).
Shapes: Mainly hydriai. Some lebetes and psykteres.

GREAT BRITAIN 9 (Plates 381-431), *OXFORD 2* (Ashmolean Museum)
SIGNIFICANT REPRESENTATION: EAST GREEK, ATTIC (B.F. AND R.F.).
Contents: Hellenic Cretan, Protocorinthian, Corinthian, Italo-Corinthian, Italian, East Greek (incl. Rhodian, Ionian, Various, Camiran Style, Late Rhodian, Naucratite Chalice Style, Fikellura Style, Plastic vases, Gorgoneion Style, Samian Style, Clazomenian, Situla Style), Attic (b.f., r.f., black, plastic, black-pattern, Hellenistic).

GREAT BRITAIN 10 (Plates 432-479), *LONDON* (British Museum 7)

SIGNIFICANT REPRESENTATION: ETRUSCAN, SOUTH ITALIAN.
Contents: Etruria and Latium (Impasto and Bucchero), Apulia (Messapian, Peucetian, Daunian), Campania (vases with painted red figures).

GREAT BRITAIN 11 (Plates 480-527), *CAMBRIDGE 2* (Fitzwilliam Museum)
SIGNIFICANT REPRESENTATION: CYPRIOT, ATTIC.
Contents: Cycladic, Minoan, Helladic, Mycenaean, Cypriot and impasto, Attic (Protogeometric, geometric, Orientalizing, b.f., r.f., w.g., black-glaze, plastic), Cretan, Boeotian, Protocorinthian, Corinthian, Italo-Corinthian, East Greek (various), "Megarian" bowls, Italiote and Etruscan black-glaze.

GREAT BRITAIN 12 (Plates 528-567), *READING 1* (University of Reading)
SIGNIFICANT REPRESENTATION: ATTIC (VARIOUS), BOEOTIAN, ETRUSCAN AND SOUTH ITALIAN.
Contents: Late Minoan, Mycenaean, Protocorinthian, Corinthian, Attic (geometric, b.f., w.g., r.f.), Boeotian (various), Euboean, East Greek, Laconian, Paestan, Campanian, Lucanian, Apulian, Black-glaze (various), Etruscan, Various, Hellenistic.

GREAT BRITAIN 13 (Plates 568-615), *LONDON 8* (British Museum)
SIGNIFICANT REPRESENTATION: EAST GREEK.
Contents: Fikellura, Clazomenian and related East Greek b.f., East Greek situlae, Miscellaneous East Greek, Attic (b.f. from Tell Defenneh), Clazomenian sarcophagi.

GREAT BRITAIN 14 (Plates 616-655), *OXFORD 3* (Ashmolean Museum)
CONTENTS: ATTIC (B.F.).
Shapes: Amphorae and hydriai.

GREAT BRITAIN 15 (Plates 656-717), *CASTLE ASHBY* (Northampton)
SIGNIFICANT REPRESENTATION: ATTIC (B.F. AND R.F.), ETRUSCAN AND SOUTH ITALIAN.
Contents: The Northampton amphora, Attic (b.f., r.f., black), Chalcidian, Corinthian, Etrusco-Corinthian, Lucanian, Etruscan, Campanian, Apulian, South Italian black vases, Figure vases, Daunian.
Shapes: Mixed, (incl. a number of b.f. amphorae, r.f. kylikes).

GREECE 1 (Plates 1-50), *ATHENS 1* (National Museum)
CONTENTS: ATTIC (GEOMETRIC, PROTOATTIC, B.F., R.F., W.G.).

GREECE 2 (Plates 51-98), *ATHENS 2* (National Museum)
CONTENTS: CORINTHIAN, ATTIC (GEOMETRIC, PROTOATTIC, B.F., R.F., W.G., OUTLINE).

ITALY 1 (Plates 1-49), *ROME* (Villa Giulia 1). (Museo Nazionale di Villa Giulia in Roma)
SIGNIFICANT REPRESENTATION: ATTIC (B.F. AND R.F.), ETRUSCAN AND SOUTH ITALIAN (VARIOUS).
Contents: Protocorinthian (Sicyonian ?), Attic (b.f. and r.f.), Bucchero, Ionic-Etruscan b.f., Faliscan, Apulian, Campanian.
Shapes: Mixed, (incl. a number of amphorae).

ITALY 2 (Plates 50-98), *ROME* (Villa Giulia 2). (Museo Nazionale di Villa Giulia in Roma)
SIGNIFICANT REPRESENTATION: ATTIC (R.F.), SOUTH ITALIAN (VARIOUS).
Contents: Attic (b.f., r.f., w.g.), Bucchero, Faliscan, Volsinian.
Shapes: Mixed, (incl. a number of krateres, kylikes).

ITALY 3 (Plates 99-147), *ROME* (Villa Giulia 3). (Museo Nazionale di Villa Giulia in Roma)
SIGNIFICANT REPRESENTATION: ATTIC (B.F.).
Contents: Ionian (b.f.), Attic (b.f., black-glaze), Etruscan (various), Lucanian.
Shapes: Mixed, (incl. a number of amphorae, kylikes).

ITALY 4 (Plates 148-197), *LECCE 1* (Museo Provinciale Castromediano di Lecce)
SIGNIFICANT REPRESENTATION: ATTIC (R.F.), SOUTH ITALIAN (VARIOUS, ESP. APULIAN).
Contents: Ionian, Attic (b.f. and r.f.), South Italian (incl. pottery of Picenum, Apulian, Daunian, Canosan, Gnathian, Lucanian).

ITALY 5 (Plates 198-247), *BOLOGNA 1* (Museo Civico)
CONTENTS: ATTIC (R.F.).
Shapes: Column krateres, kylikes.

ITALY 6 (Plates 248-299), *LECCE 2* (Museo Provinciale Castromediano di Lecce)
CONTENTS: APULIAN.

ITALY 7 (Plates 300-349), *BOLOGNA 2* (Museo Civico)
SIGNIFICANT REPRESENTATION: ATTIC (B.F.).
Contents: Attic (b.f. and w.g.).
Shapes: Mixed, (incl. many amphorae).

ITALY 8 (Plates 350-405), *FLORENCE 1* (Museo Archeologico)
SIGNIFICANT REPRESENTATION: CYPRIOT, VILLANOVAN, ATTIC (R.F.).
Contents: Cypriot (incl. Mycenaean), Villanovan, Bucchero, Etrusco-Roman, Protocorinthian, Attic (r.f.).

ITALY 9 (Plates 406-456), *RHODES 1* (Museo dello Spedale dei Cavalieri)
SIGNIFICANT REPRESENTATION: RHODIAN, ATTIC (B.F.).
Contents: Cypriot, Rhodian (various), Corinthian, Laconico-Cyrenaean, Ionian, Attic (b.f. and r.f.).

ITALY 10 (Plates 457-508), *RHODES 2* (Museo dello Spedale dei Cavalieri)
SIGNIFICANT REPRESENTATION: LATE MINOAN, RHODIAN (VARIOUS), ATTIC (VARIOUS).
Contents: Late Minoan, Rhodian (various), Corinthian, Ionian, Attico-Corinthian, Attic (b.f., w.g., r.f., plastic).

ITALY 11 (Plates 509-558), *CAPUA 1* (Museo Campano)
CONTENTS: CAMPANIAN (R.F.).
Shapes: Mixed, (incl. fishplates, skyphoi).

ITALY 12 (Plates 559-608), *BOLOGNA 3* (Museo Civico)
CONTENTS: SOUTH ITALIAN. MUCH APULIAN (INCL. GEOMETRIC) BUT ALSO GNATHIAN, CAMPANIAN, LUCANIAN.

ITALY 13 (Plates 609-657), *FLORENCE 2* (Museo Archeologico)
SIGNIFICANT REPRESENTATION: ATTIC (R.F.).
Contents: Attic (r.f. and w.g.).

ITALY 14 (Plates 658-706), *PALERMO 1* (Museo Nazionale di Palermo)
CONTENTS: ATTIC R.F. (WITH ONE OR TWO HYBRIDS).
Shapes: Mixed, (incl. kylikes, lekythoi, krateres).

ITALY 15 (Plates 707-757), *TARANTO 1* (Museo Nazionale)
SIGNIFICANT REPRESENTATION: EARLY SOUTH ITALIAN POTTERY: APULIAN AND LUCANIAN R.F.

Contents: Pottery of Bronze and Iron Ages from various local sites; Geometric from North Apulia, Canosan, Peucetian, Messapian; Apulian and Lucanian r.f.; Gnathian.
Shapes: Mixed, (incl. krateres).

ITALY 16 (Plates 758-802), *UMBRIA 1* (Musei Comunali Umbri)
(a) Orvieto (Museo dell' Opera del Duomo)
CONTENTS: ATTIC (R.F.).
(b) Spoleto (Museo Civico)
CONTENTS: IMPASTO (IRON AGE).
(c) Terni (Museo Comunale)
CONTENTS: VILLANOVAN, BUCCHERO, ETRUSCO-CAMPANIAN, VASES OF THE ROMAN PERIOD.
(d) Bettona (Museo Comunale)
CONTENTS: ETRUSCO-CAMPANIAN, ROMAN DOMESTIC.
(e) Todi (Museo Comunale)
CONTENTS: ATTIC (R.F.), ETRUSCAN (VARIOUS), ETRUSCO-CAMPANIAN, LOCAL ETRUSCO-ROMAN, ARRETINE, ROMAN DOMESTIC.

ITALY 17 (Plates 803-854), *SYRACUSE 1* (Museo Archeologico Nazionale)
CONTENTS: GEOMETRIC, ATTIC (B.F. AND R.F.), SOUTH ITALIAN.
Shapes: Mixed, (incl. a number of amphorae, hydriai, pelikai).

ITALY 18 (Plates 855-901), *TARANTO 2* (R. Museo Nazionale)
SIGNIFICANT REPRESENTATION: ATTIC (B.F.), APULIAN (R.F.).
Contents: Apulian, Tarantine (Mycenaean), Protocorinthian, Corinthian, Attic (b.f.).

ITALY 19 (Plates 902-940), *GENOA 1* (Museo Civico D'Archeologia Ligure di Genova-Pegli collezione del Castello d'Albertis di Genova)
SIGNIFICANT REPRESENTATION: ATTIC (R.F.), APULIAN, CAMPANIAN.
Contents: Corinthian, Attic (b.f., w.g., r.f., black-glaze), Hellenistic, Impasto, Italo-Corinthian, Ionic-Etruscan (b.f.), Apulian, Gnathian, Italian pottery with superimposed decoration, Paestan, Lucanian, Campanian (b.f., r.f., etc.), Terra sigillata.

ITALY 20 (Plates 941-990), *NAPLES 1* (Museo Nazionale)

CONTENTS: ATTIC (B.F.).
Shapes: Mixed, (incl. a number of amphorae, kylikes).

ITALY 21 (Plates 991-1022), *ROME* (Museo Preistorico L. Pigorini 1)
CONTENTS: CYPRIOT (BRONZE AGE), RHODIAN (BRONZE AGE), CAPENA (VARIOUS), NECROPOLI PICENE (IMPASTO), VEIO (INCL. IMPASTO, BUCCHERO NERO ETC.).

ITALY 22 (Plates 1023-1063), *NAPLES 2* (Museo Nazionale)
CONTENTS: CALENIAN.

ITALY 23 (Plates 1064-1098), *CAPUA 2* (Museo Campano)
CONTENTS: ATTIC (B.F. AND R.F.).

ITALY 24 (Plates 1099-1130), *NAPLES 3* (Museo Nazionale)
SIGNIFICANT REPRESENTATION: FASCICULE OF SOUTH ITALIAN POTTERY, FOURTH TO SECOND CENTURY B.C.
Contents: Late Italian pottery with decoration superimposed in red, Gnathian.

ITALY 25 (Plates 1131-1166), *TARQUINIA 1* (Museo Nazionale Tarquiniense)
CONTENTS: ITALIAN GEOMETRIC, ATTIC (B.F. AND R.F.), ITALIAN IMITATIONS OF GREEK R.F.
Shapes: Mixed, (incl. amphorae, kylikes).

ITALY 26 (Plates 1167-1204), *TARQUINIA 2* (Museo Nazionale Tarquiniense)
SIGNIFICANT REPRESENTATION: ATTIC (B.F. AND R.F.), ITALIAN (R.F.).
Contents: Local Italian, (Impasto and Bucchero), Tyrrhenian (b.f.), Attic (b.f. and r.f.), Italian (r.f.).
Shapes: Mixed, (incl. amphorae, kylikes).

ITALY 27 (Plates 1205-1248), *BOLOGNA 4* (Museo Civico)
Contents: Attic (r.f.) from the fifth century to early fourth century B.C.
Shapes: Volute krateres, calyx krateres, Campanian krateres, oinochoe, stamnos.

ITALY 28 (Plates 1249-1292), *ADRIA 1* (Museo Civico)
CONTENTS: ATTIC (R.F.).
Shapes: Mixed, (incl. kylikes).

ITALY 29 (Plates 1293-1332), *CAPUA 3* (Museo Campano)

CONTENTS: ITALIAN BLACK-GLAZE, CAMPANIAN (B.F.).

ITALY 30 (Plates 1337-1380), *FLORENCE 3* (Museo Archeologico)
CONTENTS: ATTIC (R.F.).
Shapes: Kylikes.

ITALY 31 (Plates 1381-1424), *MILAN 1* (Civico Museo Archeologico)
SIGNIFICANT REPRESENTATION: ATTIC (R.F.), ETRUSCAN AND SOUTH ITALIAN.
Contents: Attic (r.f.), Proto-Italiote, Etruscan, Campanian, Apulian.

ITALY 32 (Plates 1425-1468), *TURIN 1* (Museo di Antichità)
SIGNIFICANT REPRESENTATION: SOUTH ITALIAN.
Contents: Italiote with superimposed decoration, Proto-Italiote, Apulian, Lucanian, Campanian (r.f.), Paestan.
Shapes: Mixed, (incl. amphorae).

ITALY 33 (Plates 1469-1512), *BOLOGNA 5* (Museo Civico)
SIGNIFICANT REPRESENTATION: ATTIC (R.F.).
Shapes: Amphora, volute krateres, kylikes, kantharos, oinochoai, rhyton.

ITALY 34 (Plates 1513-1556), *VERONA 1* (Museo del Teatro Romano)
SIGNIFICANT REPRESENTATION: SOUTH ITALIAN.
Contents: Corinthian, Italo-Corinthian, Italo-geometric, Ionico-Etruscan, Attic (b.f., r.f., black-glaze), South Italian and Sicilian (incl. Apulian, Sicilian r.f., Daunian, Canosan, Messapian, Gnathian, Campanian).

ITALY 35 (Plates 1557-1600), *TARANTO 3* (Museo Nazionale)
SIGNIFICANT REPRESENTATION: ATTIC (B.F. AND R.F.), ITALIOTE (R.F.).
Contents: Corinthian, Attic (b.f. and r.f.), Italiote (r.f.), Gnathian.

ITALY 36 (Plates 1601-1644), *ROME, CAPITOLINE 1* (Musei Capitolini)
SIGNIFICANT REPRESENTATION: ATTIC (B.F.).
Contents: Protocorinthian, Corinthian, Attic (b.f., w.g.).
Shapes: Mixed, (incl. amphorae, kylikes, olpai).

ITALY 37 (Plates 1645-1688) *FERRARA 1* (Museo Nazionale)
CONTENTS: ATTIC (R.F.).
Shapes: Mixed, (incl. many krateres).

ITALY 38 (Plates 1689-1732), *FLORENCE 4* (Museo Archeologico)
CONTENTS: ATTIC (R.F.).
Shapes: Kylikes.

ITALY 39 (Plates 1733-1776), *ROME, CAPITOLINE 2* (Musei Capitolini)
SIGNIFICANT REPRESENTATION: ATTIC (R.F.).
Contents: East Greek, Italogeometric, Aristonothos krater, Etrusco-Corinthian, Laconian, Caeretan hydria, Attic (r.f.), Pontic, Etruscan (b.f.), Apulian, Protolucanian.

ITALY 40 (Plates 1777-1820), *TURIN 2* (Museo di Antichità)
SIGNIFICANT REPRESENTATION: ATTIC (B.F. AND R.F.).
Contents: East Greek, Protocorinthian, Corinthian, Italian (geometric), Italo-Corinthian, Attic (b.f. and r.f.), Etruscan (b.f. and r.f.).
Shapes: Mixed, (incl. lekythoi).

ITALY 41 (Plates 1821-1864), *ORVIETO 1* (Museo Claudio Faina)
SIGNIFICANT REPRESENTATION: ATTIC (B.F.).
Contents: Ionian, Attic (black-glaze, b.f., bilinguals), Corinthian.
Shapes: Ionian and Attic kylikes.

ITALY 42 (Plates 1865-1906), *FLORENCE 5* (Museo Nazionale)
CONTENTS: ATTIC (B.F.).
Shapes: Hydriai.

ITALY 43 (Plates 1907-1950), *TRIESTE 1* (Civico Museo di Storia ed Arte)
SIGNIFICANT REPRESENTATION: ATTIC (R.F.), APULIAN.
Contents: Corinthian, Italo-Corinthian, Attic (b.f. and r.f.), Apulian.

ITALY 44 (Plates 1951-1995), *CAPUA 4* (Museo Campano)
SIGNIFICANT REPRESENTATION: CORINTHIAN, ETRUSCAN AND SOUTH ITALIAN (VARIOUS).

Contents: Corinthian, Geometric and Subgeometric, Impasto, Bucchero, Plain, Messapian, Apulian, Gnathian, Campanian.

ITALY 45 (Plates 1996-2036), *PARMA 1* (Museo Nazionale di Antichità)
SIGNIFICANT REPRESENTATION: ATTIC (B.F. AND R.F.).
Contents: Cretan and Italian (geometric), Protocorinthian, Corinthian, Italo-Corinthian, Uncertain, Ionico-Etruscan, Attic (b.f., r.f., black-glaze), Egyptian, Etruscan.

ITALY 46 (Plates 2037-2080), *PARMA 2* (Museo Nazionale di Antichità)
SIGNIFICANT REPRESENTATION: ETRUSCAN AND SOUTH ITALIAN.
Contents: Bucchero, Etruscan (r.f.), Apulian, Italiote (various incl. Gnathian, Hellenistic and Roman types).

ITALY 47 (Plates 2081-2025), *COMO 1* (Civico Museo Archeologico "Giovio")
SIGNIFICANT REPRESENTATION: ETRUSCAN AND SOUTH ITALIAN (VARIOUS), ATTIC (B.F. AND R.F.).
Contents: Corinthian, Etrusco-Corinthian, Attic (b.f. and r.f.), Hellenistic, Etruscan, South Italian (various), Campanian, Lucanian, Paestan.

ITALY 48 (Plates 2126-2169), *FERRARA 2* (Museo Nazionale)
CONTENTS: ATTIC (B.F.).
Shapes: Mixed, (incl. lekythoi, oinochoai).

ITALY 49 (Plates 2170-2210), *MILAN* (Collezione "H.A." 1)
CONTENTS: APULIAN (R.F.).
Shapes: Mainly amphorae, krateres.

ITALY 50 (Plates 2211-2254), *PALERMO* (Collezione Mormino 1)
CONTENTS: ATTIC (B.F., W.G., SIX TECHNIQUE, R.F., BLACK-GLAZE).
Shapes: Lekythoi.

ITALY 51 (Plates 2255-2294), *MILAN* (Collezione "H.A." 2)
SIGNIFICANT REPRESENTATION: CORINTHIAN, ITALO-CORINTHIAN, ATTIC (R.F.).
Contents: Corinthian, Italo-Corinthian, Laconian, Attic (b.f. and r.f.), Etruscan and Italian (various).
Shapes: Mixed, (incl. krateres).

ITALY 52 (Plates 2295-2337), *GELA 1* (Museo Archeologico Nazionale)
CONTENTS: PROTOCORINTHIAN AND CORINTHIAN.
Shapes: Mixed, (incl. many alabastra and aryballoi).

ITALY 53 (Plates 2338-2377), *GELA 2* (Museo Archeologico Nazionale)
SIGNIFICANT REPRESENTATION: CORINTHIAN, EAST GREEK.
Contents: Corinthian, Vases of alabaster, Argive monochrome, East Greek, Laconian, Attic (black-glaze).
Shapes: Mixed, (incl. many alabastra).

ITALY 54 (Plates 2378-2421), *GELA* 3 (Museo Archeologico Nazionale)
CONTENTS: ATTIC (B.F., R.F., W.G.).
Shapes: Mainly lekythoi.

ITALY 55 (Plates 2422-2465), *TARQUINIA 3* (Museo Archeologico Nazionale)
SIGNIFICANT REPRESENTATION: ITALOGEOMETRIC.
Contents: Protocorinthian, Cumaean, Italogeometric, Etrusco-Corinthian, Archaic with linear decoration, Impasto.
Shapes: Mixed, (incl. many oinochoai).

ITALY 56 (Plates 2466-2512), *GELA 4* (Museo Archeologico Nazionale)
SIGNIFICANT REPRESENTATION: ATTIC (B.F.).
Contents: Attic (b.f., various incl. black-glaze, w.g.).
Shapes: Mixed (incl. many lekythoi).

ITALY 57 (Plates 2513-2560), *FIESOLE* (Collezione Costantini 1)
CONTENTS: CORINTHIAN, PLASTIC VASES, ATTIC (B.F. AND R.F.).
Shapes: Mixed, (incl. amphorae).

ITALY 58 (Plates 2561-2600), *FIESOLE* (Collezione Costantini 2)
SIGNIFICANT REPRESENTATION: DAUNIAN, APULIAN, GNATHIAN, ITALIAN BLACK-GLAZE.
Contents: Impasto, Bucchero, Etruscan, Daunian, Peucetian, Messapian, Apulian, Campanian, Gnathian, Italian (black-glaze).

JUGOSLAVIA 1 (Plates 1-48), *ZAGREB 1* (Musée National)
CONTENTS: LOCAL PREHISTORIC POTTERY FROM

VUCEDOL (DANUBE VALLEY). SOME WHOLE PIECES AND MANY SHERDS.

JUGOSLAVIA 2 (Plates 49-96), *ZAGREB 2* (Musée National)
CONTENTS: LOCAL POTTERY FROM SARVAS, DALJ AND VELIKA GORICA (DANUBE VALLEY).

JUGOSLAVIA 3 (Plates 97-127), *BELGRADE* (Musée du Prince Paul)
CONTENTS: LOCAL POTTERY FROM THE DANUBE VALLEY.

JUGOSLAVIA 4 (Plates 128-175), *SERAJEVO 1* (Musée National de la République Socialiste de Bosnie-Herzégovine)
SIGNIFICANT REPRESENTATION: CYPRIOT, ATTIC (VARIOUS), BOEOTIAN (VARIOUS).
Contents: Cypriot, Mycenaean, Attic (geometric, b.f., w.g., r.f.), Protocorinthian, Corinthian, Boeotian, Various.

NETHERLANDS 1 (Plates 1-48), *THE HAGUE 1* (Musée Scheurleer)
SIGNIFICANT REPRESENTATION: ATTIC (VARIOUS).
Contents: Egyptian, Cypriot, Rhodian, Cycladic and uncertain, Plastic vases of uncertain fabric, Mycenaean, Sicyonian and Corinthian, Euboean and Ionian, Boeotian, Attic (geometric, Attico-Corinthian, b.f., r.f.), black-glaze, Relief and plastic vases, Etruscan (Bucchero Nero), Italian (geometric and Orientalizing), Apulian (Gnathian), Campanian, (Cumae, Teanum, Cales).

NETHERLANDS 2 (Plates 49-94), *THE HAGUE 2* (Musée Scheurleer)
SIGNIFICANT REPRESENTATION: EGYPTIAN.
Contents: Egyptian, Rhodian, Uncertain, Corinthian, Attic (geometric, Protoattic and Attico-Corinthian, b.f., r.f., plastic), black-glaze, Hellenistic, Etrusco-Latin (Faliscan), Apulian, Campanian (Cumae, Cales).

NETHERLANDS 3 (Plates 95-147), *LEIDEN 1* (Rijksmuseum van Oudheden)
CONTENTS: ATTIC (B.F.).
Shapes: Many neck-amphorae, but also stamnos, column krater, amphorae Type B, Panathenaic amphorae, lids, graffiti.

NETHERLANDS 4 (Plates 148-201), *LEIDEN 2* (Rijksmuseum van Oudheden)

CONTENTS: ATTIC (B.F.).
Shapes: Mixed, (incl. many lekythoi).

NEW ZEALAND 1 (Plates 1-48) (Vases and fragments from eight collections)
SIGNIFICANT REPRESENTATION: ATTIC (GEOMETRIC AND B.F.), CORINTHIAN.
Contents: Attic (Protogeometric, geometric, b.f.), Corinthian.
Shapes: Mixed, (incl. amphorae, aryballoi, kylikes, lekythoi, oinochoai).

NORWAY 1 (Plates 1-52) (Public and Private Collections)
SIGNIFICANT REPRESENTATION: CORINTHIAN, ATTIC (B.F. AND R.F.).
Contents: Corinthian, Attic (b.f., r.f., w.g., plastic, black-glaze, patterned).
Shapes: Mixed, (incl. many b.f. lekythoi).

POLAND 1 (Plates 1-54), *GOLUCHOW 1* (Musée Czartoryski)
SIGNIFICANT REPRESENTATION: ATTIC (B.F. AND R.F.).
Contents: Egyptian, Cypriot, Corinthian, Greek Archaic plastic vases, Attic (Attico-Corinthian, b.f., r.f., w.g.), Plastic vases, Black-glaze, Etruscan, Italian r.f., Apulian, Campanian, Italian polychrome, Hellenistic (Centuripe Ware), Pottery of Arezzo, Carthaginian, Pottery of Gaul and Germany, Coarse ware (uncertain).

POLAND 2 (Plates 55-96), *CRACOW 1* (Collections de Cracovie)
SIGNIFICANT REPRESENTATION: ATTIC (B.F. AND R.F.), SOUTH ITALIAN.
Contents: Egyptian, Asia Minor, South Russian, Cypriot, Archaic Greek plastic vases, Mycenaean, Greek Bucchero, Small Greek Archaic vases, Corinthian and Italo-Corinthian, Boeotian, Attic (geometric, b.f., r.f., with net decoration, w.g., plastic, black-glaze), Hellenistic vases with paint applied to the glaze, Hellenistic relief, Early pottery of Latium, Etruscan (Bucchero, b.f.), Italo-Ionian, Italian of Attic r.f. style, Apulian, Italian polychrome, Greek black-glaze, Coarse Italian pottery (Graeco-Roman period), Roman pottery of Dalmatia, Algerian, Tunisian, Pottery of France and Germany, Graeco-Roman coarse pottery of unknown provenance.

POLAND 3 (Plates 97-129) (Collections Diverses)

SIGNIFICANT REPRESENTATION: ATTIC AND SOUTH ITALIAN.
Contents: Egyptian, Asia Minor, South Russian, Geometric of the Cyclades, Ionian, Cypriot, Coarse pottery pointed amphorae, Macedonian, Corinthian and Italo-Corinthian, Laconian, Boeotian, Attic (geometric, b.f., r.f., net decoration, w.g., black-glaze), Hellenistic with paint applied to glaze, Hellenistic decoration with barbotine or relief, Greek black-glaze, Italo-geometric, Etruscan, Italo-Ionic, Pottery of Arezzo, Italian of Attic r.f. style, Apulian, Campanian, Lucanian, Italian black-glaze, Coarse Italian pottery of Graeco-Roman period, Carthaginian, Vases of France and Germany, Coarse Graeco-Roman pottery of unknown provenance.

POLAND 4 (Plates 130-177), *WARSAW* (Musée National 1)
SIGNIFICANT REPRESENTATION: ATTIC (B.F.).
Contents: Attic (protogeometric, geometric, b.f.), Boeotian (geometric and b.f.), Laconian (b.f.).
Shapes: Mixed, (incl. amphorae, kylikes, lekythoi).

POLAND 5 (Plates 178-231), *WARSAW* (Musée National 2)
SIGNIFICANT REPRESENTATION: CYPRIOT, CORINTHIAN, PLASTIC VASES, W.G.
Contents: Mycenaean, Cypriot, Rhodian, Protocorinthian, Corinthian, Plastic vases (Ionian, Rhodian, Corinthian, Attic, Egyptian), w.g. lekythoi.

POLAND 6 (Plates 232-279), *WARSAW* (Musée National 3)
CONTENTS: ATTIC (R.F.).

POLAND 7 (Plates 280-332), *WARSAW* (Musée National 4)
SIGNIFICANT REPRESENTATION: SOUTH ITALIAN R.F. (ESP. APULIAN).
Contents: Lucanian, Paestan, Apulian.

POLAND 8 (Plates 333-376), *WARSAW* (Musée National 5)
SIGNIFICANT REPRESENTATION: SOUTH ITALIAN (R.F.).
Contents: Apulian, Paestan, Campanian, Uncertain.
Shapes: Mixed, (incl. krateres).

POLAND 9 (Plates 377-430), *WARSAW* (Musée National 6)
SIGNIFICANT REPRESENTATION: ATTIC (BLACK-GLAZE), GNATHIAN, CALENIAN, CAMPANIAN, ETRUSCAN.
Contents: East Greek, Boeotian, Corinthian, Attic (black-glaze), Gnathian, Xenon Group, Teanum and "Procolom"

Group, Calenian, Campanian (various), Impasto and Bucchero Nero, Etrusco-Corinthian, Etruscan (b.f. and r.f.), Daunian.

RUMANIA 1 (Plates 1-45), *BUCHAREST 1* (Institut d'Archéologie, Musée National des Antiquités)
SIGNIFICANT REPRESENTATION: CYPRIOT, ATTIC (GEOMETRIC, B.F., R.F., BLACK-GLAZE), BOEOTIAN, ARGIVE, CORINTHIAN, APULIAN, CAMPANIAN, ITALIOTE, GNATHIAN, ETRUSCAN BUCCHERO.
Shapes: Mixed, (incl. lekythoi).

RUMANIA 2 (Plates 46-88), *BUCHAREST 2* (Collection Dr Georges et Maria Severeanu – Musée de la Ville de Bucarest, et Collections Privées)
(a) *Collection Dr Georges et Maria Severeanu*
SIGNIFICANT REPRESENTATION: CORINTHIAN, ATTIC (B.F. AND R.F.).
Contents: Attic (protogeometric, geometric, b.f., r.f.), Corinthian and imitations, East Greek, Laconian, Daunian, Italian r.f., Etruscan, Black-glaze, Hellenistic, Roman.
(b) *Collections Privées*
Contents: Attic (geometric, b.f., r.f., black-glaze), Boeotian, Italiote.

SPAIN 1 (Plates 1-49), *MADRID 1* (Musée Archéologique National)
SIGNIFICANT REPRESENTATION: CORINTHIAN, ATTIC (B.F.).
Contents: Egyptian, Cypriot, Rhodian, East Greek, Corinthian, Boeotian (geometric), Attic (geometric, Attico-Corinthian, b.f.).

SPAIN 2 (Plates 50-98), *MADRID 2* (Musée Archéologique National)
CONTENTS: SPANISH (PREHISTORIC), ATTIC (R.F.).

SPAIN 3 (Plates 99-138), *BARCELONA 1* (Musée Archéologique de Barcelone)
SIGNIFICANT REPRESENTATION: IONIAN, ATTIC (B.F. AND R.F.).
Contents: Corinthian, East Greek, Cypriot, Ionian, Plastic, Various, Attic (b.f., r.f., net decoration).

SPAIN 4 (Plates 139-178), *BARCELONA 2* (Musée Archéologique de Barcelone)
CONTENTS: IBERIAN POTTERY (VARIOUS PREHISTORIC AND SOME HELLENISTIC BLACK-GLAZE).

SWEDEN 1 (Plates 1-25), *LUND 1* (Museum of Classical Antiquities)
SIGNIFICANT REPRESENTATION: CYPRIOT, MINOAN, MYCENAEAN.
Contents: Cypriot (Early, Middle, Late, Cypro-Geometric, Cypro-Archaic), Neolithic, Minoan (Early, Middle and Late), Middle Helladic, Mycenaean, Uncertain.

SWITZERLAND 1 (Plates 1-42), *GENEVA 1* (Musée d'art et d'histoire)
SIGNIFICANT REPRESENTATION: ATTIC (R.F.), BLACK-GLAZE POTTERY FROM VARIOUS SOURCES.
Contents: Dimini Style, Mycenaean, Attic (geometric, r.f., black-glaze), Boeotian (black-glaze), Greek uncertain, Vases with relief decoration, Etruscan (incl. black-glaze), Apulian (black-glaze), Campanian (black-glaze and Calene ware), Italian uncertain.
Shapes: Mixed, (incl. a number of kylikes).

SWITZERLAND 2 (Plates 43-98), *ZÜRICH 1* (Public Collections)
SIGNIFICANT REPRESENTATION: ATTIC (B.F.), APULIAN (R.F.).
Contents: Attic (geometric, b.f., r.f., w.g., black-glaze), Corinthian (incl. geometric), Argive, Laconian, Parian, Naxian, Eretrian, Boeotian (geometric, archaic, r.f., black-glaze), Ionian, West Slope and related, Eastern Greyware, Megarian Bowls, Lucanian (r.f.), Campanian (black-decorated, r.f., black-glaze), Apulian (r.f., black-glaze), South Italian (red-painted), Sicilian (white-painted, r.f.), Gnathia Ware, Canosan, Daunian, South Italian uncertain (black-glaze).
Shapes: Mixed, (incl. a number of Attic b.f. amphorae and lekythoi; Apulian amphorae, krateres, pelikai).

SWITZERLAND 3 (Plates 99-146), *GENEVA 2* (Musée d'art et d'histoire)
SIGNIFICANT REPRESENTATION: ATTIC (B.F.).
Contents: Attic (b.f., w.g., r.f.), Boeotian, Ionian incl. "Rhodian", Melian, Etruscan.
Shapes: Mixed, (incl. a number of amphorae, lekythoi).

SWITZERLAND 4 (Plates 147-202), *BASEL 1*
SIGNIFICANT REPRESENTATION: CORINTHIAN, ATTIC (B.F.).
Contents: Mycenaean, Attic (geometric, b.f., Six's Technique), Non-Attic geometric, Cretan relief amphorae, Corinthian, Plastic vases, East Greek, Various (incl. Laconian, Boeotian, uncertain), "Chalcidian".

SWITZERLAND 5 (Plates 203-256), *OSTSCHWEIZ TICINO* (Chur, St. Gallen, Winterthur, Bellinzona Museo Civico, Bellinzona Collezione Lombardi, Locarno Collezione Rossi)
SIGNIFICANT REPRESENTATION: ITALIAN, (ESP. APULIAN).
(a) *Chur*
Contents: East Greek, Cypriot, Corinthian, Boeotian, Attic, Italo-Corinthian, Etruscan, Paestan, Campanian, Various, Apulian, Gnathian, Punic.
(b) *St. Gallen*
Contents: Corinthian, Italo-Corinthian, Daunian, Peucetian, Campanian, Apulian, Gnathian, Attic, Unguentaria, Various.
(c) *Winterthur*
Contents: Cypriot, Italo-Corinthian, Corinthian, East Greek, Boeotian, Uncertain, Attic, Etruscan, Daunian, Campanian, Lucanian, Apulian, Gnathian, Italian (various).
(d) *Bellinzona, Museo Civico*
Contents: Attic, Etruscan, Impasto and Bucchero, Italian, Punic, Campanian, Apulian, Gnathian.
(e) *Bellinzona, Collezione Lombardi*
Contents: Daunian, Canosan, Peucetian, Apulian, Unguentaria, Gnathian.
(f) *Locarno, Collezione Rossi*
Contents: Daunian, Canosan, Apulian, Unguentaria, Gnathian.
U.S.A. 1 (Plates 1-53), *HOPPIN & GALLATIN*
(a) *HOPPIN*
SIGNIFICANT REPRESENTATION: ATTIC (R.F.).
Contents: Corinthian, Uncertain, Laconian, Chalcidian, Boeotian, Attic (geometric, b.f., r.f., w.g.), Campanian.
(b) *GALLATIN*
SIGNIFICANT REPRESENTATION: ATTIC (R.F.).
Contents: Cypriot (incl. Mycenaean), Ionian, Corinthian, Attic (b.f., r.f., w.g.), Plastic vases, Campanian, Italian uncertain, Apulian, Messapian).

U.S.A. 2 (Plates 54-85), *PROVIDENCE 1* (Museum of the Rhode Is. School of Design)
SIGNIFICANT REPRESENTATION: ATTIC (R.F.).
Contents: Egyptian, Mesopotamian, Cypriot, LH III (Mycenaean), Rhodian, Greek Is. Ware, Corinthian, Chalcidian, Attic (geometric, b.f., r.f., w.g., plastic, undecorated), Apulian, Gnathia Ware, Campanian, Faliscan, Lucanian, Etruscan, Greek unclassified, Roman Ware, "Megarian" Bowls, Arretine Ware, Gallo-Roman Ware.

U.S.A. 3 (Plates 86-133), *MICHIGAN* (University of Michigan 1)
SIGNIFICANT REPRESENTATION: ATTIC, ETRUSCAN, SOUTH ITALIAN.
Contents: Egyptian, Mysian, Palestinian, Cypriot, LH III, Corinthian, Laconian, Boeotian, Attic (geometric, b.f., r.f., w.g.), Black-glaze (various), Imitations of Protocorinthian and Corinthian, Etruscan, Apulian, Campanian, Italian (miscellaneous), Roman, North African, Gallic, German, graffiti.

U.S.A. 4 (Plates 134-181), *BALTIMORE* (The Robinson Collection 1)
SIGNIFICANT REPRESENTATION: ATTIC (B.F. AND W.G.).
Contents: Asia Minor, Minoan, Cypriot, Thessalian (Prehistoric, Dimini Ware), Helladic (incl. Mycenaean), Attic (geometric, b.f., w.g.), Protocorinthian, Corinthian, Italiote, Boeotian, Etruscan, or Italiote, East Greek, Fikellura, Rhodian, Chalcidian, Kabeiric (Boeotian).
Shapes: Mixed, (incl. amphorae, kylikes, w.g. lekythoi).

U.S.A. 5 (Plates 182-243), *CALIFORNIA* (University of California 1)
SIGNIFICANT REPRESENTATION: CORINTHIAN, ATTIC (B.F. AND R.F.).
Contents: Mycenaean, Attic (geometric, b.f., r.f., plastic, w.g.), Protocorinthian and Cumaean geometric, Corinthian, Boeotian (incl. "Andrian"), Chalcidian, Etruscan, Apulian, East Greek and Boeotian plastic.
Shapes: Mixed, (incl. Corinthian alabastra, Attic amphorae, kylikes, lekythoi, hydriai, pelikai, krateres).

U.S.A. 6 (Plates 244-294), *BALTIMORE* (The Robinson Collection 2)
CONTENTS: ATTIC (R.F.).
Shapes: Kylikes, plates, amphorae, column krateres, hydriai, calyx krater, lekythos, oinochoe, skyphoi, bell krateres, loutrophoros, lebetes gamikoi.

U.S.A. 7 (Plates 295-338), *BALTIMORE* (The Robinson Collection 3)
SIGNIFICANT REPRESENTATION: ATTIC (R.F., ESP. KERTSCH STYLE), SOUTH ITALIAN.
Contents: Attic (b.f. and r.f.), South Italian (incl. Apulian, Campanian, Gnathian), Etruscan, Various Italian, Terra sigillata, Arretine Ware.
Shapes: Mixed, (incl. krateres).

U.S.A. 8 (Plates 339-412) (Fogg Museum & Gallatin Collection)
(a) *FOGG MUSEUM*
SIGNIFICANT REPRESENTATION: ATTIC (B.F. AND R.F.).
Contents: Lydian, Cypriot, Cycladic, Boeotian, Attic (Protoattic, geometric, b.f., r.f., w.g., black-glaze, Hellenistic), Corinthian, Italo-Corinthian, Boeotian, Greek black-glaze, Etruscan and South Italian, (incl. Apulian, Gnathian, Campanian, Lucanian), Gallo-Roman, graffiti.
Shapes: Mixed, (incl. amphorae).
(b) *GALLATIN COLLECTION*
SIGNIFICANT REPRESENTATION: ATTIC (B.F. AND R.F.).
Contents: Egyptian, Cypriot, East Greek, Uncertain, Corinthian, Attic (b.f., w.g., r.f., "Six's Technique", black-glaze, plastic, Hellenistic), Boeotian?, Etruscan Bucchero, Apulian, South Italian, Campanian, Sicilian.
Shapes: Mixed, (incl. amphorae).

U.S.A. 9 (Plates 413-460), *NEW YORK 1* (Metropolitan Museum of Art)
CONTENTS: ARRETINE RELIEF WARE – STAMPS, MOULDS, VASES, COVERS AND FRAGMENTS.

U.S.A. 10 (Plates 461-489), *SAN FRANCISCO* (San Francisco Collections 1)
SIGNIFICANT REPRESENTATION: ATTIC (B.F. AND R.F.).
Contents: Mycenaean, Cypriot, Attic (geometric, b.f., w.g., r.f.), Corinthian, "Halys Theriomorphic", Etruscan (b.f.), Unclassified, Italiote (r.f.).
Shapes: Mixed, (incl. amphorae, lekythoi).

U.S.A. 11 (Plates 490-532), *NEW YORK 2* (Metropolitan Museum of Art)
SIGNIFICANT REPRESENTATION: ATTIC (B.F.).
Contents: Attic (b.f., black, red-and-black).
Shapes: Kylikes (Komast, Siana, Little Master, Band, Cassel, Droop, Forerunners of Eye Cups, Eye cups, Hybrids, Miscellaneous Late b.f., Black, Red-and-black stemless).

U.S.A. 12 (Plates 533-580), *NEW YORK 3* (Metropolitan Museum of Art)
CONTENTS: ATTIC (B.F.) VASES AND SHERDS.
Shapes: Panel amphorae (Types A, B and C), Panathenaic amphorae (prize vases, uninscribed Panathenaics, miniature Panathenaics).

U.S.A. 13 (Plates 581-622), *BRYN MAWR 1* (Bryn Mawr College)
CONTENTS: ATTIC (R.F.). A FEW WHOLE VASES (PINAKES AND KYLIKES) AND MANY FRAGMENTS. LATE 6TH TO EARLY 4TH CENTURY B.C.
Shapes: pinakes, kylikes, rhyta, skyphoi, amphorae, pelikai, krateres, hydriai, oinochoai, lekythoi, askoi.

U.S.A. 14 (Plates 623-680), *BOSTON 1* (Museum of Fine Arts)
CONTENTS: ATTIC (B.F.).
Shapes: One-piece amphorae, neck amphorae and fragments.

U.S.A. 15 (Plates 681-728), *CLEVELAND 1* (Museum of Art)
SIGNIFICANT REPRESENTATION: ATTIC (B.F., R.F., W.G.), APULIAN.
Contents: Attic (geometric, b.f, r.f., w.g., black-glaze), Protocorinthian, Corinthian, Etruscan, South Italian (Apulian, Campanian, Gnathia Ware, Calenian), Arretine.
Shapes: Mixed, (incl. a number of b.f. amphorae, r.f. krateres, w.g. lekythoi).

U.S.A. 16 (Plates 729-780), *NEW YORK 4* (Metropolitan Museum of Art)
CONTENTS: ATTIC (B.F.).
Shapes: Neck amphorae (incl. ovoid, early panel, Panathenaic, figures on shoulder and body, standard, special types, small and late, doubleens, Botkin Class, Class of Cabinet des Médailles 218).

U.S.A. 17 (Plates 781-840), *TOLEDO 1* (Museum of Art)
SIGNIFICANT REPRESENTATION: ATTIC (B.F. AND R.F.).
Contents: Attic (geometric, Protoattic, b.f., r.f., w.g., head vase, black-glaze).
Shapes: Mixed, (incl. a number of b.f. amphorae, and b.f. and r.f. kylikes).

U.S.A. 18 (Plates 841-892), *LOS ANGELES 1* (County Museum of Art)
SIGNIFICANT REPRESENTATION: ATTIC (B.F. AND R.F.), SOUTH ITALIAN.
Contents: Corinthian, Attic (b.f., r.f., w.g.), Early South Italian, Apulian (incl. Gnathian), Lucanian, Campanian, Paestan.
Shapes: Mixed, (incl. a number of b.f. and r.f. amphorae, r.f. krateres, pelikai.

U.S.A. 19 (Plates 893-943), *BOSTON 2* (Museum of Fine Arts)
CONTENTS: ATTIC (B.F.).
Shapes: Mixed, (esp. hydriai, kylikes).

List of CVA fascicules according to museum location

ADRIA 1	ITALY 28
ALTENBURG 1	GERMANY 17
ALTENBURG 2	GERMANY 18
ALTENBURG 3	GERMANY 19
ATHENS 1	GREECE 1
ATHENS 2	GREECE 2
BALTIMORE 1	U.S.A. 4
BALTIMORE 2	U.S.A. 6
BALTIMORE 3	U.S.A. 7
BARCELONA 1	SPAIN 3
BARCELONA 2	SPAIN 4
BASEL 1	SWITZERLAND 4
BELGRADE 1	JUGOSLAVIA 3
BERKELEY 1	U.S.A. 5
BERLIN 1	GERMANY 2
BERLIN 2	W. GERMANY 21
BERLIN 3	W. GERMANY 22
BERLIN 4	W. GERMANY 33
BERLIN 5	W. GERMANY 45
BOLOGNA 1	ITALY 5
BOLOGNA 2	ITALY 7
BOLOGNA 3	ITALY 12
BOLOGNA 4	ITALY 27
BOLOGNA 5	ITALY 33
BONN 1	GERMANY 1
BONN 2	W. GERMANY 40
BOSTON 1	U.S.A. 14
BOSTON 2	U.S.A. 19
BOURGES and TOURS 1	FRANCE 30
BRAUNSCHWEIG 1	GERMANY 4
BRUSSELS 1	BELGIUM 1

GOLUCHOW 1	POLAND 1
GOTHA 1	GERMANY 24
GOTHA 2	GERMANY 29
THE HAGUE 1	NETHERLANDS 1
THE HAGUE 2	NETHERLANDS 2
HAMBURG 1	W. GERMANY 41
HANNOVER 1	W. GERMANY 34
HEIDELBERG 1	W. GERMANY 10
HEIDELBERG 2	W. GERMANY 23
HEIDELBERG 3	W. GERMANY 27
HEIDELBERG 4	W. GERMANY 31
HOPPIN and GALLATIN	U.S.A. 1
KARLSRUHE 1	W. GERMANY 7
KARLSRUHE 2	W. GERMANY 8
KASSEL 1	W. GERMANY 35
KASSEL 2	W. GERMANY 38
LAON 1	FRANCE 20
LECCE 1	ITALY 4
LECCE 2	ITALY 6
LEIDEN 1	NETHERLANDS 3
LEIDEN 2	NETHERLANDS 4
LEIPZIG 1	GERMANY 14
LEIPZIG 2	E. GERMANY 2
LIMOGES and VANNES 1	FRANCE 24
LONDON 1	GREAT BRITAIN 1
LONDON 2	GREAT BRITAIN 2
LONDON 3	GREAT BRITAIN 4
LONDON 4	GREAT BRITAIN 5
LONDON 5	GREAT BRITAIN 7
LONDON 6	GREAT BRITAIN 8
LONDON 7	GREAT BRITAIN 10
LONDON 8	GREAT BRITAIN 13
LOS ANGELES 1	U.S.A. 18
LUND 1	SWEDEN 1
MADRID 1	SPAIN 1
MADRID 2	SPAIN 2
MAINZ (Röm. Germ. Zentralmuseum 1)	W. GERMANY 42
MAINZ (Röm. Germ. Zentralmuseum 2)	W. GERMANY 43
MAINZ (University 1)	W. GERMANY 15
MANNHEIM 1	W. GERMANY 13
MICHIGAN 1	U.S.A. 3
MILAN (Civico Museo Archeologico 1)	ITALY 31
MILAN (Collezione 'H.A.' 1)	ITALY 49
MILAN (Collezione 'H.A.' 2)	ITALY 51
MUNICH 1	GERMANY 3

MUNICH 2	GERMANY 6
MUNICH 3	W. GERMANY 9
MUNICH 4	W. GERMANY 12
MUNICH 5	W. GERMANY 20
MUNICH 6	W. GERMANY 28
MUNICH 7	W. GERMANY 32
MUNICH 8	W. GERMANY 37
NAPLES 1	ITALY 20
NAPLES 2	ITALY 22
NAPLES 3	ITALY 24
NEW YORK 1	U.S.A. 9
NEW YORK 2	U.S.A. 11
NEW YORK 3	U.S.A. 12
NEW YORK 4	U.S.A. 16
NEW ZEALAND 1	NEW ZEALAND 1
NORWAY 1	NORWAY 1
ORVIETO 1	ITALY 41
OSTSCHWEIZ TICINO 1	SWITZERLAND 5
OXFORD 1	GREAT BRITAIN 3
OXFORD 2	GREAT BRITAIN 9
OXFORD 3	GREAT BRITAIN 14
PALERMO 1	ITALY 14
PALERMO 2	ITALY 50
PARMA 1	ITALY 45
PARMA 2	ITALY 46
PARIS (Bibliothèque Nationale 1)	FRANCE 7
PARIS (Bibliothèque Nationale 2)	FRANCE 10
PARIS (Louvre 1)	FRANCE 1
PARIS (Louvre 2)	FRANCE 2
PARIS (Louvre 3)	FRANCE 4
PARIS (Louvre 4)	FRANCE 5
PARIS (Louvre 5)	FRANCE 8
PARIS (Louvre 6)	FRANCE 9
PARIS (Louvre 7)	FRANCE 11 (Index)
PARIS (Louvre 8)	FRANCE 12
PARIS (Louvre 9)	FRANCE 14
PARIS (Louvre 10)	FRANCE 17
PARIS (Louvre 11)	FRANCE 18
PARIS (Louvre 12)	FRANCE 19
PARIS (Louvre 13)	FRANCE 21
PARIS (Louvre 14)	FRANCE 22 (Index)
PARIS (Louvre 15)	FRANCE 23
PARIS (Louvre 16)	FRANCE 25
PARIS (Louvre 17)	FRANCE 26
PARIS (Louvre 18)	FRANCE 27
PARIS (Louvre 19)	FRANCE 28

VIENNA (Kunsthistorisches Museum 2)	AUSTRIA 2
VIENNA (Kunsthistorisches Museum 3)	AUSTRIA 3
WARSAW 1	POLAND 4
WARSAW 2	POLAND 5
WARSAW 3	POLAND 6
WARSAW 4	POLAND 7
WARSAW 5	POLAND 8
WARSAW 6	POLAND 9
WÜRZBURG 1	W. GERMANY 39
WÜRZBURG 2	W. GERMANY 46
ZAGREB 1	JUGOSLAVIA 1
ZAGREB 2	JUGOSLAVIA 2
ZÜRICH 1	SWITZERLAND 2

List of Abbreviations

AA	*Archäologischer Anzeiger*
ABV	John D. Beazley, *Attic Black-figure Vase-Painters*. Oxford: Clarendon, 1956.
Agora	Brian A. Sparkes and Lucy Talcott, *The Athenian Agora XII: Black and Plain Pottery*. Princeton: American School of Classical Studies at Athens, 1970.
AJA	*American Journal of Archaeology*
AK	*Antike Kunst*
AM	*(Athenische Mitteilungen) Mitteilungen des Deutschen Archäologischen Instituts, Athenische Abteilung.*
Amyx	D.A. Amyx, "The Attic Stelai" *Hesperia* 27 (1958): 169-310.
Arias, Hirmer, and Shefton	P.E. Arias, M. Hirmer and B.B. Shefton, *A History of Greek Vase-Painting*. London: Thames and Hudson, 1962.
ARV	John D. Beazley, *Attic Red-figure Vase-Painters* 2nd ed. Oxford: Clarendon, 1963.
ASCS	American Society for Classical Studies
BCH	*Bulletin de la Correspondance Hellénique*
b.f.	black-figure
BM	British Museum
Beazley, *V. Amer.*	John D. Beazley, *Attic Red-figured Vases in American Museums.* Cambridge, Mass.: Harvard University Press, 1918.
Boardman, *BF*	John Boardman, *Athenian Black Figure Vases.* London: Thames and Hudson, 1974.

Boardman, *RF*	John Boardman, *Athenian Red Figure Vases. The Archaic Period.* London: Thames and Hudson, 1975.
Boston I, II, & III	L.D. Caskey and J.D. Beazley, *Attic Vase Paintings in the Museum of Fine Arts, Boston.* London: Oxford University Press, 1931- .
BSA	Annual of the British School at Athens
Coldstream	J.N. Coldstream, *Greek Geometric Pottery.* London: Methuen, 1968.
Cook	R.M. Cook, *Greek Painted Pottery.* 2nd ed. London: Methuen, 1972.
CIG	*Corpus Inscriptionum Graecarum*
CVA	*Corpus Vasorum Antiquorum*
Daremberg and Saglio	C. Daremberg and E. Saglio, *Dictionnaire des antiquités grecques et romaines.* Paris.
Delt.	*Archaiologikon Deltion*
esp.	especially
Frisk	H. Frisk, *Griechisches etymologisches Wörterbuch.* Heidelberg, 1967.
Gericke	Helga Gericke, *Gefässdarstellungen auf Griechischen Vasen.* Berlin: Verlag Bruno Hessling, 1970.
HASB	*Hefte des archäologischen Seminars der Universität Bern*
Hesp.	*Hesperia*
incl.	including
IG	*Inscriptiones Graecae*
JdI	*Jahrbuch des Deutschen Archäologischen Instituts*
JHS	*Journal of Hellenic Studies*
Lazzarini	Maria L. Lazzarini, "I Nomi dei Vasi Greci nelle Iscrizioni dei Vasi Stessi". *Archeologica Classica* XXV-XXVI (1973–4): 341-375.
Mon. Ant.	*Monumenti Antichi per cura della R. Accademia dei Lincei*
Payne	Humfry Payne, *Necrocorinthia.* Oxford: Clarendon, 1931.
RE	*Paulys Realencyclopädie der classischen Altertumswissenschaft*
REA	*Revue des Études Anciennes*
r.f.	red-figure
RM	*(Römische Mitteilungen) Mitteilungen des Deutschen Archäologischen Instituts, Römische Abteilung*
Richter and Milne	Gisela M.A. Richter and Marjorie J.

	Milne, *Shapes and Names of Athenian Vases.* New York: Metropolitan Museum of Art, 1935.
Schiering	Wolfgang Schiering, *Griechische Tongefässe: Gestalt, Bestimmung und Formenwandel.* Berlin: Gebr. Mann Verlag, 1967.
Simon	Erika Simon, *Die Griechischen Vasen.* Munich: Hirmer Verlag, 1976.
Webb	Virginia Webb, *Archaic Greek Faience: Miniature scent bottles and related objects from East Greece, 650*–500 B.C. Warminister, England: Aris and Phillips, 1978.
w.g.	white-ground

Index

Y0-CAM-822

The Series on Social Emotional Learning

School–Family Partnerships for Children's Success
Evanthia N. Patrikakou, Roger P. Weissberg, Sam Redding, and Herbert J. Walberg, editors

Building Academic Success on Social and Emotional Learning: What Does the Research Say?
Joseph E. Zins, Roger P. Weissberg, Margaret C. Wang, and Herbert J. Walberg, editors

How Social and Emotional Development Add Up: Getting Results in Math and Science Education
Norris M. Haynes, Michael Ben-Avie, and Jacque Ensign, editors

Higher Expectations: Promoting Social Emotional Learning and Academic Achievement in Your School
Raymond J. Pasi

Caring Classrooms/Intelligent Schools: The Social Emotional Education of Young Children
Jonathan Cohen, editor

Educating Minds and Hearts: Social Emotional Learning and the Passage into Adolescence
Jonathan Cohen, editor

Social emotional learning is now recognized as an essential aspect of children's education and a necessary feature of all successful school reform efforts. The books in this series will present perspectives and exemplary programs that foster social and emotional learning for children and adolescents in our schools, including interdisciplinary, developmental, curricular, and instructional contributions. The three levels of service that constitute social emotional learning programs will be critically presented: (1) curriculum-based programs directed to all children to enhance social and emotional competencies, (2) programs and perspectives intended for special needs children, and (3) programs and perspectives that seek to promote the social and emotional awareness and skills of educators and other school personnel.

School–Family Partnerships for Children's Success

EDITED BY

Evanthia N. Patrikakou

Roger P. Weissberg

Sam Redding

Herbert J. Walberg

FOREWORD BY

Joyce L. Epstein

Teachers College
Columbia University
New York and London

To my parents, who shaped my spirit. To my teachers, who enriched my thinking. To George and Ioanna, my partnerships in life.

—Evanthia N. Patrikakou

To my nuclear family, who taught me about family partnerships: Ned, Snooks, Kenny, Helen, Ellen, Lew, Stephanie, Elizabeth, and Ted.

—Roger P. Weissberg

Raising four children and watching them raise what today totals seven grandchildren has kept at least one of my feet planted on terra firma. For that, I thank them.

—Sam Redding

To Madoka B. Walberg and Herbert J. Walberg III.

—Herbert J. Walberg

Published by Teachers College Press, 1234 Amsterdam Avenue, New York, NY 10027

Library of Congress Cataloging-in-Publication Data

School-family partnerships for children's success / edited by Evanthia N. Patrikakou . . . [et al.]; foreword by Joyce L. Epstein.
p. cm.—(The series on social emotional learning)
Includes bibliographical references and index.
ISBN 0-8077-4601-0 (cloth : alk. paper)—ISBN 0-8077-4600-2 (pbk. : alk. paper)
1. Education—Parent participation. I. Patrikakou, Evanthia N. II. Series.

LB1048.5.S45 2005
371.19'2—dc22

2005043974

ISBN 0-8077-4600-2 (paper)
ISBN 0-8077-4601-0 (cloth)

Printed on acid-free paper

Manufactured in the United States of America

12 11 10 09 08 07 06 05 8 7 6 5 4 3 2 1

Contents

Foreword

Children learn and grow at home, at school, and in the community. Their experiences may be positive or negative, but it is clear that the people in these three contexts influence student learning and development from infancy on. Despite this certainty, there still are many questions that must be addressed to enable more families, ideally all families, to become and remain involved in their children's education in positive ways across the grades.

This book, *School–Family Partnerships for Children's Success*, presents useful research summaries, reports, perspectives, and recommendations on aspects of family involvement and on the work that educators and parents may do together to help more students succeed. The collection makes four important contributions to strengthen the knowledge base on school, family, and community partnerships.

1. *The book confirms that parental involvement—or, more broadly, school, family, and community partnerships—is a multidimensional concept.* Across chapters, the authors' discussions of home and school connections confirm the usefulness of the research-generated framework of six types of involvement (Epstein, 2001). Their lists of involvement activities include examples for all of the framework's components of *parenting, communicating, volunteering, learning at home, decision making, and collaborating with the community* (see the outline in the Introduction).

Although some authors call for common measures of involvement to be used in research studies, that is unlikely to occur in this increasingly complex field. Specific studies, such as those in this book, explore parents' experiences, teachers' practices to involve families at different school levels, and/or students' experiences with family and community involvement and results. The disparate studies will not, and need not, measure the same involvement activities nor address the complete framework, just as useful studies of reading programs do not, and need not, measure every component of a complete reading curriculum. Nevertheless, the sum total of results of diverse studies of family and community involvement will continue to increase researchers' and educators' understanding of the nature, results, interconnections, and multidimensionality of the six types of involvement.

2. *The book deepens an understanding that student success in school is a multidimensional concept.* Successful learning and development are defined not only by achievement test scores, but also by many other social and emotional skills, attitudes, and behaviors that students develop and improve across the grades. Although every parent and educator knows that this is true (e.g., expecting children to become good citizens, with good jobs, and happy, fair, and honest people), it is important, in this age of ultranarrow school accountability focused mainly on student achievement test scores, to see strong support for the argument that student success must be considered more broadly.

The authors measure and discuss numerous valued traits and behaviors. The cumulative list, across chapters, includes *academic outcomes* such as readiness for school, participation in class, completion of homework, report card grades, test scores, course credits, high school completion, enrollment in postsecondary education, and other academically related behaviors. Each chapter also discusses important *social and emotional outcomes* for students' personal development, such as motivation to learn; attitudes toward schoolwork; dimensions of self-esteem; positive relationships with teachers, peers, friends, and other adults in school; attendance; good behavior; perseverance on tasks; postsecondary planning; and other indicators of citizenship, good character, and academic commitment.

Some may consider these personal qualities "soft," but the authors show repeatedly that social and emotional characteristics are influenced by school and family partnership practices *and* that they are linked to student achievement and other indicators of success. (See Chapter 6 for a summary of an impressive body of work by Reynolds and his colleagues on the short-term and long-term effects of parental involvement in and after early childhood education programs on students' academic, social, and emotional outcomes.) The authors state the need for better measures of social and emotional attitudes and behaviors. Their work should encourage others to take the challenge to explore and advance this essential argument about success in school.

It is very clear that neither the quality of schools nor the success of students can be understood by measuring and monitoring only achievement test scores. I recall an interesting finding from one of my early studies of students' satisfaction with school. Student attitudes about school were only slightly correlated with achievement test scores, more highly correlated with report card grades, and most highly correlated with student participation in class. The more actively engaged the student, the more positive the attitude about school. Longitudinally, students who liked school, liked their teachers, and were committed to learning did better in school over time and stayed in school, even if they were barely passing their courses. This book opens questions about whether more students would develop the kinds of positive attitudes and behaviors that lead to success in school and high school completion if educators, families, and community partners worked together to promote these characteristics.

3. *The book increases attention to the need for schools and school districts to develop and implement effective and inclusive programs of school, family, and community partnerships.* All of the authors offer strong recommendations to apply research results, known to date, to improve school programs that will involve more families in productive ways. Several remind readers that parents' jobs, family resources, neighborhoods, and other social and economic variables at home and in the community affect students' achievement and social and emotional development (see, for example, Chapter 5). Still, the call is clear for schools to create programs and opportunities that will involve all families and to strengthen and sustain their influences on student success in school.

The recommendations for developing programs, policies, and practice focus on different partners and results. For example, some authors suggest that school programs of partnership should strengthen parents' beliefs about the importance of involvement and increase parents' feelings of competence in communicating with teachers and working with their children on schoolwork (see Chapters 1, 2, and 3). Others suggest that partnership programs should improve teachers' practices to reach more families and to incorporate families' cultural strengths in the classroom and in involvement activities (see Chapter 4). Still others recommend that programs of family and community involvement should focus on the "bottom line" of producing measurable results for students. Clearly, varied measures and methods will be needed to understand program effects on parents, teachers, and students.

Several authors also recommended that teachers and principals should receive more and better preservice and advanced education and professional development on school, family, and community partnerships (see Chapter 9, and information for courses in Epstein, 2001). Research will be needed on how college courses on family and community involvement affect school practice, and how both preservice and inservice professional development affect the organization of partnership programs, outreach and responses of parents, and effects on students.

In order for research to be useful in practice, researchers and educators must work together to make complex structures and processes understandable and applicable in elementary, middle, and high schools. Readers may be interested to know that many of the authors' recommendations about developing partnership programs are being implemented by schools, districts, and states that are working with the National Network of Partnership Schools at Johns Hopkins University (Epstein et al., 2002). Over 1,000 sites are presently partners in applying research-based tools and materials to design, develop, implement, and evaluate programs to involve families of their students in ways that will help more students improve academics, attitudes, and behaviors. Examples of partnership programs and practices in diverse communities across the country may be found on the Web site *www.partnershipschools.org* (2002), in the section "In the Spotlight."

Increasingly, educators are being pressed to implement research-based approaches to improve school programs and practices. The recommendations to apply research on school, family, and community partnerships to improve school programs, strengthen college courses, and develop more informed policies on family involvement (see Chapters 7 and 8) are timely. To date, responses of educators and their applications of research on this topic in practice and policy have been encouraging.

4. *The book suggests many topics for new research on the dimensions of family and community involvement and the effects of partnership programs and practices on students' academic, social, and emotional development across the grades.* It is, of course, always important to address new questions with increasingly better measures and methods to improve any field of study. The topics suggested throughout the book need to be studied intensively, longitudinally, and in diverse communities.

Several authors provide complex conceptual models to help readers visualize how parents and the community may affect student learning and development. Readers may also be interested in an additional theory, *overlapping spheres of influence,* which integrates sociological and social-psychological perspectives (Epstein, 2001). That theory posits that children learn more and better when home, school, and community share some responsibilities for student learning and development. It includes an external model of contextual effects and an internal model of interpersonal interactions and networking to represent the varied connections that may occur as parents, teachers, and others are drawn together or pushed apart in their support for students' learning and development. The conceptual and theoretical models suggest many paths of partnerships, which should be tested in new research to increase knowledge about how school programs and specific practices of involvement affect diverse families and their children.

Among the topics in need of more research, two are particularly pertinent. There are too few studies and too little understanding of the effects of theoretically linked involvement activities and outcomes. As one example, studies are needed on the effects of family and community involvement activities in reading on students' reading achievement, attitudes about reading, self-concept as a reader, and other academic, social, and emotional indicators of students' enjoyment, commitment, and success in reading. Similarly, studies are needed on the effects of theoretically linked involvement activities in math, science, other academic subjects, and the social and emotional outcomes for student success that are discussed in this book. Another neglected topic, discussed by several authors, is the students' role in the conduct of school, family, and community partnerships for differential effects on academic, social, and emotional outcomes across the grades.

The book has other good qualities. The chapters are comprehensive and readable. Authors cross-reference one another's chapters to inform readers of the connections between and among aspects of involvement. Combined, the bibliographies document the history, diversity, and growth in research on school, family, and community partnerships in psychology, sociology, and education.

Across chapters, the authors show by example that it is possible to unpack the complexities of the multidimensional concept of school, family, and community partnerships and the diverse dimensions of student success in academic, social, and emotional learning. The issues that are raised are compelling and should entice researchers to join this exciting and important agenda.

Joyce L. Epstein, Ph.D.,
Director, Center on School,
Family and Community Partnerships,
and the National Network of Partnership Schools,
Johns Hopkins University, Baltimore, Maryland

REFERENCES

Epstein, J. L. (2001). *School, family, and community partnerships: Preparing educators and improving schools.* Boulder, CO: Westview Press.

Epstein, J. L., Sanders, M. G., Simon, B. S., Salinas, K. C., Jansorn, N. R., & Van Voorhis, F. L. (2002). *School, family, and community partnerships: Your handbook for action, second edition.* Thousand Oaks, CA: Corwin Press.

www.partnershipschools.org. (2002). Web site for the National Network of Partnership Schools at Johns Hopkins University.

Preface: A Book in the Making . . .

After a decade of collaboration on researching school–family partnerships and working with educators and policymakers to make these partnerships happen, we dreamt of having the opportunity to bring together colleagues, leaders in the fields of home–school relations and human development, to prepare a book offering a comprehensive view of the topic. This book has made this dream come true. Over the years, we refined the original concept and set a framework to embrace the different dimensions of school–family partnerships and the academic, social, and emotional effects they have on children and adolescents.

The collaborative work of the editorial team and the high-quality chapters the gifted authors developed made this effort truly enjoyable and stimulating. The enthusiasm of all who participated has been inspirational, and it was their cooperation that made this endeavor feasible.

We are grateful to the Mid-Atlantic Regional Educational Laboratory for Student Success (LSS) at Temple University for supporting this dream. LSS sponsored a conference on school–family partnerships as part of their Signature Series of National Invitational Conferences in December 2002. These Signature Series are part of the LSS mission to disseminate information supporting student success. The gathering was supported by a grant from the Institute of Education Sciences at the U.S. Department of Education to LSS. Marilyn Murphy, LSS Co-Director and Director of Outreach and Dissemination, coordinated the conference and continues to assist in promoting constructive home–school relations. Kent McGuire, Executive Director of the Center for Research in Human Development and Education at Temple University, has provided enthusiastic, ongoing support for our school–family partnership efforts. Their support is very much appreciated.

Initial versions of the chapters of this book were featured at the conference, where experts in home–school relations and human development engaged in an extremely stimulating and productive discussion. More than

100 researchers; administrators; teachers; local, state, and federal education officials; parents; and top experts in school–family relations from many disciplines participated in large- and small-group presentations and discussions. The conference proceedings placed the spotlight on the crucial role that school–family partnerships play in children's education and development, and emphasized that such partnerships should be an important factor in national education reform.

Happy readings!

Eva, Roger, Sam, and Herb
Spring 2005

Introduction

School–Family Partnerships: Enhancing the Academic, Social, and Emotional Learning of Children

Evanthia N. Patrikakou, Roger P. Weissberg, Sam Redding & Herbert J. Walberg

The realization of children's potential depends, to a great degree, on the contexts within which they develop and learn, as well as on interconnections between those contexts. From the onset of a child's life, the family and relationships formed among family members are profound catalysts of social, emotional, and cognitive development. From the critical attachment of the child to the caregiver to the later years of adolescent development and citizenship, families are the first context, the first socialization system in which a child's individual skills interact with the immediate environment and result in individual growth. Once children enter the stage of formal schooling, schools become another important context influencing their development. However, it is not only single settings that contribute to a child's development; most importantly, *interrelationships* among contexts play decisive roles in human development. More supportive links among settings produce greater potential for healthy development (Bronfenbrenner, 1979). This introduction and the book as a whole will focus on the multidimensional links between the school and home environments, what influences them, and how they contribute to a child's academic, social, and emotional learning.

 ISBN 0-8077-4600-2 (paper), ISBN 0-8077-4601-0 (cloth).

IN SEARCH OF A DEFINITION OF SCHOOL–FAMILY PARTNERSHIPS

One challenging issue in reviewing this literature is the way school–family relations have been defined, measured, studied, and evaluated. Research studies usually operationally define the construct of home–school connections, but a broadly accepted definition is far from being used. When studies and policies do not offer well-defined definitions or well-developed frameworks for school–family partnerships, the challenge to measure the nature and outcomes of such partnerships becomes even more pronounced. Indicators seem to vary from study to study, often making it difficult to compare across studies or synthesize the literature (Fan & Chen, 2001).

The broad term "parent involvement" has been mostly used intuitively, but its operational use has not been consistent or clear (Redding, 2000). Terms such as "parent involvement," "parent participation," "home–school connections," and "school–family partnerships" are used to describe a broad array of parent beliefs, behaviors, and practices.

Recently, studies in the broader area of parent involvement programming have focused on issues of design, implementation, and evaluation. Such efforts have raised the need for a concrete conceptual framework and also point to the importance of using rigorous research methods to provide robust evidence of program effectiveness (Mattingly, Prislin, McKenzie, Rodriguez, & Kayzar, 2002).

The Multidimensional Nature of School–Family Partnerships

The lack of a common framework or definition lies primarily in the *multidimensional* nature of parent and teacher influences on children, as well as the complexity of home–school connections (see Figure I.1). *Child development* considerations (Eccles, 1999; Elias, Bryan, Patrikakou, & Weissberg, 2003; Steinberg, 1992), the *beliefs and expectations* held by all individuals involved in the educational process (Eccles & Harold, 1996; Patrikakou, 1997, 2004; Reynolds & Walberg, 1992), the *different roles* that parents, students, and teachers play (Christenson & Sheridan, 2001; Hoover-Dempsey & Sandler, 1997), *cultural perspectives* (Laosa, 1997; Taylor, Casten, & Flickinger, 1993), and the *policies* that outline or mandate schools to forge relationships with families all contribute to what is known as school–family partnerships (Chavkin & Williams, 1988; Moles, 1997; Redding, 2000; Weiss, 1996).

FIGURE I.1. The Multidimensionality of School–Family Partnerships

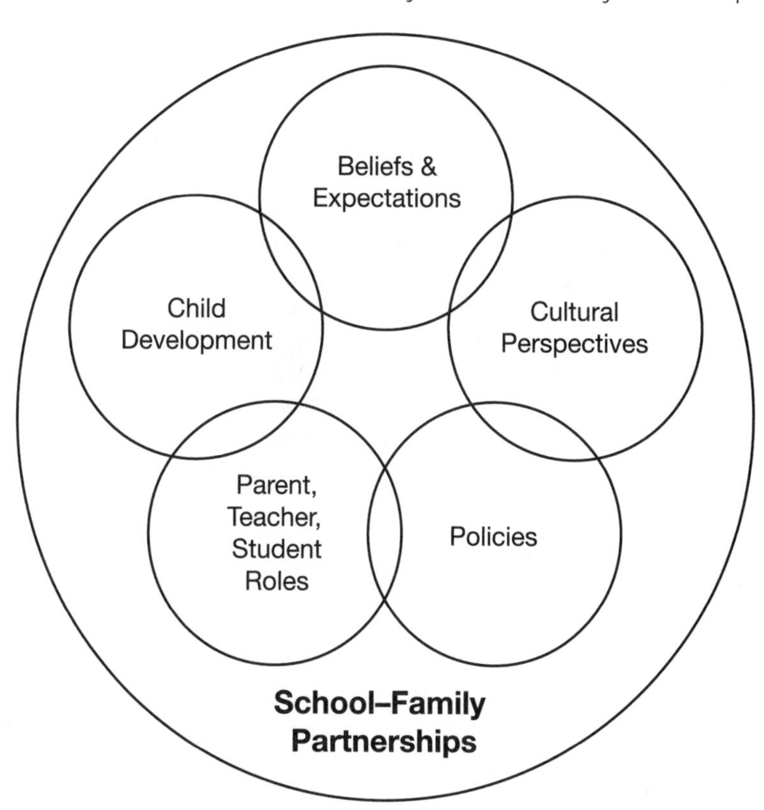

Partnership Orientation. We believe that the term "school–family partnerships" reflects the *multidimensional* nature of home–school interactions, and indicates a *shared responsibility* that both home and school have in children's education. The *partnership orientation*, which is emphasized throughout this volume, differs from the fragmented offering of isolated parent involvement activities in that it reflects a relationship between home and school rather than an occasional interface of the two institutions. Analyzing the multidimensional nature of school–family partnerships is one of the main purposes of this book. Presenting the relationship between school–family partnerships and children's academic, social, and emotional learning is the other central objective.

In the next section we offer an overview of the field of school–family partnerships. Following that discussion, we review the relationship between school–family partnerships and school success. We then offer conceptual organizers and a broader conceptual framework that encompasses the multidimensional nature of school–family partnerships. Finally, we provide an overview of chapter contents.

SCHOOL–FAMILY PARTNERSHIPS FRONT AND CENTER

School–family partnerships have increasingly been the focus of research, policy, and practice efforts in recent years. Part of the interest in the topic has been the finding that children benefit when schools and families work closely and cooperatively. Such benefits include higher grades and test scores, as well as better attendance, improved behavior at home and at school, better interpersonal skills, and more responsible decision making (Henderson & Mapp, 2002).

Acknowledged in Educational Policies

Such strong findings have been reflected in major legislation implemented by the U.S. Department of Education. For example, in 1990 the National Education Goals Panel ambitiously proclaimed increased parental participation in education as a key goal: "By the year 2000, every school will promote partnerships that will increase parental involvement in promoting the social, emotional, and academic growth of children" (National Education Goals Panel, 1995, p. 13). The panel proposed that state and local education agencies work together to develop partnership programs that would meet the varying needs of bilingual, disabled, or disadvantaged children and their parents. Programs would support the academic work of children at home, promote shared decision making at school, and hold schools and teachers accountable for high standards of achievement. In the years since the panel called for greater collaboration between schools and families, more attention has been placed on the importance of parent involvement, but the goal has not been fully met.

Most recently, the No Child Left Behind Act of 2001 (NCLB, 2002) became the centerpiece of the department's education strategy. NCLB acknowledges that parents play an integral role in their children's learning, and that they should be given the opportunity to act as full partners in their child's education. However, the precise roles that parents can assume within the various NCLB programs are not clearly defined. Therefore, states, districts, school personnel, and parents alike will benefit from theoretically and evidence-based approaches to enhance school–family ties.

As a Means of Preventing Risky Behaviors

Outreach to parents is not the only critical element for student success that may get sidetracked. While many educators, schools, and districts are concerned with the social and emotional development of students, in addition to their academic performance, they tend to focus their efforts almost exclusively on academic progress and high-stakes assessment. This becomes even more alarming considering all the risks, including drug use and violence, facing schools, families, and students today.

Preventing such risky behaviors involves many factors. One is the involvement of parents. It is imperative that through their relationship with schools, parents become aware of their great positive influence and preventive power. Major prevention efforts such as the National Youth Anti-Drug Media Campaign (2004), a historic initiative to educate and empower all youth to reject illegal drugs launched in 1998, underline the important role parents play in prevention that works. The campaign's motto, "Parents: The Anti-Drug" reflects the critical importance of parent involvement in children's and youth's healthy development (http://theantidrug.com).

Enhancing Social and Emotional Learning. Another important ingredient in effective prevention initiatives is the enhancement of social and emotional learning. Social and emotional learning is the process of acquiring the skills to recognize and manage emotions, demonstrate caring and concern for others, make responsible decisions, establish positive relationships, and handle challenging situations effectively. Social and emotional learning is fundamental to children's academic learning, social and emotional development, health and mental well-being, motivation to achieve, and citizenship (CASEL, 2003; Elias et al., 1997). The resolution passed by the National Conference of State Legislatures in 2002 underlines the importance of such competencies for school and later success in life, and calls for parents, schools, and policymakers to pay attention to and better integrate social and emotional learning into the educational process: "Legislators have an obligation to help our children acquire the skills they need to become productive and contributing members of our society. As part of this responsibility, we must encourage our schools to ensure that children are well trained in academic subjects and also given the social-emotional skills that build character and lay the foundation of good citizenship" (p. 1).

SCHOOL–FAMILY PARTNERSHIPS AND SCHOOL SUCCESS

What is school success? Is it good grades and mastery of reading, writing, mathematics, and science? Does an emphasis on academics suffice to foster a child's positive health, character, social and emotional development, and civic engagement?

Educators, parents, and policymakers have become increasingly aware that in addition to the mastery of academic subjects, school success encompasses a broad array of competencies and behaviors. Successful students not only perform academically and are committed to lifelong learning, they also display responsible and respectful social behaviors, practice safe and positive health habits, and are engaged participants in their families, schools, and communities (Greenberg et al., 2003).

In order for children to succeed in school there needs to be a synergy of many factors and a collaboration of all the people and systems involved in a child's education. In addition to developing the basic academic skills, children need to develop competencies including awareness of self and others, self-management, relationship skills, and responsible decision making (CASEL, 2003). It has been indicated that schools will be most successful in their educational mission when they integrate efforts to promote children's academic, social, and emotional learning (Zins, Weissberg, Wang, & Walberg, 2004). "Scholastic achievement must go hand-in-hand with the acquisition of traits such as honesty, cooperation, fairness, respect for others, kindness, trustworthiness, the ability to resolve conflict, and the insight to understand why such character traits are important" (National Conference of State Legislatures, 2002). It has also been shown that the more interconnections are established between the major social institutions and contexts in a child's life, the more long-term benefits can be achieved (Reynolds, Temple, Robertson, & Mann, 2001).

In light of recent incidents of school violence, and increasing concerns from parents and teachers regarding children's social and emotional growth, school–family partnerships should be considered a powerful force for children's success in general, and not just academics. Strong school–family partnerships that involve coordinated efforts between teachers and parents must expand their framework to encompass social and emotional learning, and therefore help children develop and apply the necessary skills to succeed academically, socially, and emotionally at school and in life.

Evidence of Academic, Social, and Emotional Benefits

Positive academic outcomes stemming from parent involvement range from gains in early childhood to adolescence and beyond. Most common measures used to define academic achievement are report card grades, grade point averages, standardized test scores, teacher ratings, cognitive test scores, grade retention, and dropout rates (Henderson & Mapp, 2002). Research evidence on home–school relations and academic achievement has indicated that children whose parents are involved with early childhood programs such as Head Start score higher on cognitive development scales, use a richer vocabulary, and speak using more complex sentences than do children whose families are not part of such programs (Mathematica Policy Research and the Center for Children and Families at Teachers College, Columbia University, 2001). Also, participants in early childhood programs that had a family collaboration and support component are more likely to score at or above national norms on scholastic readiness tests at school entry. Most importantly, these gains continue to be prominent in later educational performance, with fewer grade retentions and increased high school completion rates (Henderson & Mapp, 2002). Home–school communication and parent involvement remain strong predictors of academic achievement even in high school. Students whose parents stayed well informed and held high expectations for them had higher grades, completed more academic credits, and were more likely to plan for college (Catsambis, 2002; Patrikakou, 1997, 2004).

In several studies investigating academic achievement, definitions were broadened to include measures on improved behavior and healthy development. Even in a somewhat fragmented way, researchers examining home–school relations have been acknowledging for a long time the value that social and emotional factors have on learning and academic achievement. More recently, these factors have been recognized for their own merits and for the important role they play in academic success (Zins et al., 2004).

Children who participated in early childhood programs with a family component displayed much less delinquent behavior later in life than those who did not (e.g., 40% fewer arrests for violent offenses; Reynolds, 2000). Adolescents who are supported at home and at school display more positive attitudes about school, better attendance and behavior, and increased class preparation (Henderson & Mapp, 2002). In turn, researchers have found that positive behavior at school is linked to positive learning outcomes (Haynes, Ben-Avie, & Ensign, 2003). Specifically, social and emotional learning has been shown to increase mastery of subject material and motivation to learn; reduce anxiety, enhance attention, and improve study skills; and increase commitment to school and the time devoted to schoolwork. Social and emotional learning has also been shown to improve attendance and graduation rates, as well as constructive employment, while it reduces suspensions, expulsions, and grade retention (CASEL, 2003; Zins et al., 2004).

CONCEPTUAL ORGANIZERS OF PARENT INVOLVEMENT

In recent years, informative conceptual organizers, also called "typologies," have been developed to describe different roles that parents can play at home and at school. These typologies have stimulated more specific definitions and more precise indicators.

Three Central Types of Parent Involvement

There are two broad types often used to categorize parent involvement: "at home" and "at school" (Henderson & Mapp, 2002). At-home involvement may include activities such as helping with homework, whereas at-school involvement may involve attending school events. An enriched version of this framework includes "communication" as a third, separate type of home–school relations, and actually sets communication as the catalyst of parent involvement activities at home and at school (Patrikakou & Weissberg, 1999; Rubenstein, Patrikakou, Weissberg, & Armstrong, 1999). Two-way communication between home and school reflects the reciprocity that a collaborative relationship must have in order to be most effective. Establishing a positive, proactive, persistent, and personalized communication channel between parents and educators increases parent participation in learning activities at home and at school (Patrikakou & Weissberg, 1999, 2000).

Epstein's Types of Parent Involvement. A more detailed typology that has been widely used by researchers and practitioners alike is that of Epstein and her colleagues (1987, 1995). Epstein's typology fleshes out the dimensions of communication, parent involvement at home, and parent involvement at school and offers specific ways parents can be involved under each type. It offers six types of parent involvement, six ways through which families and schools can work together to maximize student benefits. These six types of involvement are:

Type 1: Parenting. The basic obligations of the family, such as establishing positive home conditions to encourage school success.
Type 2: Communicating. The communication between home and school that facilitates the flow of information about school curriculum and the child's progress.
Type 3: Volunteering. The recruitment of parents to act as volunteers in order to help and support school initiatives and functions.
Type 4: Learning at Home. Parental involvement at home, such as help-

ing children with homework and other learning activities, based on the information schools provide.

Type 5: Decision Making. Active parent involvement in school decisions and advocacy to lobbying for school improvements.

Type 6: Collaborating with the Community. The identification and dissemination of a network of available resources and services in the community in order to assist parents and schools in their efforts to better their children's education.

The next section offers a broader conceptual framework that encompasses the multidimensional nature and different types of school–family partnerships, as well as the complex processes of academic, social, and emotional learning.

AN INTEGRATIVE FRAMEWORK OF SCHOOL–FAMILY PARTNERSHIPS

We believe that the best framework to encompass the *multidimensional* nature of school family partnerships is an ecological, developmental framework. Such a framework allows for the inclusion of the effects that school–family partnerships have both on child development, as well as on partnerships themselves (see Figure I.2). There are four levels of contexts and influences ranging from the most immediate *microsystem* (such as the family or the school environment) to the more distant *macrosystem* (such as broad cultural contexts).

The center of the system is the individual with her abilities and developmental characteristics. All influences experienced in any single setting, such as the family, school, or neighborhood, are part of the *microsystem*. Influences and interrelations among microsystems are part of the *mesosystem*, the second level of influences. It is the mesosystem where school–family partnerships are formed and maintained, contributing to the student's academic, social, and emotional learning. As we discussed earlier in this chapter, the existence of strong reciprocal links between home and school has a strong positive influence on a child's development, whereas the lack of such links becomes a risk factor in development (Weissberg & Greenberg, 1998). The more supportive links there are between settings, the more potential there is for healthy development. All chapters of the book involve the mesosystem and investigate the different forms of interactions that are developed and fostered within this structure.

The next system is the *exosystem*, which includes influences from broader social groups and institutions (e.g., school board or community

FIGURE I.2. An Ecological Perspective of School–Family Partnerships

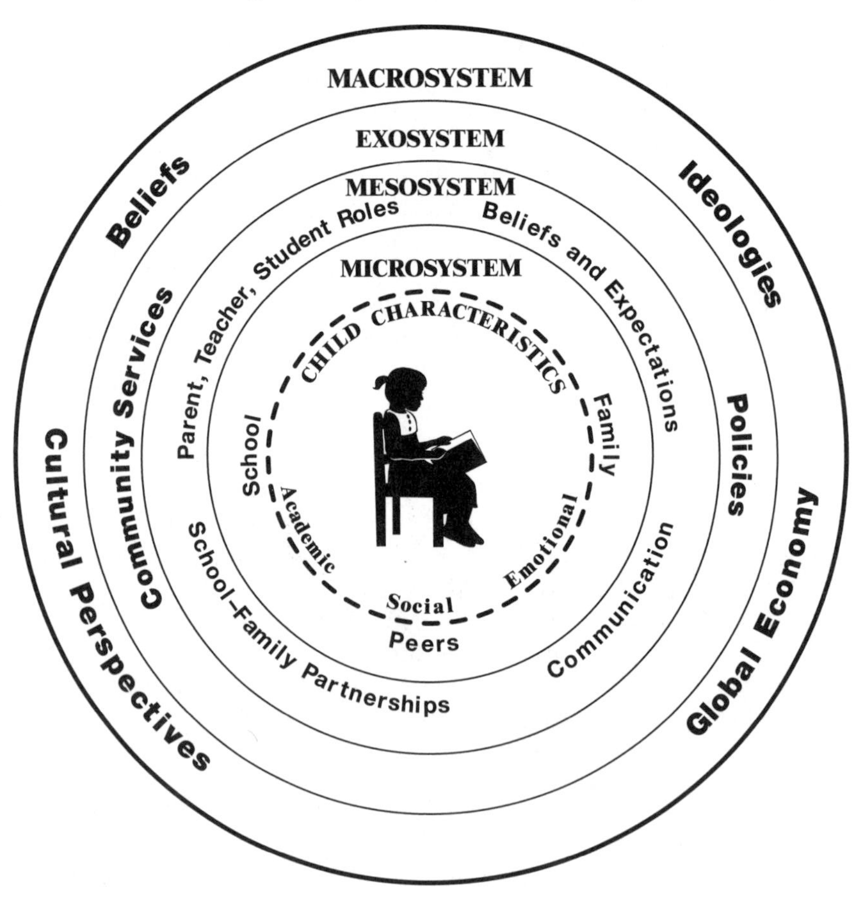

centers). While these settings may not include the individual directly, events occurring within them can have a great effect on the child. For example, school board decisions regarding changes in school policies or school closures will affect the child's schooling and alter interactions among microsystems, such as home and school.

The final level is the *macrosystem,* which encompasses the broader traditions, ideologies, and cultural characteristics. Influences within this system include effects of major historical events (e.g., wars) or more gradual socioeconomic changes such as the globalization of the economy. These broader issues can have a great impact on lower systems. For example, if a global economic crisis trickles down to the U.S. economy, it will result in loss of jobs and budget cuts for school programs. Given the rapidly changing demographics indicating that we are becoming a more diverse nation, the impact of cultural beliefs and broad ideological differences within the macrosystem can have a huge impact on school–family partnerships and student success (Forum on Child and Family Statistics, 2002).

Chapters in this book do not limit school–family partnerships to the mesosystem, a limitation that represents the traditional review of the area. Authors expand their review and investigation of the multilevel nature of home-school relationships to systems above and beyond the obvious mesosystem. The following section provides an overview of the chapters' content.

OVERVIEW OF THE BOOK

The contributors to this book are experts in the field of school–family partnerships and school success and have spent years investigating aspects of the multidimensional nature of school–family partnerships. The authors provide a rich description of issues influencing school–family partnerships; discuss benefits for the academic, social, and behavioral learning of children and youth; and offer recommendations for parents, educators and other school personnel to better inform research, policy, and practice.

There are three major parts in this book, and each consists of three chapters.

Part I: Conceptual Frameworks of School–Family Partnerships

Part I contains Chapters 1, 2, and 3, which identify the broad issues and conceptual frameworks of school–family partnerships and children's success. The authors provide a comprehensive review of influences stemming from the microsystem to the macrosystem. They also discuss the importance that home–school interconnections have in children's development; provide the scientific base for such connections to academic, social, and emotional learning; and offer recommendations for educators, parents, and policymakers.

In Chapter 1, Sandra Christenson, Yvonne Godber, and Amy Anderson provide an overview of the critical issues facing families and schools today and why a partnership orientation can contribute significantly to student success. They present the benefits of school–family connections and acknowledge the structural and psychological influences affecting parents, educators, and their relationship. They explore avenues for strengthening school–family partnerships and offer recommendations to enhance the partnership orientation in children's schooling.

In Chapter 2, Kathleen Hoover-Dempsey, Joan Walker, and Howard Sandler present a thorough picture of the factors that motivate parents to get involved in their children's education and foster academic, social, and emotional development. They offer a multilevel picture of the parent involvement process and discuss three important factors that motivate parents to become involved in the educational process: (a) parental role construction for involvement, (b) parental efficacy for helping the child learn, and (c) parental perception of school outreach. Based on their model, they offer a number of concrete steps that schools may take to increase all three factors and expand their collaboration with parents.

Finally, in Chapter 3, Pam Davis-Kean and Jacquelynne Eccles discuss challenges in forming meaningful school–family partnerships and offer a model that can serve as a framework to understand and overcome those challenges. The "social executive functioning model" is an analogy drawn from the coordination and interaction of the multiple systems in the human body and brain. It is used to conceptualize the multidimensional and multilevel nature of interactions between oneself and the social contexts, including school–family partnerships. The authors also provide recommendations for better coordination between the social contexts of home and schools.

Part II: Cultural and Empirical Perspectives

Part II (Chapters 4, 5, and 6) explores different cultural and empirical perspectives. Cultural issues are central when discussing school–family partnerships for many reasons. Broader beliefs, ideologies, and stereotypes, influences found in the macrosystem, significantly impact home–school relations. In addition, each partnership involves the distinct cultures of home and the school, which should not be assumed to target the same goals or hold similar beliefs and expectations. Finally, the rapidly changing demographic trends are transforming the ethnic composition of the U.S. population, and many questions arise regarding the relationship between families and schools.

Under Part II, authors discuss the cultural dimension of home–school relations and how it affects learning and school success in a multicultural society. The need of rigorous evaluation of parent programming, including its cultural aspects, is also underlined, and empirical evidence of programs that include a parent involvement component are discussed.

In Chapter 4, Luis Laosa discusses the major issues involved in intercultural interactions and the impact that these have on children's lives and schooling. He examines the experiences that children and their families may face in intercultural encounters that occur at the intersection between the family and the school. He also presents the ensuing changes that this contact between cultures may bring about in individuals, families, and schools. Awareness of such processes and their outcomes can foster positive relations among the nation's diverse cultural groups. Laosa offers ways to better integrate cultural issues when forming school–family partnerships for the benefit of students' academic, social, and emotional learning.

In Chapter 5, Ron Taylor closely examines the academic, social, and emotional development of African-American students, and what benefits can occur by forming bonds between families, schools, and other formal and informal social institutions. The author provides empirical evidence to support the determinants of psychological functioning of both parents and children and how they can lead to well-being and school and life success. Recommendations are offered on ways in which schools can enhance the social and emotional well-being of children and families from economically disadvantaged environments.

Chapter 6 by Arthur Reynolds and Melissa Clements offers additional empirical evidence to support the value of school–family partnerships, especially in economically disadvantaged urban areas. The authors provide an overview of research evidence on parent involvement indicators that support children's learning and development. Most importantly, the chapter offers meticulous research evidence stemming from a longitudinal study in a low-income area of Chicago. Results support the importance of school–family partnerships to enhance the academic achievement of children and foster the social and emotional well-being of youth. Based on the longitudinal benefits that school–family partnerships can have for the academic, social, and emotional development of children, the authors offer recommendations for enhancing home–school connections, especially in early education settings.

Part III: Policy Issues

Part III (Chapters 7, 8, and 9) examines policy issues. Chapters in this part of the book examine local (exosystem) and broader (macrosystem) policy initiatives. Authors shift the focus of the multidimensional nature of home–school relations to policy-related issues, and how broader-level issues in policy decisions on the federal and state levels affect school–family partnership practices, which in turn have a great impact on children's lives. In addition to federal and state perspectives, teacher preparation implications are also discussed in depth.

Part III begins with Chapter 7, a description of federal initiatives by Oliver Moles. This chapter offers a comprehensive picture of federal legislation and programs that over the years have targeted home–school collaborations and social and emotional learning, in addition to academics. All relevant parts of the No Child Left Behind Act of 2001 (2002), the centerpiece of educational policy, are closely examined and discussed. The author offers a detailed account of federal programs and the learning areas they target (i.e., academic, social, and emotional), as well as who is responsible for implementation, monitoring, and evaluation. Specific recommendations for action regarding parents are offered for policymakers, school and school districts, and parents.

Next, in Chapter 8, Sam Redding and Pam Sheley review a broad array of state and local initiatives and discuss important issues involved in school governance and local control. They also offer a detailed description of model programs that have been implemented in three states and draw the critical points common in successful state and local efforts. Finally, they offer recommendations for establishing effective state policies.

In Chapter 9, Nancy Chavkin discusses the important issue of teacher preparation. If one considers all the evidence supporting school–family partnerships and their benefits for the academic, social, and emotional learning of children and youth, it is dumbfounding that only a few states require teachers to study and develop skills in parent involvement and that therefore most higher education institutions do not prepare teachers to work with and outreach to parents. The author provides a thorough analysis of the existing data, offers frameworks and models for teacher training, presents promising current initiatives, and offers a comprehensive approach to increasing educator preparation for school–family partnerships.

Finally, the conclusion summarizes the common themes identified in the three parts of the book. We highlight the cross-cutting dimensions and recommendations that were raised by the authors and underline the factors that have been identified as essential ingredients for establishing and maintaining school–family partnerships.

Other books in the Series on Social and Emotional Learning (SEL) have put the spotlight on the importance of SEL, have provided the scientific base linking it to school success, and have offered a variety of suggestions to incorporate it into formal schooling. This book adds to the series by placing emphasis on one crucial factor of school success and academic, social, and emotional learning: school–family partnerships. This book contributes to the series and to the broader field of school–family partnerships and school success by: (a) creating a multidimensional framework that supports a comprehensive review of the field; (b) reviewing home–school interconnections from different angles, including the conceptual, cultural, and policy perspectives; and (c) offering recommendations for educators, parents, policymakers, and all those involved in children's development and schooling to support a partnership orientation for children's school and life success. Each author offers a different facet in a prism that reflects the importance of establishing and maintaining school–family partnerships. We hope that in these pages you will find the information and recommendations you need to improve the nature and quality of school–family partnerships for the benefit of all children.

REFERENCES

Bronfenbrenner, U. (1979). *The ecology of human development.* Cambridge, MA: Harvard University Press.

CASEL [Collaborative for Academic, Social, and Emotional Learning]. (2003). *Safe and sound: An educational leader's guide to evidence-based social and emotional learning programs.* Chicago: Author.

Catsambis, S. (2002). Expanding knowledge of parental involvement in children's secondary education: Connections with high school seniors' academic success. *Social Psychology of Education, 5*(2), 149–177.

Chavkin, N., & Williams, D. (1988). Critical issues in teacher training for parent involvement. *Educational Horizons, 66*(2), 87–89.

Christenson, S. L., & Sheridan, S. M. (2001). *School and families: Creating essential connections for learning.* New York: Guilford Press.

Eccles, J. (1999). The development of children ages 6 to 14. *The Future of Children, 9*(2), 30–44.

Eccles, J. S., & Harold, R. D. (1996). Family involvement in children's and adolescents' schooling. In A. Booth & J. F. Dunn (Eds.), *Family-school links* (pp. 3–34). Mahwah, NJ: Erlbaum.

Elias, M. J., Bryan, K., Patrikakou, E. N., & Weissberg, R. P. (2003). Challenges in creating effective home–school partnerships in adolescence: Promising paths of collaboration. *School-Community Journal, 13*(1), 133–153.

Elias, M. J., Zins, J. E., Weissberg, R. P., Frey, K. S., Greenberg, M. T., Haynes, N. M., Kessler, R., Schwab-Stone, M. E., & Shriver, T. P. (1997). *Promoting social and emotional learning: Guidelines for educators.* Alexandria, VA: Association for Supervision and Curriculum Development.

Epstein, J. L. (1987). Parent involvement: What research says to administrators. *Education and Urban Society, 19*(2), 119–135.

Epstein, J. L. (1995). School/family/community partnerships—Caring for the children we share. *Phi Delta Kappan, 76*(9), 701–712.

Fan, X., & Chen, M. (2001). Parental involvement and students' academic achievement: A meta-analysis. *Educational Psychology Review, 13*(1), 1–22.

Forum on Child and Family Statistics. (2002). *America's children: Key national indicators of well-being 2002*. Retrieved from www.nichd.nih.gov/publications/pubs/childstats/report2002.pdf

Greenberg, M. T., Weissberg, R. P., O'Brien, M. U., Zins, J. E., Fredericks, L., Resnik, H., & Elias, M. J. (2003). Enhancing school-based prevention and youth development through coordinated social, emotional, and academic learning. *American Psychologist, 58*, 466–474.

Haynes, N. M., Ben-Avie, M., & Ensign, J. (Eds.). (2003). *How social and emotional development add up: Getting results in math and science education*. New York: Teachers College Press.

Henderson, A. T., & Mapp, K. L. (2002). *A new wave of evidence*. Austin, TX: National Center for Family and Community Connections with Schools.

Hoover-Dempsey, K. V., & Sandler, H. M. (1997). Why do parents become involved in their children's education? *Review of Educational Research, 67*(1), 3–42.

Laosa, L. M. (1997). Research perspectives on constructs of change: Intercultural migration and developmental transitions. In A. Booth, A. C. Crouter, & N. Landale (Eds.), *Immigration and the family: Research and policy on U.S. immigrants* (pp. 133–148). Mahwah, NJ: Erlbaum.

Mathematica Policy Research, Inc. and the Center for Children and Families at Teachers College, Columbia University. (2001). *Building their futures: How early Head Start programs are enhancing the lives of infants and toddlers in low-income families*. Washington, DC: Administration on Children, Youth, and Families, Department of Health and Human Services. Available: http://www.acf.dhhs.gov/programs/core/ongoing_research/ehs/ehs_reports.html

Mattingly, D. J., Prislin, R., McKenzie, T. L., Rodriguez, J. L., & Kayzar, B. (2002). Evaluating evaluations: The case of parent involvement programs. *Review of Educational Research, 72*(4), 549–576.

Moles, O. (1997). *Reaching all families: The federal initiative in family–school partnerships*. Washington, DC: U.S. Department of Education. Office of Educational Research and Improvement. (ERIC Document Reproduction Service No. ED413072).

National Conference of State Legislatures. (2002, August). *Resolution of character education and social and emotional learning.* Washington, DC: Author. Available: http://www.casel.org/downloads/Safe%20and%20Sound/3D_Endorsements.pdf

National Education Goals Panel. (1995). *The national education goals report: Building a nation of learners.* Washington, DC: Government Printing Office.

National Youth Anti-Drug Media Campaign (2004). *Parents: The antidrug.* Available: http://theantidrug.com.

No Child Left Behind Act of 2001. 20 U.S.C. §6301 (2002).

Patrikakou, E. N. (1997). A model of parental attitudes and the academic achievement of adolescents. *Journal of Research and Development in Education, 31*(1), 7–26.

Patrikakou, E. N. (2004). Adolescence: Are parents relevant? Invited article. *The Harvard Family Research Project. Family Involvement Network of Educators: Research*

Digests. Available: http://www.gse.harvard.edu/hfrp/projects/fine/resources/digest/adolescence.html

Patrikakou, E. N., & Weissberg, R. P. (1999, February 3). Seven P's to promote school–family partnership efforts. *Education Week*, pp. 34, 36.

Patrikakou, E. N., & Weissberg, R. P. (2000). Parents' perceptions of teacher outreach and parent involvement in children's education. *Journal of Prevention and Intervention in the Community, 20*(1/2), 103–119.

Redding, S. (2000). *Parents and learning*. Educational practices series—2. Brussels, Belgium: International Academy of Education; Geneva, Switzerland: The International Bureau of Education.

Reynolds, A. J. (2000). *Success in early intervention: The Chicago child–parent centers*. Lincoln, NE: University of Nebraska Press.

Reynolds, A. J., Temple, J. A., Robertson, D. L., & Mann, E. A. (2001). Long-term effects of an early childhood intervention on educational achievement and juvenile arrest: A 15-year follow-up of low-income children in public schools. *Journal of the American Medical Association, 285*(18), 2339–2346.

Reynolds, A. J., & Walberg, H. J. (1992). A structural model of science achievement and attitude: An extension to high school. *Journal of Educational Psychology, 84*(3), 371–382.

Rubenstein, M., Patrikakou, E. N., Weissberg, R. P., & Armstrong, M. (1999). *Enhancing school–family partnerships: A teacher's guide*. Chicago: The University of Illinois at Chicago.

Steinberg, L. (1992). Impact of parenting practices on adolescent achievement: Authoritative parenting, school involvement, and encouragement to succeed. *Child Development, 63*(5), 1266–1281.

Taylor, R. D., Casten, R., & Flickinger, S. (1993). The influence of kinship social support on the parenting experiences and psychosocial adjustment of African-American adolescents. *Developmental Psychology, 29*, 382–388.

Weiss, H. (1996). *Preparing teachers for family involvement*. (ERIC Document Reproduction Service No. ED396823).

Weissberg, R. P., & Greenberg, M. T. (1998). School and community competence—enhancement and prevention programs. In W. Damon (Series Ed.) & I. E. Sigel & K. A. Renninger (Vol. Eds.), *Handbook of child psychology: Vol. 4. Child psychology in practice* (5th ed., pp. 877–954). New York: Wiley.

Zins, J. E., Weissberg, R. P., Wang, M. C., & Walberg, H. J. (Eds.). (2004). *Building school success on social and emotional learning: What does the research say?* New York: Teachers College Press.

PART I

CONCEPTUAL FRAMEWORKS OF SCHOOL–FAMILY PARTNERSHIPS

1

Critical Issues Facing Families and Educators

Sandra L. Christenson, Yvonne Godber
& Amy R. Anderson

There might well be no other time in history that indicates better the overwhelming importance of constructive family–school connections for the academic, social, and emotional learning of youth. Consider, for example, the criticality of these connections for educational accountability, including high-stakes assessment required by the No Child Left Behind act (U.S. Department of Education, 2002). Schools are now being held accountable for the improvement of *all* students, including those with learning and behavioral challenges, English language learners, the highly mobile, and the homeless. Standards and accountability are excellent; however, in the absence of family- *and* school-based opportunities and supports for children to learn, relate effectively with others, and self-manage their emotions, they portend disastrous consequences for some students (e.g., retention, social promotion, dropout). Making education a priority requires that we expand the definition of learning to include social and emotional learning, which are essential in fostering students' academic success.

Although school–family partnerships should be front and center in our efforts to promote desired learning outcomes, the primary issue facing educators and families is moving these partnerships from rhetoric to reality. Namely, we suggest that the overarching issue for improving educational outcomes is to promote *shared responsibility between families and educators*. Currently, the failure to explain children's learning progress and needs as a function of contributions from multiple contexts is an impediment.

 ISBN 0-8077-4600-2 (paper), ISBN 0-8077-4601-0 (cloth).

We provide four illustrative examples of the importance and impact of thinking systemically about fostering student learning.

THE PRIMARY CHALLENGE FOR THE PARTNERSHIP

The theoretical underpinnings of current school practices for interacting with families are not based in systems-ecological theory, which, as described in the Introduction, provides a conceptual framework for organizing the reciprocal influences on children's learning. The mesosystem, which includes school–family partnerships, emphasizes the congruent socialization practices for students as learners (Bronfenbrenner, 1992). The goal of family–school connections for children's learning must be to create a culture of success—one that enhances learning experiences and competencies across home and school and underscores that partnership means shared goals, contributions, and accountability (Fantuzzo, Tighe, & Childs, 2000).

Pianta and Walsh (1996) described a necessary belief system for educators, one wherein educators understand that children develop and learn in the context of the family, and that the child/family system must interface in a positive way with the school system and schooling issues for children's educational performance to be optimal. The quality of the family–school relationship, represented in the pattern of interactions over time or the shared meaning that is created for the purpose of supporting children's learning, is the focal point when a systemic orientation is adopted. Accordingly, Pianta and Walsh have denoted a clearer understanding of risk for school failure by extending the discussion beyond status characteristics (e.g., poverty) to include the quality of school–family partnerships as a primary contributing factor to the level of child risk. For example, they theorized that children are educated in low-risk circumstances if the child/family and schooling systems are functional, where home and school communicate to provide children with congruent messages about their learning. In contrast, high-risk circumstances occur when children derive meanings from messages delivered at home or school that result in conflicting values, motivations, or goals about learning for the student. For example, students may feel less need to comply with a teacher's request to complete an assignment or follow a rule if the students do not like the assignment or rule and have been told by their parents that the teacher does not know what he or she is doing.

Risk and resilience are not properties of children, but are distributed across systems and reside in the interactions, transactions, and relationships among the multiple systems that envelop children. Using the analogy of a three-legged stool, Pianta and Walsh (1996) argue that children's level of academic, social, and emotional competence cannot be understood or fostered by locating problems in child, family, or school contexts in the absence of a focus on the dynamic influence of relationships among the systems. When problems arise, schools often blame parents for student failure when, in fact, the fault actually lies with the interface between home and school (Weiss & Edwards, 1992). Therefore, the kind of question that will advance our knowledge of students' academic, social, and emotional outcomes is, How are resources of the child and the learning context (family and school) organized to respond to problems or help the child meet developmental demands or demands of assigned tasks in school over time?

Thinking systemically underscores not only the critical nature of continuity across and the cumulative effect of positive transactions between socializing systems for educating children, but also provides clarity with respect to the question: Who is the client? Services are not directed only to students, but consideration is given to how school *and* family practices affect student learning. Hence, priority is given to the effect of the family–school interface on explanations for learning outcomes.

Accordingly, Christenson and Sheridan (2001) defined four features of partnering between families and schools: 1) a student-focused philosophy wherein educators and families collaborate to enhance learning opportunities, progress, and success for students in four domains: academic, social, emotional, and behavioral; 2) a belief in shared responsibility for educating and socializing children—both families and educators are essential and provide resources for children's learning; 3) an emphasis on the quality of the interface and ongoing connection between families and schools; and 4) a preventive, solution-oriented focus in which the partners strive to create conditions that facilitate student success.

Similarly, McWilliam, Tocci, and Harbin (1998) argued that family-centered practices originating in the early intervention field have much to offer the view typically taken toward families by the K-12 system. It should be noted that many families experience a "disconnect" between pre-K and the K-12 systems. Rather than viewing work with families as necessary for educating children, a family-centered approach considers the needs, wishes, and skills of the family as important as the needs of the child. Truly ecological in nature, family-centered practices illustrate how welcoming family–school interactions can be achieved. These practices include a focus on the whole family (rather than simply the child), positiveness (assuming the best about families), sensitivity (empathizing with families), responsiveness (doing whatever is necessary to respond to personal concerns), and friendliness (viewing and treating families as "professional" friends). When schools make a concerted effort to include families as partners, families respond by becoming more involved (Bempechat, 1998).

A sound partnership orientation provides students with supports for learning in terms of shared expectations, consistent structure, cross-setting opportunity to learn, mutual support, trusting relationships, and modeling (Ysseldyke & Christenson, 2002). In Figure 1.1, examples of contextual influences that foster students' academic, social, and emotional learning outcomes when family and school partner are described. We now turn to relevant illustrations in school practices of the importance and impact of thinking systemically to improve educational outcomes.

Viewing Students Within a System

As presented in the Introduction, the benefits of school–family partnerships for students involve social and emotional learning in addition to improved academic achievement. We know that family involvement in education has been established as a significant, positive correlate of student learning (Henderson & Mapp, 2002). It has been demonstrated that when parents are involved, students show improvement in grades; test scores, including reading and math achievement; attitude toward schoolwork; behavior; self-esteem; completion of homework; academic perseverance; and participation in classroom learning activities. Benefits also include fewer placements in special education, greater enrollment in postsecondary education, higher attendance rates, lower dropout rates, fewer suspensions, and greater realization of exceptional talents. Furthermore, findings from the National Longitudinal Study on Adolescent Health reveal that adolescents have a higher probability of avoiding high-risk behavior when family and school connectedness is strong (Resnick et al., 1997).

The concept of student engagement is a very direct way to illustrate a link to students' academic, social, and emotional learning and that the goals of schooling for students must be viewed systemically. For example, although educators have focused primarily on rates of *academic engagement* (i.e., time on task, accrual of credits), there is renewed interest in ways to enhance students' *cognitive engagement* (e.g., taking responsibility for one's decisions, actions guided by future goals, confidence in one's ability, persisting in the face of challenge), *behavioral engagement* (i.e., asking for help when needed, attendance), and *psychological engagement* (i.e., sense of belonging, relationship skills). The engagement subtypes are positive, significant low-to-moderate correlates of academic achievement (Christenson & Anderson, 2002), and the interplay among social, emotional, behavioral, and academic learning is being increasingly recognized. For example, evidence that social skills are positively predictive and problem behaviors are

FIGURE 1.1. Outcomes Associated with School–Family Partnerships

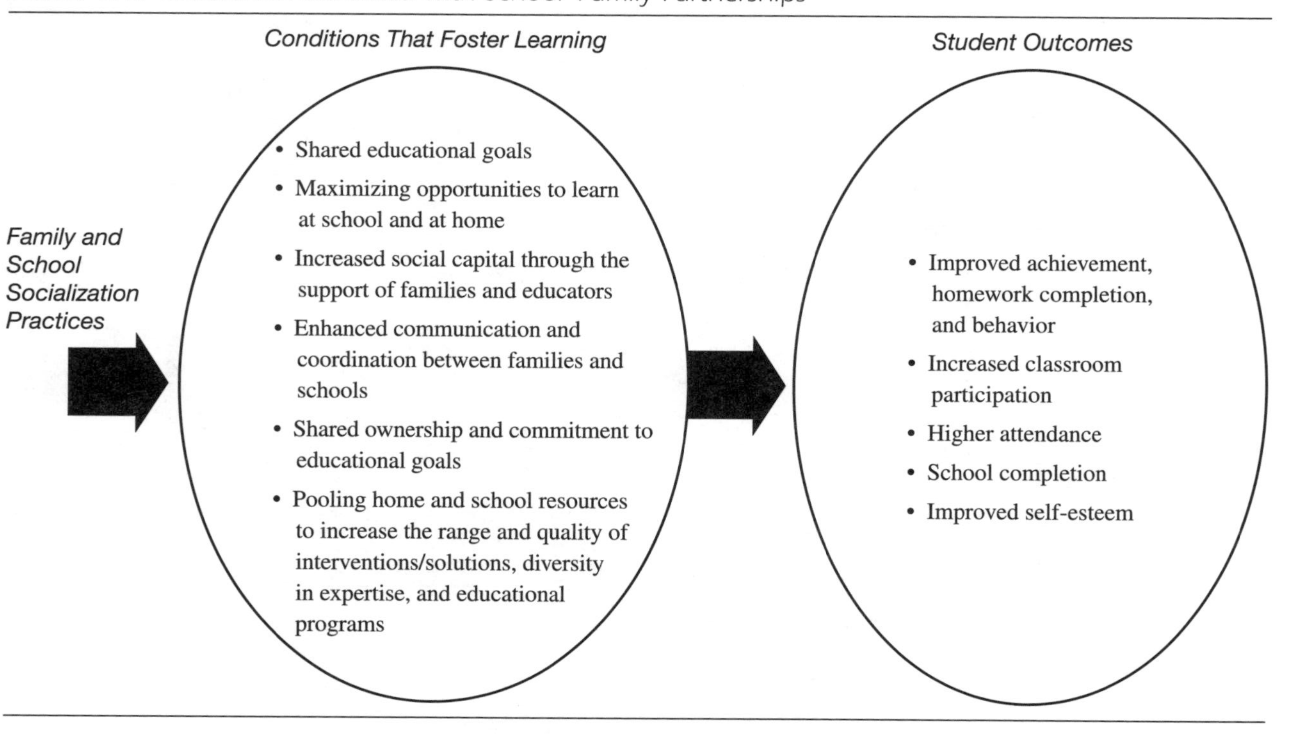

negatively predictive of concurrent academic achievement is amassing (Malecki & Elliott, 2002). All forms of engagement are necessary to meet the demands of schooling, a primary developmental task for youth. And engaged learners are supported by constructive school–family partnerships.

Being Responsive to Life Conditions of Youth

Selected statistics elucidate challenges faced by school–family partnerships to meet the needs and adequately educate all students who cross the school door. The conditions of children's lives and constraints faced by parents and educators must be understood so that the partnership can be appropriately responsive. For example, some parents need information only about the essential nature of their role and home support for learning strategies. Certain families need information and attention to a specific situation (e.g., need for reading resources), whereas others require information, attention, and ongoing support to maintain a connection with educators.

Tremendous variability among youth in the United States in terms of their access to health care and housing, economic security, opportunities to learn, and social capital is apparent from data collected by the Children's Defense Fund (2002) and in a recent publication on *America's Children: Key National Indicators of Well-Being 2002* (Federal Interagency Forum on Child and Family Statistics, 2002). For example, in 2000, 36% of children in the United States were non-white, and in 2001, 19% of children had at least one foreign-born parent. Children in the latter group were more likely to have a parent who had not earned a high school diploma, greatly affecting their health, economic security, and educational opportunities. Many children who are United States citizens do not receive benefits because their parents, who are immigrants, are unaware of the social system.

We are increasingly diverse in terms of family composition and parental working arrangements. In 2001, 69% of American children lived with two parents; 22% resided with only their mothers, 4% with only their fathers, and another 4% with neither parent (Council on Contemporary Families, 2001). The number of unmarried partners grew by 71% during the 1990s, while 33% of all births in the year 2000 were to unmarried women. Children who live in a family with only one parent are substantially more likely to have incomes below the poverty line than children who live in households with two parents (biological, step, or adoptive). Furthermore, children today are more likely to have their custodial parent or both parents in the workforce, affecting parental availability to monitor child behavior, discuss personal matters, and support education (i.e., social capital).

The environmental conditions in which children live directly affect their learning and physical and emotional health. In 1999, 35% of households with children experienced problems with physically inadequate, crowded, and/or expensive housing (i.e., costs more than 30% of household income), while 16% of children lived in areas that did not meet clean air quality standards, and 18% of children lived in households with food resource insecurity. Homelessness, including high mobility, continues to be a problem for school-age children.

To complicate matters, the additional support and resources needed for all children to achieve academically and develop into healthy, productive adults are often underfunded or unavailable to those who need them most. For example, Head Start serves only three out of every five eligible children, and Early Head Start serves less than 5%. In addition, hundreds of thousands of children are on waiting lists for child care assistance programs; only one in seven children eligible for assistance currently receives it. The need for child care does not end when children enter school. As many parents are in the workforce, countless children are left unattended after school, a time when they are more likely to engage in risky behaviors such as smoking, drinking, sex, or crime (e.g., violent juvenile crime peaks between the hours of 3:00 P.M. and 7:00 P.M.).

No picture of the condition of children's lives is complete without considering their education. Unfortunately, the educational outlook for some students in our schools is sobering. One in three youth are behind a year or more in school, two in five youth never complete a single year of college, one in eight never graduates from high school, and one in twelve has a disability.

Embracing Diversity

All too often, diversity, including ethnicity, socioeconomic status, language, culture, sexual orientation, and life experiences, is viewed as a complicating factor or as an attribution for why some students fail. Edwards (1992) cautioned against a target group orientation wherein there is false labeling of some non-white groups as not interested or unwilling to support children's learning; overemphasizing child performance within a group ignores our knowledge of heterogeneity within a group and the importance of treating the child/family system individually. Viewing diversity as an opportunity for educators, in collaboration with others, to make a difference and to help children and youth develop learning competencies offers promise to address the gaps in educational performance for some students. Families and educators from different cultures who engage in information- and resource-sharing about what works for a student illustrate thinking systemically. They recognize that it is seldom the event, but rather interpretations of the event, about which families and educators must dialogue (Thorp, 1997).

For educators to embrace diversity as an opportunity to build capacity for families and educators to partner, they must understand and attend to ethnic and cultural differences with respect to purposes of education (Bempechat, 1998), fostering parents' role construction in education and parental level of self-efficacy (Hoover-Dempsey & Sandler, 1997), ways to enhance nonblaming interactions (Weiss & Edwards, 1992), and shared decision making that creates a posture of reciprocity (Harry, 1993). As will be discussed in Chapter 5, parents who have cultural capital—or knowledge about the school's expectations and ways in which the school operates—are better advocates for their children. Often the difference between parents who participate and those who do not is that those who do have knowledge about school practices and recognize their critical role.

Bempechat (1998) has championed the need to look for culturally universal and specific factors for the academic success of poor and minority students. She found that high-achieving ethnically diverse students linked their school success to ability through the notion of sustained effort; learning was perceived as a process that required persistence, diligence, and delay of gratification in the face of challenges, as well as a means to improving oneself. Her research elevates the crucial nature of family–school socialization practices to foster students' identity as learners. Specifically, there is a need to understand the kinds of messages students are provided about their success and failures, and how culture, context, and motivation play critical and unique roles for students to maintain positive attitudes and high expectations for success and engagement with learning (see Chapter 4 for an in-depth discussion of cultural issues). Although different cultures understand the purpose of education and their role vis-à-vis education differently, Bempechat cautioned, appropriately so, about dichotomies for parents' beliefs and behaviors. She recommended thinking of the many and varied ways parents adapt their cultural beliefs/values to parenting goals and the goals of schooling. She implicitly raised the question: How can educators work effectively with families if they do not understand parental beliefs and perspectives?

Prioritizing Education

While the value of education is sometimes lacking in our society and in business and industry, it is also not a priority in every home. Home and school provide different inputs for the socialization process of children; opportunities, demands, and rewards come from schools, whereas attitudes, effort, and conception of self come from the social environment of the household (Coleman, 1987). Educational outcomes result from the interaction of the qualities that the child brings from home and experiences in school. Schools do make a difference for children; however, they do not have an equal effect on all children. There is greater variation in family resources than school resources for children's learning (Christenson, 2003, 2004). For example, the power of out-of-school learning time helps to explain school performance differences; home learning resources and opportunities, especially during summer, are a differentiating factor between low and high achievers; and low school performance often reflects discontinuity in expectations and support for learning between school and home.

Bempechat's (1998) distinction of parental roles; namely, academic and motivational support for learning is invaluable for enhancing students' learning. *Academic support* refers to the ways in which parents foster their children's intellectual and cognitive development. It is what parents do that is *directly* related to their children's experience in school. In contrast, *motivational support* refers to the ways in which parents foster the development of attitudes and habits of learning that are essential for school success. It is what parents do that is more *indirectly* related to school success, yet still obviously key. The emotional support—how parents get their children to believe that they are capable, competent, important individuals—is critical in setting the stage not only for children's confidence in school, but in all of life. Motivational support roles may be underutilized by parents when, in fact, they may be the most important for the child to be prepared for learning in the classroom (Bempechat, 1998). These roles, often forgotten by educators, can be highlighted in coordinated home–school interventions.

Considering Macrosystemic Influences

As was presented in the Introduction, influences from more distal systems, such as the macrosystem, must also be recognized when discussing school–family partnerships. For example, legislation, school reform and funding, poverty, and supports for working families have a great impact on daily student performance as well as on the concerted effort to create and sustain partnerships. The value and importance of education should be apparent in every aspect of our society if all children are to succeed. One way of demonstrating the value of education is through public funding, whether for teacher salaries, classroom materials and supplies, technology, or safe physical structures. Another way of demonstrating the value of education is through employer support for working parents. Family-friendly policies, such as allowing for flexible work hours and offering affordable child care and health care benefits, provide families with opportunities to support their children's education. Finally, a change in preservice education of school personnel that infuses a focus on shared responsibility for educational outcomes—what families and educators do to support student learning—places a priority on education.

AVENUES FOR STRENGTHENING SCHOOL–FAMILY PARTNERSHIPS

Adopting a Partnership Orientation

Traditional school-based parent involvement practices have been activity-driven, illustrated by the commonly asked question: How can we involve families? The emphasis on activities seems to persist in many schools despite the fact that we know achieving collaboration with parents is not primarily a function of the activities provided. Offering involvement activities devoid of a healthy family–school relationship has yielded less than desired levels of active parental engagement, particularly for families most alienated by traditional school practices (Christenson & Sheridan, 2001). Ironically, those are often the families that programs hope to reach. While schools are continuously improving upon the manner in which they work with families, those who tend to become engaged with activities offered by educators often match the culture of the school. Much remains to be known about how to strengthen relationships with families who have multiple needs, or who exist outside the expectations and norms held by schools.

Time to Move Beyond "Activities"

Although activities provide good ideas, not all good ideas work in every school–student–family context. It is more important to focus on a process for partnering, especially since student concerns are not always resolved in one contact. A goal in selecting activities to enhance academic, social, and emotional learning of youth is to achieve a goodness of fit for parents, teachers, and students.

Christenson and Sheridan (2001) proposed that "Four A's"—Approach, Attitudes, Atmosphere, and Actions—represent a process to develop quality partnerships. The approach taken (i.e., families and educators are essential socialization agents), the attitudes parents and educators hold about each other, and an atmosphere conducive for collaboration are all prerequisite conditions that set the stage for school and family actions to enhance learning outcomes. As presented in Figure 1.1, adopting a partnership orientation attends to critical contextual influences on student learning (e.g., maximizing opportunities to learn at school and at home). It also requires that school-based practices change. Drawn from the work of Christenson (2003, 2004), examples of actions aimed at enhancing shared responsibility for educational outcomes appear in Figure 1.2.

Structure and Psychology—The Essential School–Family Partnership Scaffolding

The strength of school–family partnerships is in large part determined by both structural and psychological variables. *Structure* refers not only to how well the roads leading families into schools are "paved" (e.g., frequency and quality of interactions, roles for families in schools), but also macrosystemic influences exerting pressure on schools, families, and communities (e.g., poverty, child care, time constraints, lack of resources). Intrinsic psychological influences (e.g., attitudes, stereotypes, efficacy, and culture) similarly direct the course partnerships take. Influenced by the work of Christenson and Sheridan (2001), Liontos (1992), and Weiss and Edwards (1992), we have provided examples in Figure 1.3 of both structural and psychological variables—the essential scaffolding for healthy school–family partnerships—that influence the extent to which families and educators are able to partner to promote optimal academic, social, and emotional learning outcomes. Although the issues are presented for each partner and the relationship, it would be a mistake to view an issue in isolation. For example, many issues for families become evident after they have transacted with a school policy and/or practice; issues must be understood systemically or within the family–student–school context.

Discussions in the literature on how schools find the appropriate level of access for the community call into question whether school-defined passageways and prescribed mechanisms for parents to "enter" the school can satisfy families' desire to be involved in their children's learning and schooling. Parents want to be included as equal partners—they want to know more about how the school functions, have access to the school, and obtain information on their children's performance relative to others and how to help their children succeed. Families require additional information on the curriculum, including the content and delivery, and their role in making sure their children's learning is enhanced to the greatest extent possible. Unfortunately, families are not often invited to this decision-making table with educators. For instance, although parents are concerned that schools provide safe learning environments for their children, teachers

Figure 1.2. Recommendations to Enhance Shared Responsibility for Learning

Approach	• Communicate how parents are integral to attaining optimal educational goals for students. • Ensure that the philosophy and policies of both the school *and* the district (i.e., mission statement) explicitly articulate that the family–school relationship (e.g., bidirectional communication) impacts students' academic, social, behavioral, and emotional learning. • Modify assessment practices to ensure that adequate information about the child's academic and social behavior across settings is used to make decisions. "Co-construct" the bigger picture. • Communicate in print and nonprint forms that positive habits of learning are maximized when there is congruence across home and school about the value of education, expectations for performance, and support for educational programming. • Model the relevance of a quality school–family partnership by fostering bidirectional communication, enhancing problem solving across home and school, encouraging shared decision making, and reinforcing congruent home–school support for students' learning.
Attitudes	• Encourage families and school personnel to engage in perspective-taking across home and school—modeling that there are no problematic individuals (parents, teachers, or students), only a problematic situation that requires the attention of the student, home, and school. • Promote active parental engagement in decision making; educators should inform, invite, and include parents to help address concerns for students' learning.
Atmosphere	• Ensure that there are ample opportunities for trust-building events (e.g., multicultural potlucks/student celebrations, principal's hour, family fun nights, committees designed to address home–school issues, workshops where parents and school personnel learn together), and that these events occur before serious decisions are made with respect to educational programming (e.g., special education placement, grade placement).
Actions	• Include parents as policymakers with respect to home–school practices and policies by establishing a family–school team. • Ensure that all families, even those with limited contact with schools and/or negative personal experience with schools, understand how schools function. • Conduct staff development that infuses a focus on partnering with families.

FIGURE 1.2. (continued)

- Ensure that goals and understanding of child behavior are mutually determined in intervention planning.
- Keep and sustain a focus on the salience of education by negotiating a consistent, feasible way for families to support students' reading and learning.
- Underscore all communication with shared responsibility between families and schools (e.g., discussing co-roles, partnership agreements).
- Invite parental assistance to resolve school-based concerns, foster parental goals for children and youth, and find out what parents desire to fulfill their commitment to their children's educational success.
- Maximize the power of students' use of in- and out-of-school time with home and school interventions.
- Learn from and be responsive to the needs of parents whose children are not successful in school and/or parents of groups of children who represent gaps in educational outcomes at a national level.
- Design, implement, and evaluate problem-solving structures that include perspective-taking, learning from each other, and sharing constraints of each system.
- Reach out to parents who are identified as "hard to reach" for the purpose of learning what would help them foster their children's learning competence.
- Help parents understand policies and practices, be a resource for their questions, provide regular information on their child's progress and resources to address the gaps in learning, and foster a positive learning environment at home.
- Communicate the desire to develop a working partnership with families, the crucial nature of family input for children's educational progress, the importance of working together to identify a mutually advantageous solution in light of concerns, and clarity about conditions that foster students' academic, social, behavioral, and emotional learning.
- Involve students in parent–teacher conferences and intervention planning, problem-solving conferences, or IEP meetings.
- Conduct workshops with parents *and* educators together, which enhance information sharing about students.
- Promote research to identify effective family–school partnership practices.

FIGURE 1.3. Issues for Families, Educators, and the Relationship

Families	*Structural Issues* • Lack of role models, information, and knowledge about resources • Lack of supportive environment and resources (e.g., poverty, limited access to services) • Economic, emotional, and time constraints • Child care and transportation issues
	Psychological Issues • Feelings of inadequacy • Adopting a passive role by leaving education to schools • Linguistic and cultural differences, resulting in less how-to knowledge about how schools function and their role • Suspicion about treatment from educators • Experiencing a lack of responsiveness to parental needs
Educators	*Structural Issues* • Lack of funding for family outreach programs • Lack of training for educators on how to maintain a partnership with families • Time constraints
	Psychological Issues • Ambiguous commitment at both school and district level to parent involvement • Use of negative communication about students' school performance and productivity • Use of stereotypes about families, such as dwelling on family problems as an explanation for students' performance • Stereotypic views of people, events, conditions, or actions that are not descriptive of behavior, but portray a casual orientation • Doubts about the abilities of families to address schooling concerns • Fear of conflict with families • Narrow conception of the roles families can play

FIGURE 1.3. (continued)

Family–School Relationship	*Structural Issues* • Limited time for communication and meaningful dialogue • Communication primarily during crises • Limited contact for building trust within the family–school relationship • Limited skills and knowledge about how to collaborate • Lack of a routine communication system • Limited understanding of the constraints faced by the other partner *Psychological Issues* • Partial resistance toward increasing home–school cooperation • Lack of belief in a partnership orientation to enhance student learning/development negatively influences interactions • A blaming and labeling attitude that permeates the home–school atmosphere • A win-lose rather than a win-win attitude in the presence of conflict • Tendencies to personalize anger-provoking behaviors by the other individual • Misunderstanding differences in parent–educator perspectives about children's performance • Psychological and cultural differences that lead to assumptions and building walls • Limited use of perspective-taking or empathizing with the other person • Limiting impressions of child to observations in only one environment • Assumption that parents and teachers must hold identical values and expectations • Failure to view differences as a strength • Previous negative interactions and experiences between families and schools • Failure to recognize the importance of preserving the family–school relationship across time

vary on the degree to which they feel parents should share decision-making power in conjunction with discipline policies. According to a study conducted by Metropolitan Life (Binns, Steinberg, & Amorosi, 1997), 60% of all participating teachers believe parents should be actively consulted about changes in the discipline policy, whereas 40% believe that they need only be kept informed. On the other hand, 74% of minority teachers believe parents need to be actively consulted about disciplinary policy. The more schools can take into account parent preferences and make arrangements to elicit and incorporate family needs and wishes, the better students will fare.

How, when, and why core parental beliefs influence their decisions to become engaged with their children's learning provide one important—yet relatively unexplored and underutilized—example of intrinsic variables that affect the relationship between families and schools. As will be discussed in detail in Chapter 2, parental efficacy encompasses parents' beliefs in their own ability to help their children succeed and solve school-related problems, and whether outcomes are attributed to luck, effort, and/or ability levels. Fan and Chen's (2001) meta-analysis on parent involvement points to the need to incorporate more intrinsic variables when seeking to understand when it is that parents become involved. Their study revealed that the strongest aspect of parent involvement had to do with the hopes and expectations parents hold for their children's academic success, and yet educators seldom query parents about this. In fact, educators have erred by emphasizing the structural rather than the psychological aspects that influence the quality of relationships. While both frequency and type of contact between educators and families foster positive relationships, it is the quality, rather than the quantity, of contacts that makes the largest difference in how interactions between families and educators are viewed (Patrikakou & Weissberg, 1999).

Relationships between families and school personnel—with each partner holding different priorities, goals, roles, and expectations—is a sensitive area that can easily become conflict-ridden. Attempting to engage parents requires that schools ask for and heed parental suggestions and concerns. The literature indicates that parents become involved when very satisfied *and* when very dissatisfied (Zellman & Waterman, 1998). Considering how school policies and practices influence whether parents enter the school building as friend or foe has tremendous implications for parents and teachers, not to mention students, who are often caught in the middle of both parties. Questions for school personnel to ask include: What can schools do to make sure that parents are satisfied, aware, and presented with numerous opportunities? What can schools do to make sure parents feel comfortable, confident, and eager to be involved? How can events and activities be arranged to take advantage of what families feel they can offer? What information and resources do parents desire in order to feel empowered?

CONCLUSION

Consensus is clearly emerging that a new social contract between families and schools—one that is partnership-oriented—is needed. To develop strong home–school bonds, it is essential to apply systems theory to synchronize home and school practices. The difference of looking at families as "essential partners" and as "desirable extras" can be explained in part by the degree to which both home and school are viewed as contexts for children's learning. Specific examples of thinking systemically—being responsive to the conditions of children's lives; viewing student outcomes as comprised of academic, social, and emotional competencies; embracing and celebrating the strengths of an increasingly diverse culture; bolstering efforts aimed at prioritizing education; and orienting attention toward broader macrosystemic influences surrounding and molding the partnership—illustrate the urgency of adopting a partnership approach to improve educational outcomes.

Far too often, activities and agendas are developed without attention to process variables influencing the quality of family–school connections for children's learning. One word captures the essence of a partnership orientation, and provides a framework for families, educators, and policymakers to consider when uniting their efforts on students' behalf: *relationships*. The needs, wishes, capabilities, and beliefs of each partner—including oft-neglected and misunderstood intrinsic psychological variables—must not be viewed in isolation, but rather as part of the complex school–student–family system. Establishing partnerships by simply removing structural barriers or developing "activities" will only go so far in benefiting students. Taking into account and nurturing personal qualities and preferences has the potential to move the partnership to an entirely different plane. Relationship-building is integral to the success of this new social contract.

Creating and strengthening school–family partnerships is no easy task given the time, leadership, and concerted effort required for them to succeed. At the same time, we hope educators and policymakers will consider and commit the time and resources necessary to this exciting opportunity for improving students' academic, social, and emotional learning. Strong family–school connections and supports help students meet established educational standards. This is particularly true for those facing the greatest social and educational barriers. As different family–school policies and practices are implemented, a clearer delineation and understanding of parents', teachers', and students' rights, responsibilities, and resources for enhancing learning outcomes should emerge. By partnering, families and schools can achieve the goal of helping *all* children succeed.

REFERENCES

Bempechat, J. (1998). *Against the odds: How "at-risk" students EXCEED expectations.* San Francisco: Jossey Bass.

Binns, K., Steinberg, A., & Amorosi, S. (1997). *The Metropolitan Life survey of the American teacher 1998: Building family–school partnerships: Views of teachers and students.* New York: Louis Harris and Associates.

Bronfenbrenner, U. (1992). Ecological systems theory. In R. Vasta (Ed.), *Annals of child development. Six theories of child development: Revised formulations and current issues* (pp. 187–249). London: Jessica Kingsley.

Children's Defense Fund. (2002). *National data.* Retrieved October 25, 2002, from http://www.childrensdefense.org/data/default.asp

Christenson, S. L. (2003). The family–school partnership: An opportunity to promote the learning competence of *all* students. *School Psychology Quarterly, 18*(4), 454–482.

Christenson, S. L. (2004). The family–school partnership: An opportunity to promote the learning competence of *all* students. *School Psychology Review, 33*(1), 83–104.

Christenson, S. L., & Anderson, A. R. (2002). Commentary: The centrality of the learning context for students' academic enabler skills. *School Psychology Review, 31*(3), 378–393.

Christenson, S. L., & Sheridan, S. M. (2001). *School and families: Creating essential connections for learning.* New York: Guilford Press.

Coleman, J. (1987, August–September). Families and schools. *Educational Researcher,* 32–38.

Council on Contemporary Families. (2001). *America's changing families.* Retrieved October 10, 2002, from http://www.contemporaryfamilies.org/public/families.php

Edwards, P. A. (1992). Strategies and techniques for establishing home–school partnerships with minority parents. In A. Barona & E. Garcia (Eds.), *Children at-risk: Poverty, minority status, and other issues in educational equity* (pp. 217–236). Silver Spring, MD: National Association of School Psychologists.

Fan, X., & Chen, M. (2001). Parental involvement and students' academic achievement: A meta-analysis. *Educational Psychology Review, 13,* 1–22.

Fantuzzo, J., Tighe, E., & Childs, S. (2000). Family involvement questionnaire: A multivariate assessment of family participation in early childhood education. *Journal of Educational Psychology, 92*(2), 367–376.

Federal Interagency Forum on Child and Family Statistics. (2002). *America's children: Key national indicators of well-being 2002.* Retrieved October 25, 2002, from http://www.childstats.gov/ac2002/index.asp

Harry, B. (1993). *Cultural diversity, families, and the special education system: Communication and empowerment.* New York: Teachers College Press.

Henderson, A. T., & Mapp, K. L. (2002). *A new wave of evidence: The impact of school, family, and community connections on student achievement.* Austin, TX: Southwest Educational Development Laboratory.

Hoover-Dempsey, K. V., & Sandler, H. M. (1997). Why do parents become involved in their children's education? *Review of Educational Research, 67,* 3–42.

Liontos, L. B. (1992). *At-risk families and schools: Becoming partners.* Eugene, OR: University of Oregon, College of Education, ERIC Clearinghouse on Educational Management.

Malecki, C. K., & Elliott, S. N. (2002). Children's social behaviors as predictors of academic achievement: A longitudinal analysis. *School Psychology Review, 17*(1), 1–23.

McWilliam, R. A., Tocci, L., & Harbin, G. L. (1998). Family-centered services: Service providers' discourse and behavior. *Topics in Early Childhood Special Education, 18,* 206–221.

Patrikakou, E. N., & Weissberg, R. P. (1999). The seven P's of school–family partnerships: A philosophy built on research. *Education Week, 18*(21), 34, 36.

Pianta, R., & Walsh, D. B. (1996). *High-risk children in schools: Constructing sustaining relationships.* New York: Routledge.

Resnick, M. D., Bearman, P. S., Blum, R. W., Bauman, K. E., Harris, K. M., Jones, J., Tabor, J., Beuhring, T., Sieving, R. E., Shew, M., Ireland, M., Bearinger, L. H., & Udry, J. (1997). Protecting adolescents from harm: Findings from the National Longitudinal Study on adolescent health. *The Journal of the American Medical Association, 278*(10), 823–832.

Thorp, E. K. (1997). Increasing opportunities for partnerships with culturally and linguistically diverse families. *Intervention in School and Clinic, 32*(2), 261–269.

U.S. Department of Education. (2002). *No Child Left Behind.* Retrieved October 11, 2004, from http://www.nclb.gov

Weiss, H. M., & Edwards, M. E. (1992). The family–school collaboration project: Systemic interventions for school improvement. In S. L. Christenson & J. C. Conoley (Eds.), *Home–school collaboration: Enhancing children's academic and social competence* (pp. 215–243). Silver Spring, MD: National Association of School Psychologists.

Ysseldyke, J. E., & Christenson, S. L. (2002). *FAAB: Functional assessment of academic behavior: Creating successful learning environments.* Longmont, CO: Sopris West.

Zellman, G. L., & Waterman, J. M. (1998). Understanding the impact of parent school involvement on children's educational outcomes. *Journal of Educational Research, 91*(6), 370–380.

2

Parents' Motivations for Involvement in Their Children's Education

Kathleen V. Hoover-Dempsey,
Joan M. T. Walker & Howard M. Sandler

Educators, policymakers, and researchers have long focused on parental involvement as a complement to the fundamental importance of strong teaching and curricula to student achievement. Despite agreement that parental involvement has positive consequences (e.g., Comer, 1993; Epstein, 1996; Fan & Chen, 2001; Henderson & Mapp, 2002; Meidel & Reynolds, 1999), educators and parents have reported barriers to involvement. For example, parents have reported that teachers do not really welcome their involvement; they have also reported an unmet need for *specific* suggestions about *how* to help their children (e.g., Epstein & Dauber, 1991; Harry, 1992; Pena, 2000). Teachers, for their part, have reported poor administrative support for involvement efforts and very limited training in effective involvement strategies (e.g., Hoover-Dempsey, Walker, Jones, & Reed, 2002). Thus, schools' and parents' involvement efforts often "miss the mark."

We and others have suggested that school efforts to increase the effectiveness of parental involvement must be grounded in an understanding of psychological variables underlying parents' decisions to become involved in their children's education (e.g., Grolnick, Benjet, Kurowski, & Apostoleris, 1997; Hoover-Dempsey & Sandler, 1995, 1997). Our model of involvement (Hoover-Dempsey & Sandler, 1995, 1997) is focused on three major questions: Why do parents choose to become involved? How does their involvement, once engaged, influence student outcomes? What

 ISBN 0-8077-4600-2 (paper), ISBN 0-8077-4601-0 (cloth).

student outcomes are associated with parental involvement efforts? The model thus describes a full and dynamic involvement process.

In this chapter we focus on the first question: Why do parents become involved? While we examine the motivators of involvement in some detail, the question is important because it enables a process of parental contributions to students' school learning. We review the full model briefly as context for understanding parents' motivations for involvement before examining each motivation in turn.

THE PARENTAL INVOLVEMENT PROCESS: THE MODEL AS CONTEXT FOR UNDERSTANDING PARENTS' MOTIVATIONS

Our model is constructed in five levels (Figure 2.1). As shown in the first level of our model, we suggest that parents become involved in their children's education for four major reasons: parental *role construction* for involvement (Do parents believe they *should* be involved?), parental *efficacy* for helping the child learn (Do parents believe that their involvement *will make a difference*?), parental perception of *invitations* to involvement from the school (Do parents believe that the school *wants* their involvement?) and from the child (Do parents believe that the child *wants or needs* their involvement?).

Level 2 of the model describes factors that are likely to influence the forms of involvement they choose (e.g., home-based and/or school-based activities). These variables include parents' perceptions of their own *skills and knowledge* (e.g., Do they believe they have the knowledge to help the child with math homework?) and competing *demands on parents' time* from other family and employment responsibilities (e.g., Do work hours allow time to help with homework? Do the demands of infant care allow participation in school-based conferences?). At this level, specific *invitations to involvement* from the child, teacher(s), and school also influence parents' involvement forms. For instance, many parents are likely, if asked and given suggestions, to help the student review spelling words and so forth.

Level 3 of the model suggests that parents' involvement, once engaged, influences students' school outcomes through specific mechanisms. Among the most important are parents' *modeling* of appropriate school-related skills (e.g., "One way you can use to think about solving this problem is to . . ."), *reinforcement* of learning and attributes related to learning (e.g., "You really worked hard to organize that report, and you've done a great job"), and *instruction* (e.g., offering teaching and related help with specific school assignments and needs).

Figure 2.1. Model of the Parental Involvement Process

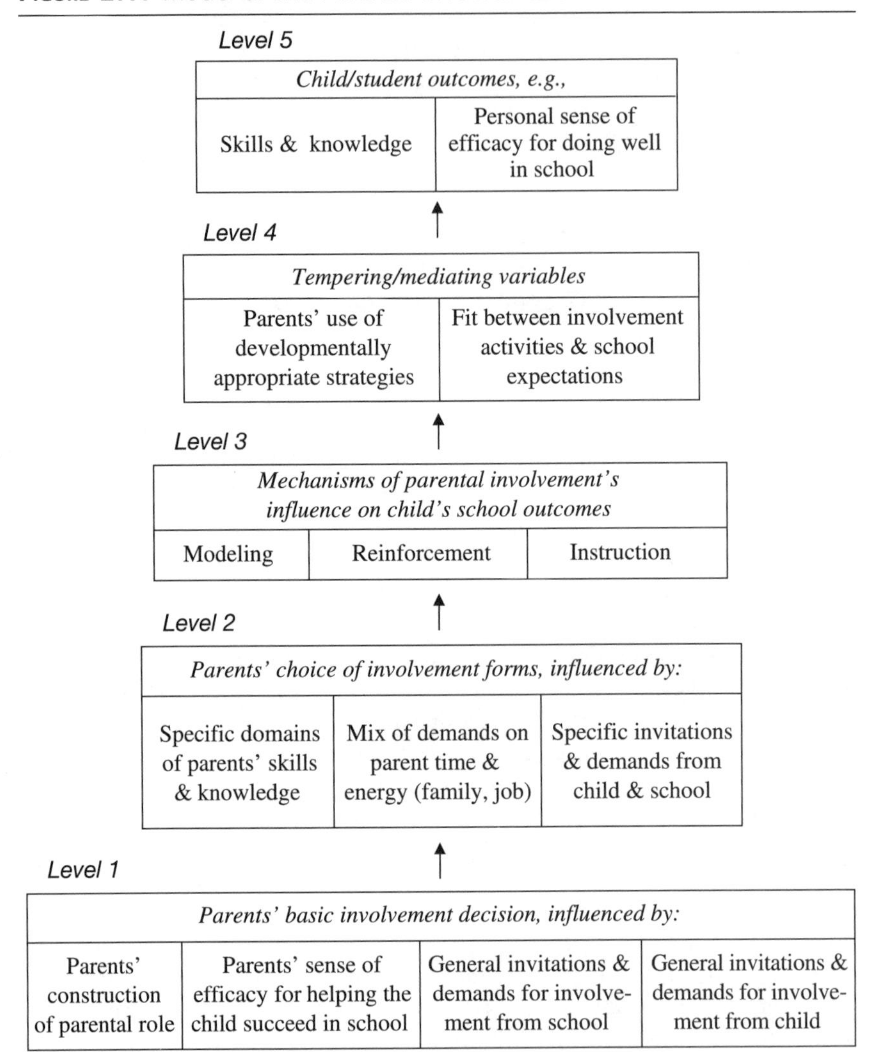

Adapted from Hoover-Dempsey & Sandler, 1995, 1997.

At Level 4, the model suggests that parents' involvement activities influence student outcomes to the extent that parents use developmentally appropriate strategies (e.g., close monitoring of time on homework may help younger children complete tasks, but the same strategy used with adolescents may be perceived as intrusive). The model also suggests that the fit between parents' choice of activities and the school's expectations for involvement will also influence parents' ability to support learning outcomes through involvement (e.g., if the parent and school make similar assumptions about appropriate student learning tasks and habits, parents and school are likely to have positive influence on learning).

The final level of the model (Level 5) focuses on specific student outcomes influenced by parental involvement. These outcomes include, but are not limited to, achievement, skills and knowledge, students' personal sense of efficacy for doing well in school, and other attributes that facilitate student learning (e.g., self-regulation, motivation; e.g., Hoover-Dempsey et al., 2001).

WHAT MOTIVATES PARENTS TO BECOME INVOLVED?

Having situated parents' motivations for involvement within the full model, we turn now to more detailed consideration of each identified motivation. Following this section, we offer suggestions for teachers and schools wishing to increase the incidence and effectiveness of parental involvement. We conclude with brief consideration of critical links among school invitations to parents, parents' involvement decisions and choices, and student school outcomes.

Parental Role Construction for Involvement

Roles are composed of sets of expectations or beliefs held by individuals and groups for the behavior of individual members and beliefs about the behaviors characteristic of group members (Hoover-Dempsey & Sandler, 1997). These expectations or beliefs include patterns of ideas that guide the individual's choice of behaviors within specific contexts, as well as the individual's interpretation of others' behaviors within those contexts. Roles also reflect the individual's understanding of personal responsibility for behaviors within a context, and understanding of the range of behaviors that are appropriate in the context. Thus, roles are socially constructed by individuals in the context of their social groups. Because contributors to role construction (individuals and groups) are subject to change, individuals' role constructions are also subject to change.

These observations on role are complemented by theory and research in developmental psychology. For example, parents' ideas about appropriate roles in children's education are shaped not only by relevant social contexts (e.g., family, school, culture), but are also guided by parents' beliefs and ideas about child-rearing (how children develop, effective child-rearing practices: e.g., Stevenson, Chen, & Uttal, 1990). In general, parents' role construction for involvement in children's education is created in the interaction of: (a) beliefs about appropriate and desired child outcomes; (b) beliefs about who is responsible for these outcomes; (c) perceptions of important group members' (e.g., family, teachers, other parents) expectations for parental behaviors; and (d) parental behaviors related to those beliefs and expectations. These sources of role construction underscore its grounding in families' cultural context, and suggest the importance of attending well to variations across families and school communities in understanding role construction for involvement.

Work that we have conducted on parental role construction (e.g., Hoover-Dempsey & Jones, 1997; Hoover-Dempsey & O'Connor, 2002; Hoover-Dempsey, Wilkins, Sandler, & O'Connor, 2004; Walker, Wilkins, Dallaire, Sandler, & Hoover-Dempsey, in press) suggests that role construction has two major manifestations: *active* and *passive*. An active parental role construction reflects: (a) belief that primary responsibility for the child's educational outcomes belongs to the parent; (b) belief that the parent should be active in meeting these responsibilities; and (c) personal behavior that includes active support for the child's school learning. We have observed two forms of active role construction: parent-focused (parent is ultimately responsible) and partnership-focused (parent and school ultimately share responsibility). A passive parental role construction (usually a school-focused orientation) reflects: (a) belief that the school holds primary responsibility for the child's educational outcomes; (b) belief that the parent need act only when the school initiates contact; and (c) behaviors that include reliance on the school and general acceptance of school decisions.

Parents' accounts of involvement may include beliefs and behaviors compatible with both active and passive role construction, even as they reflect the predominance of one. Although this has not been tested yet, we suspect that the relative predominance of active or passive role construction for a parent at a given time is linked to: (a) the context of the child's current school year (provided by the teacher and school); (b) the parent's perceptions of the child's developmental status and learning needs; and (c) the parent's judgments about personal ability to meet the child's needs in the current context.

While productive family–school partnerships are likely enhanced by an active role construction (whether parent- or partnership-focused), a parent's ability to enact a partnership role depends on a school context that enables partnership. Because roles are socially constructed, a partnership focus appears possible only when parents hold active role beliefs *and* the social context (the school) invites and enables partnership. In the "Recommendations for School Action" section of the chapter, we address the power of the social context to shape parents' role construction and offer suggestions for increasing active role construction.

Parental Sense of Efficacy for Helping the Child Succeed in School

Self-efficacy refers to beliefs in one's capability to act in ways that will produce desired outcomes (Bandura, 1997). Self-efficacy beliefs are a significant factor in personal decisions about one's goals, the effort one puts into those goals, one's persistence in the face of obstacles, and the accomplishment of those goals. Bandura identified four primary sources of self-efficacy beliefs: personal mastery experiences, vicarious experiences, verbal persuasion, and physiological or affective arousal. Applied to the domain of parental involvement, this work suggests that parental self-efficacy for helping the child succeed in school requires experiences of success in helping the child learn (personal mastery), opportunities to observe similar parents or adults successfully helping children in school-related tasks (vicarious experience), encouragement from important others (verbal persuasion), and support for the positive feelings that come with success as well as realistic encouragement from others when doubts emerge (affective arousal).

Self-efficacy theory thus suggests that parents make involvement decisions in part by thinking about the likely outcomes of their actions, and that they develop goals for involvement based on their appraisal of their capabilities in the situation (Bandura, 1989). If a parent believes that her help with homework will make a difference for the child, she is likely to become involved in helping with homework; if she believes that her involvement in parent–teacher conferences will not make any difference for the child, she is likely to avoid them. This theory also suggests that a parent with strong self-efficacy for helping the child succeed is likely to persist in the face of challenges and work through difficulties to successful outcomes. Low self-efficacy, on the other hand (i.e., low expectations for personal success in helping the child learn), may be associated with low persistence or simply quitting if it appears that involvement is not making any difference.

Thus, personal sense of efficacy for helping the child succeed in school is important to positive family–school relationships. Parents who believe that their involvement will make a difference for the child are more likely than parents who doubt it to engage in activities that might help the child learn. Because sense of efficacy influences parental goals for involvement and persistence when confronted with obstacles, it also shapes parents' involvement choices in uncertain or discouraging situations.

Self-efficacy and role construction are also linked. A parent with high self-efficacy who believes his actions will really help his child learn is likely to receive support from this belief for a more active role construction. At the same time, a parent with low self-efficacy who may try but perceives little benefit from his attempts is likely to adopt a more passive role construction (e.g., "My help probably won't do much good, so there's not much point in being involved"). Weak efficacy or low outcome expectancies combined with a passive role construction may lead to very low or no involvement.

In sum, sense of efficacy for helping the child succeed in school is an important contributor to parents' involvement decisions. Further, like parental role construction, efficacy is influenced by interactions with others in the social context (teachers, school, family, other parents). We address implications of the dynamic nature of self-efficacy in more detail in the "Recommendations for School Action" section of the chapter and offer suggestions for increasing parental self-efficacy.

Parental Perceptions of Invitations to Involvement

Our model suggests that the "others" who are most influential in parents' decisions about involvement are the school, the child's teacher(s), and the child. Looking first at the school, a relatively strong body of evidence suggests that school invitations, manifested in a positive school climate, and consistent invitations to be involved in learning at home influence parents' decisions about becoming involved in elementary, middle, and high school students' education (e.g., Dauber & Epstein, 1993). General school invitations include broad school attributes or activities that convey to all parents that involvement is welcome and that it is a valuable resource for supporting student learning and success. A welcoming school climate, user-friendly newsletters describing students' grade-level work, and clear suggestions for home-based support of learning all indicate that parental involvement is expected and wanted by the school, that the parent's active engagement in the child's schooling is important to the child's learning, and that the parent's activities at home or at school are important to the child's success. Such invitations are particularly important for families who feel uncertain about involvement, such as those whose adult members did not do well in school themselves, those from cultures different from the middle-class norm, and those who feel on the margins of society in general and schools in particular; they are also often critically important to non-English-speaking families (e.g., Collignon, Men, & Tan, 2001; Comer, 1993; Garcia Coll et al., 2002; Lopez, Sanchez, & Hamilton, 2000).

General invitations to involvement from the child may come from child attributes (e.g., grade level, general school performance) and characteristic behaviors (e.g., difficulty with schoolwork or valuing of parental help: Hoover-Dempsey, Bassler, & Burow, 1995; Walker & Hoover-Dempsey, 2001). General invitations from the child may be explicit (e.g., "I don't understand this problem!" "It helps when you explain things") or implicit (e.g., the parent's assessment of recent test grades suggests that the child needs help). Because most parents respond to their children's general characteristics and try to respond to their explicit needs, varied invitations from the child are a potentially important contextual motivator of parents' involvement.

While general invitations from the school and child often stand on their own as contributors to parental involvement, they may also contribute to active role construction and relatively strong sense of efficacy because the latter are subject to influence from the social context. Thus invitations to involvement often convey school valuing of an active parental role in children's school success. Further, because such invitations may act as verbal encouragement to become involved—and, if accepted, offer opportunities for personal success in helping the child—they also have the potential to strengthen the sense of self-efficacy.

Overall, our model and related research (e.g., Griffith, 1998; Hoover-Dempsey et al., 1995; Kohl, Lengua, & McMahon, 2002) suggest that invitations from school and child will influence parents' decisions about involvement. Because these invitations emerge from two important contexts for parents (school and child), they are susceptible to intervention and change (e.g., the school's efforts to become more welcoming of parental involvement; the school's efforts to encourage children to share and seek parents' responses to homework tasks). In the next section, we address specific steps schools can take to increase the incidence and power of invitations to involvement, as well as specific suggestions for school support of more active parental role construction and a more positive sense of efficacy for helping children succeed in school.

RECOMMENDATIONS FOR SCHOOL ACTION

Here we offer examples of actions schools may take to increase parents' active role construction, sense of efficacy for helping the child succeed, and positive perceptions of invitations to involvement. These suggestions, summarized in Figure 2.2, do not offer magic solutions to perennial issues in parental involvement, but, rather, illustrate ways in which schools might explicitly support parents' motivations to become involved. They are intended to stimulate discussion among teachers, parents, administrators, and policymakers about strategies likely to be effective in different school settings and cultures. We believe that these and related strategies offer critical support as teachers and parents work to create the family–school partnerships central to students' success.

Steps to Increase Active Parental Role Construction

Create Positive School Assumptions About the Importance of Parents' Contributions to Student Success. This strategy requires that schools believe and communicate to parents and school staff that parents do in fact have an important role to play in their children's educational success. Schools may create and convey these assumptions in many ways, for example, by offering specific information about how parents' active engagement supports student learning, and creating and maintaining an environment that actively values and respects parents' suggestions and presence.

Communicate Clearly With Parents About the Importance of Their Active Involvement in Children's Schooling. This can be accomplished through two-way communications—communications in which the school and teachers not only offer information, but listen and take into account the parent's ideas and contributions. Creating more effective parental involvement may require that schools develop strong listening skills as well as habits of adaptation and flexibility. Such skills may be particularly critical when parents and schools represent different cultures. Creating varied opportunities for consistent and accessible interactive communications is also vital; this can be accomplished, for example, by installing phones in teachers' classrooms, maintaining adequate Internet connections to support e-mail, creating opportunities for relaxed and enjoyable parent–teacher interactions, and allocating resources to support parent–teacher communication (e.g., preparation of classroom newsletters for parents, creating a parent–teacher room for conversations at school, hiring community members deeply familiar with the culture and language[s] of the school's families).

Steps to Increase Parents' Sense of Efficacy

Communicate Clearly That Many Involvement Behaviors Influence School Success. This means helping parents understand that their involve-

FIGURE 2.2. Steps Schools Might Take to Enhance Parental Involvement

Supporting parents' active role construction and sense of partnership with the school	Enact the values, expectations, and behaviors that enable parent–school partnership: • Believe and communicate that parents have an important role to play in their children's educational success. • Offer specific information about ways in which parents' active engagement in children's schooling supports student learning conducive to achievement (e.g., more positive attitudes about school, increased time on homework, greater persistence in learning tasks, improved knowledge and use of effective learning strategies, stronger beliefs that effort is important to learning success). • Create and maintain a school environment that overtly values and respects parental presence and input.
Supporting parents' sense of efficacy for helping the child succeed in school	Offer opportunities to help children learn successfully, to observe other parents or adults who are successfully helping students learn, to receive encouragement from numerous sources, and to receive support for the positive feelings that come with success and encouragement when doubts or obstacles arise: • Communicate with parents about the positive influence of varied parental involvement behaviors on student attributes that lead to student school success. • Use existing parent–teacher structures to enhance involvement (e.g., create opportunities for sharing stories of similar parents' success in helping their children learn). • Give parents specific feedback on the positive influence of their involvement activities (e.g., "Her spelling test results have really improved since you've been quizzing her two nights a week").
Supporting parents' perceptions of general invitations to involvement from the school	Convey that parental involvement is expected and wanted, that involvement is important to the child's success, and that parents' activities at home and at school are all important contributors to the child's learning and success: • Develop overtly welcoming practices that respect and build on family culture and strengths. • Seek, develop, and welcome involvement that engages parents and teachers in two-way communications. • Offer a full range of home-based and school-based involvement opportunities; advertise them clearly, attractively, and repeatedly.
Building parents' perceptions of general invitations to involvement from the child	Help students convey that parental involvement is expected, wanted, and valuable: • Develop interactive homework assignments (e.g., require students to ask for their parents' help or listening; focus particularly on time- or task-limited activities). • Encourage student–parent conversation about school activities.

ment (e.g., through encouraging children, modeling school-appropriate behaviors, reinforcing good work) contributes to student characteristics associated with higher achievement. Among these student attributes are increased time on homework, greater persistence in learning tasks, improved knowledge and use of effective learning strategies, better classroom behavior, and stronger beliefs about the importance of effort in learning success (Glasgow, Dornbusch, Troyer, Steinberg, & Ritter, 1997; Hoover-Dempsey et al., 2001; Sanders, 1998; Xu & Corno, 1998). Information from schools, such as specific suggestions that parents might enact to support these learning-enhancing attributes, may be especially helpful.

Give Parents Specific Feedback on the Positive Influence of Their Involvement. Teachers can give feedback with reference to specific parental behaviors and specific child progress (e.g., "Her spelling test results have really improved since you've been quizzing her two nights a week"). They can also relate stories of similar parents' success. Finally, teachers and schools can also focus their feedback on parents' more general involvement (e.g., how the parent's presence at school communicates to the student that school is important, how parental encouragement of student effort supports a sense of personal control over learning outcomes).

Steps to Increase Parents' Perceptions of Invitations to Involvement

Develop Welcoming Practices That Clearly Respect and Build on Family Culture and Strengths. This strategy can be implemented in a variety of ways, including: visual displays in school entry areas and hallways; the development of schoolwide positive responses to parents (e.g., courteous and clear responses to parent phone calls and walk-in visitors, a "customer-friendly" attitude in responding to parents' or family members' questions and concerns). Schools may also use existing structures that offer points of regular contact between parents and school personnel (e.g., PTO, after-school programs) as a means of conveying that parents' involvement is consistently welcome, expected, and valuable.

Offer a Full Range of Home-Based and School-Based Involvement Opportunities. Standard approaches to involvement (e.g., open house, scheduled parent–teacher conferences, progress reports) should be included, but the involvement activities that schools routinely invite can be expanded to include other, less traditional approaches to involvement (e.g., Denham & Weissberg, 2003; Greenberg et al., 2003; Webster-Stratton, Reid, & Hammond, 2001). Such initiatives might include educational programs to help parents and teachers advocate more effectively for student needs, and parent–teacher partnerships focused on enhancing social-emotional learning—a domain of development shared by families and schools. Expanding the range of involvement activities is often critical to engaging more parents in children's education, especially in communities where families may feel estranged from traditional parent–school relationships.

Develop Interactive Homework Assignments. To get parents more actively involved, students might be asked to request parental or other family member input and opinion (these requests are often most effective if short and easy to implement). Schools may also offer specific, time-limited, developmentally appropriate suggestions for involvement (e.g., "Read together for 10 minutes three nights this week"; "Go over one of tonight's homework problems together"), and might consider using a programmatic approach to increasing parental involvement in student homework, such as Epstein and colleagues' Teachers Involving Parents in Schoolwork (TIPS) program (e.g., Epstein & Van Voorhis, 2001; Van Voorhis, 2001). Adoption of such suggestions may have the dual benefit of increasing parent–student interactions related to schoolwork and increasing parent opportunities to see what students are learning in school.

Encourage Parents to Talk With Students About General or Specific Aspects of School. This suggestion follows findings that parents' at-home discussion of school-related activities is strongly related to student achievement (e.g., Sui-Chu & Willms, 1996). While developmental differences in children across the grades are important to the specifics of such discussions (e.g., form, timing, content), the reality that parental involvement in students' education continues to be influential throughout the secondary school years (e.g., Steinberg, Elmen, & Mounts, 1989) supports the ongoing usefulness of this recommendation.

Recommendations in Sum

This set of suggestions serves the broad goal of supporting school and family efforts to create, support, and maintain effective parental involvement in students' education. As described in the next and final section, the benefits of these and related suggestions—especially when developed by school faculty, staff, and parents as appropriate to the needs of specific schools and school communities—are to be found in enhanced parental motivation for involvement, increases in productive family–school partnerships, and gains in student attributes that underlie achievement.

WHY SHOULD SCHOOLS TAKE THESE STEPS?

School efforts to enhance parents' motivations for involvement are important because they are likely to increase the incidence and effectiveness of parental involvement with children. For instance, parents who are encouraged to develop an *active role construction* are likely to model the importance of consistent and often cooperative communications about school tasks (e.g., talking with the student about school activities; talking with the teacher about the student's learning). When students observe and experience such interactions, they have further access to some of the cognitive and social skills necessary for effective classroom learning (e.g., reflecting on learning in order to communicate about it; social self-efficacy for interacting with and asking questions of the teacher). Similarly, parents whose *sense of efficacy for involvement* is well supported are likely to notice and reinforce students for learning effort, skill development, and accomplishments. Parents who receive invitations for involvement from the school, especially when invitations include specific suggestions for helping the student learn, are likely to offer instructionally related involvement that supports learning goals for students.

Schools should also take these steps because they and parents make unique and often complementary contributions to student achievement and outcomes that lead to achievement. Such outcomes that support student achievement include motivation to learn, ability to regulate one's emotions and learning strategy use, personal sense of efficacy for doing well in school, and sense of efficacy for relating well to teachers and peers (e.g., Elias et al., 1997; Epstein & Van Voorhis, 2001; Grolnick & Slowiaczek, 1994; Gutman & Midgley, 2000; Hoover-Dempsey et al., 2001; Walker & Hoover-Dempsey, in press). Thus, when parents and schools focus on building the student's cognitive, social, and emotional skills essential to effective learning, schools, families, teachers, and students benefit.

In fact, we argue that recognizing parents' contributions to the cognitive, social, and emotional *precursors* of effective learning is perhaps more important than focusing on parents' direct influence on student achievement. This is because achievement ultimately depends not on the parent's beliefs, thinking, or behaviors but rather on the *student's* beliefs, thinking, and behaviors. Student behaviors are often outside parents' direct control (e.g., parents generally do not do their children's homework or take their tests). Moreover, subject-matter knowledge and problem-solving skills in various content areas are often beyond the expertise of many parents, especially as students enter secondary school.

Our view of involvement as a critical support system for school efforts to improve student achievement is supported by theoretical perspectives on learning (e.g., Rogoff, 1998; Vygotsky, 1978), and research investigating links among students' beliefs about what they should and can do in their own education, their perceptions of school, and their school learning (e.g., Roeser, Midgeley, & Urdan, 1996; Skinner, Wellborn, & Connell, 1990) supports this suggestion. Just as our model suggests that parents' decisions to become involved in their children's education are influenced by interactions with others (e.g., teachers, school staff), so, too, is student learning influenced by the beliefs and behaviors of important adults (parents and teachers) in the social contexts of schooling.

CONCLUSION

By attending to the major motivations underlying parents' involvement in children's education, schools are more likely to hit the mark of supporting increasingly effective parental contributions to learning. Developing and implementing strategies such as we have suggested here will require the allocation of resources (e.g., time, place, or funds); however, the implementation of such strategies offers important school support for parents' unique and complementary contributions to student learning.

NOTE

We offer many thanks to the members of the Peabody/Vanderbilt Family–School Partnership Lab. A portion of the work reported here was supported by a grant from the Institute for Education Sciences (formerly the Office of Educational Research and Innovation), U.S. Department of Education: *Parental Involvement: A Path to Enhanced Achievement,* #R305T010673.

REFERENCES

Bandura, A. (1989). Human agency in social cognitive theory. *American Psychologist, 44,* 1175–1184.

Bandura, A. (1997). *Self-efficacy: The exercise of control.* New York: W. H. Freeman.

Collignon, F. F., Men, M., & Tan, S. (2001). Finding ways in: Community-based perspectives on southeast Asian family involvement with schools in a New England state. *Journal of Education for Students Placed at Risk, 6*(1&2), 27–44.

Comer, J. P. (1993). *School power: Implications of an intervention project.* New York: Free Press.

Dauber, S. L., & Epstein, J. L. (1993). Parents' attitudes and practices of involvement in inner-city elementary and middle schools. In N. F. Chavkin (Ed.),

Families and schools in a pluralistic society (pp. 53–71). Albany, NY: State University of New York.

Denham, S., & Weissberg, R. (2003). In M. Bloom & T. P. Gullotta (Eds.), *A blueprint for the promotion of prosocial behavior in early childhood* (pp. 13–50). New York: Kluwer/Academic Publishers.

Elias, M. J., Zins, J. E., Weissberg, R. P., Frey, K. S., Greenberg, M. T., Haynes, N. M., et al. (1997). *Promoting social and emotional learning.* Alexandria, VA: Association for Supervision and Curriculum Development.

Epstein, J. L. (1996). Perspectives and previews on research and policy for school, family, and community partnerships. In A. Booth & J. F. Dunn (Eds.), *Family–school links: How do they affect educational outcomes?* (pp. 209–246). Hillsdale, NJ: Erlbaum.

Epstein, J. L., & Dauber, S. L. (1991). School programs and teacher practices of involvement in inner-city elementary and middle schools. *Elementary School Journal, 91,* 291–305.

Epstein, J., & Van Voorhis, F. L. (2001). More than minutes: Teachers' roles in designing homework. *Educational Psychologist, 36*(3), 181–193.

Fan, X., & Chen, M. (2001). Parental involvement and students' academic achievement: A meta-analysis. *Educational Psychology Review, 13*(1), 1–22.

Garcia Coll, C., Akiba, D., Palacios, N., Bailey, B., Silver, R., DiMartino, L., & Chin, C. (2002). Parental involvement in children's education: Lessons from three immigrant groups. *Parenting: Science and Practice, 2*(3), 303–324.

Glasgow, K. L., Dornbusch, S. M., Troyer, L., Steinberg, L., & Ritter, P. L. (1997). Parenting styles, adolescents' attributions, and educational outcomes in nine heterogeneous high schools. *Child Development, 68,* 507–529.

Greenberg, M. T., Weissberg, R. P., O'Brien, M. U., Zins, J. E., Fredericks, L., Resnick, H., & Elias, M. J. (2003). Enhancing school-based prevention and youth development through coordinated social, emotional, and academic learning. *American Psychologist, 58*(6/7), 466–474.

Griffith, J. (1998). The relation of school structure and social environment to parent involvement in elementary schools. *Elementary School Journal, 99*(1), 53–80.

Grolnick, W. S., Benjet, C., Kurowski, C. O., & Apostoleris, N. H. (1997). Predictors of parent involvement in children's schooling. *Journal of Educational Psychology, 89*(3), 538–548.

Grolnick, W. S., & Slowiaczek, M. L. (1994). Parents' involvement in children's schooling: A multidimensional conceptualization and motivational model. *Child Development, 64,* 237–252.

Gutman, L. M., & Midgley, C. (2000). The role of protective factors in supporting the academic achievement of poor African American students during the middle school transition. *Journal of Youth and Adolescence, 29*(2), 223–248.

Harry, B. (1992). An ethnographic study of cross-cultural communication with Puerto Rican-American families in the special education system. *American Educational Research Journal, 29,* 471–494.

Henderson, A. T., & Mapp, K. L. (2002). *A new wave of evidence: The impact of school, family, and community connections on student achievement.* Austin, TX: Southwest Educational Development Laboratory.

Hoover-Dempsey, K. V., Bassler, O. C., & Burow, R. (1995). Parent involvement in elementary children's homework: Parameters of reported strategy and practice. *Elementary School Journal, 95,* 435–450.

Hoover-Dempsey, K. V., Battiato, A. C., Walker, J. M. T., Reed, R. P., DeJong, J. M., & Jones, K. P. (2001). Parental involvement in homework. *Educational Psychologist, 36*(3), 195–209.

Hoover-Dempsey, K. V., & Jones, K. (1997, March). *Parental role construction and parental involvement in children's education.* Paper presented at the annual meeting of American Educational Research Association, Chicago, IL.

Hoover-Dempsey, K. V., & O'Connor, K. J. (2002, April). *Parental involvement in children's education: Exploring the structure and functions of parental role construction.* Paper presented at the annual meeting of the American Educational Research Association, New Orleans, LA.

Hoover-Dempsey, K. V., & Sandler, H. M. (1995). Parental involvement in children's education: Why does it make a difference? *Teachers College Record, 95,* 310–331.

Hoover-Dempsey, K. V., & Sandler, H. M. (1997). Why do parents become involved in their children's education? *Review of Educational Research, 67*(1), 3–42.

Hoover-Dempsey, K. V., Walker, J. M. T., Jones, K. P., & Reed, R. P. (2002). Teachers Involving Parents (TIP): Results of an in-service teacher education program for enhancing parental involvement. *Teaching and Teacher Education, 18*(7), 1–25.

Hoover-Dempsey, K. V., Wilkins, A. S., Sandler, H. M., & O'Connor, K. P. J. (2004, April). *Parental role construction for involvement: Theoretical, measurement and pragmatic issues in instrument development.* Paper presented at the annual meeting of the American Educational Research Association, San Diego, CA.

Kohl, G. W., Lengua, L. J., & McMahon, R. J. (2002). Parent involvement in school: Conceptualizing multiple dimensions and their relations with family and demographic risk factors. *Journal of School Psychology, 38*(6), 501–523.

Lopez, L. C., Sanchez, V. V., & Hamilton, M. (2000). Immigrant and native-born Mexican-American parents' involvement in a public school: A preliminary study. *Psychological Reports, 86,* 521–525.

Meidel, W. T., & Reynolds, A. J. (1999). Parent involvement in early intervention for disadvantaged children: Does it matter? *Journal of School Psychology, 37*(4), 379–402.

Pena, D. C. (2000). Parent involvement: Influencing factors and implications. *Journal of Educational Research, 94*(1), 42–54.

Roeser, R. W., Midgley, C., & Urdan, T. (1996). Perceptions of the school psychological environment and early adolescents' psychological and behavioral functioning in school: The mediating role of goals and belonging. *Journal of Educational Psychology, 88*(3), 408–422.

Rogoff, B. (1998). Cognition as a collaborative process. In D. Kuhn & R. S. Siegler (Eds.), *Handbook of child psychology: Vol. 2. Cognition, perception, and language* (pp. 679–744). New York: Wiley.

Sanders, M. (1998). The effects of school, family, and community support on the academic achievement of African American adolescents. *Urban Education, 33,* 385–409.

Skinner, E. A., Wellborn, J. G., & Connell, J. P. (1990). What it takes to do well in school and whether I've got it: A process model of perceived control and children's engagement and achievement in school. *Journal of Educational Psychology, 82*(1), 22–32.

Steinberg, L., Elmen, J. D., & Mounts, N. S. (1989). Authoritative parenting, psychosocial maturity, and academic success among adolescents. *Child Development, 60,* 1424–1436.

Stevenson, H. W., Chen, C., & Uttal, D. H. (1990). Beliefs and achievement: A study of Black, White, and Hispanic children. *Child Development, 61,* 508–523.

Sui-Chu, E. H., & Willms, J. D. (1996). Effects of parental involvement on eighth-grade achievement. *Sociology of Education, 69,* 126–141.

Van Voorhis, F. L. (2001). Interactive science homework: An experiment in home and school connection. *National Association of Secondary School Principals' Bulletin, 85*(627), 20–32.

Vygotsky, L. S. (1978). *Mind in society*. Cambridge, MA: Harvard University Press.

Walker, J. M., & Hoover-Dempsey, K. V. (2001, April). *Age-related patterns in students' invitations to parental involvement in homework.* Symposium paper presented at the annual meeting of the American Educational Research Association, Seattle, WA.

Walker, J. M., & Hoover-Dempsey, K. V. (in press). Parental involvement and classroom management. In C. M. Evertson & C. S. Weinstein (Eds.), *Handbook for classroom management: Research, practice, and contemporary issues.* Hillsdale, NJ: Erlbaum.

Walker, J. M., Wilkins, A. S., Dallaire, J., Sandler, H. M., & Hoover-Dempsey, K. V. (in press). Parental involvement: Model revision through scale development. *Elementary School Journal.*

Webster-Stratton, C., Reid, J., & Hammond, M. (2001). Preventing conduct problems, promoting social competence: A parent and teacher training partnership in Head Start. *Journal of Clinical Child Psychology, 30,* 283–302.

Xu, J., & Corno, L. (1998). Case studies of families doing third-grade homework. *Teachers College Record, 100,* 402–436.

3

Influences and Challenges to Better Parent–School Collaborations

Pamela E. Davis-Kean &
Jacquelynne S. Eccles

As previous chapters have pointed out, schools and families are partners in the healthy academic, social, and emotional development of the child. During the elementary years, and often earlier, children spend more time in the school and the home than anywhere else. Thus, both need to communicate and interact often in order to coordinate and manage the healthy development of a child. The research, however, suggests that this type of coordination and communication is not common and that if it does occur, it is based on specific features of parents, teachers, and schools (Comer & Haynes, 1991; Epstein, 1990; Hoover-Dempsey & Sandler, 1997). There is evidence to suggest that both parents and teachers would like more interaction, but it is often difficult to achieve, and even if it initially exists it declines rather rapidly as children make their transition into secondary school (Dauber & Epstein, 1989; Eccles & Harold, 1993; Epstein, 1986). The research also suggests that having high-quality links (e.g., frequent telephone and personal communication) between parents and teachers promotes positive school success for children (Comer, 1980; Comer & Haynes, 1991; Epstein, 1990). So the question is: What are the challenges to creating these links, and how can those challenges be reduced so that the links can be retained over time? In this chapter we review the basic literature on challenges to forming partnerships between schools and families and then introduce a developmental model that presents a way that schools and families can work together in their important roles in the academic, social, and emotional development of children. We end with recommendations on how families and schools can make changes to support the academic, social, and emotional development of children in the varying contexts where they spend their time.

CHALLENGES TO PARENTAL INVOLVEMENT

Eccles and Harold (1993) introduced a model that examined a wide range of variables across multiple contexts (family, school, community) that likely influence parental involvement. In this model, they proposed how distal influences such as neighborhood and school characteristics and more proximal characteristics of teacher and parental beliefs influence whether or not parents get involved in the schools and have subsequent effects on child outcomes. Recent reviews of research on parent involvement (Henderson & Mapp, 2002) have supported the existence of these various influences and challenges to better interactions and communication between parents and the schools. These major influences and challenges are reviewed and discussed below.

Parent/Family Characteristics

This category encompasses the general socioeconomic status of the family as well as the mental health of the parents. Here it is noted that parents' education and workforce participation are important predictors of involvement, with more highly educated parents being more involved and those in the workforce having limited time for involvement (Hoover-Dempsey & Sandler, 1997). Other important parent/family characteristics likely to be important to involvement are social and psychological resources of the parents, parents' perceptions of their child, parents' beliefs about their role in their child's education, parents' attitude toward school, parents' ethnic identity, socialization practices, and parents' prior involvement in children's education. As was discussed in Chapter 2, these factors are important in determining how willing parents are to participate in the schools. Parents may not, for example, have confidence that they can influence their child's achievement by participating in school (parental efficacy). They may place more emphasis on getting the child to school on time with the appropriate materials and not on assisting in the classroom. They may have language difficulties and feel strange about participating with adults and children whom they cannot understand, or have difficulty understanding the materials that are sent home explaining participation opportunities (Garcia Coll et al., 2002). Dornbusch (1994) and Romo (1999), for example, found that parents and adolescents, especially from minority populations, do not know what courses are required for getting admitted into the university system. The adolescents believed they were taking, and doing quite well in, the right courses; however, they found out too late that the courses would not meet the requirements and that they were not being taught the information needed to do well on the entrance exams. As a consequence, they had to change their post–high school plans despite their parents' best efforts. Similarly, Garcia-Coll et al. (2002) found that parents' reports of language difficulty were one of the strongest predictors of lack of involvement in schools for Portuguese, Dominican, and Cambodian immigrant groups regardless of socioeconomic status. They also found that the Dominican group was impacted the least by the language discomfort and had more interaction with the schools. They believe that this is due to the strong emphasis in the schools on bilingual education and materials that kept parents informed and helped the teachers and parents coordinate educational challenges such as homework.

Thus, it is very important that schools be aware that parents, like their children, do not enter the school environment with the same characteristics, background, or general understanding of how they might interact or be a part of the relationship with the school. The schools need to provide very specific assistance with materials sent home with the schools and make no assumptions about basic literacy or language comprehension.

Community Characteristics

Similar to the parental characteristics that influence families' involvement in schools are the characteristics of the communities in which the families are members. Parents develop parenting behaviors that are consistent with the demands of their environments. In high-risk neighborhoods, for example, parents are more likely to concentrate on protecting their children from the negative influences and consequences of that environment than on other issues of development. These parents are not necessarily less involved with their children; instead they have chosen to focus their parenting influence on protection rather than on talent or achievement development. In lower-risk neighborhoods, danger is less of a concern and there are generally more resources (e.g., after-school programs, recreation programs, enrichment activities for dance and music) that parents can use to help in developing their children. There is also an interaction between the parent characteristics and the communities. Parents with lower education, lower-status jobs, low social support, lower emotional intelligence or social skills, and lower financial resources are often residing in high-risk environments. These parents face additional challenges in assisting their children with education or school-related activities. Schools need to use appropriate strategies and innovations to assist these parents to get involved and stay involved (Eccles & Harold, 1993). For these children and families, schools may be the one safe and stable place they can go to experience positive involvement with the community. Schools could capitalize more on this fact by providing programs that serve both parents and children (see Chapter 5 for an extended discussion of this issue).

Child Characteristics

Characteristics of the child also influence the participation of parents in school settings, especially when children enter adolescence, which is a time of tremendous exploration and growth (for a discussion of the longitudinal effects of school–family partnership programs, see Chapter 7). Adolescents' lives are changing as they transition first from elementary to middle school and then to high school. These transitions to new environments present risks to self-esteem and motivation (Eccles et al., 1993). Schools and families must work together not only to prevent negative outcomes, but also to promote positive, healthy, successful development. This requires open and ongoing communication and collaboration.

Parents may feel that as children get older they want less participation from their parents. Some parents may also feel less adequate at helping or participating with adolescents because the schoolwork becomes more complicated over the school years. Parents who were initially active in elementary schools might find fewer avenues to contribute in middle and junior high schools as their children spend more time with peers and the schools have fewer activities requiring or encouraging parent involvement. One reason for this decline in involvement might be parent response to the growing autonomy of their children. It becomes increasingly difficult as children grow older for parents to know what their role should be in the education of their children when issues such as college selection and occupation in the workforce begin to take prominence in children's lives. However, parent involvement is not only necessary as adolescents enter a transition that can lead to a healthy or risky developmental trajectory (Eccles & Harold, 1993), it is actually what many adolescents want.

School and Teacher Characteristics and Practices

Schools and teachers also play a strong role in initiating and sustaining involvement with parents (Dauber & Epstein, 1989). Factors such as teacher beliefs about the role of the parent in the classroom, the opportunities for involvement that are made available in the classroom and school, and the schools' policies (or lack thereof) regarding parental involvement are crucial in creating an atmosphere that is conducive to parents' involvement at school (Eccles & Harold, 1993). There are two major challenges that schools need to overcome if they want to increase parents' involvement: the organization and physical structure of the school, and the school personnel's beliefs and attitudes regarding parent involvement across all grades. The physical structure of the school may make it difficult for parents to easily find classrooms or to locate personnel who can help them with issues. The school may make no allowances for space for parents, for example, a parent room or an area where they can get refreshments or meet with teachers. Teacher attitudes and beliefs are also important predictors of how accommodating a classroom is to the parent and whether or not a parent feels welcome and invited to participate in the classroom or school. The more positive teacher attitudes are, the more welcome parents feel, the more likely they are to be involved.

In Chapter 2, Hoover-Dempsey and her colleagues outlined a model of parents' involvement in which parents' choices and behaviors were stressed as the key determinant of parents' involvement. This model is especially useful for identifying potential challenges to parents' involvement. They presented a five-level model that ascends from general parental beliefs and characteristics to positive outcomes for children. The barriers are similar to the Eccles and Harold model (1993) in two ways: (1) parents need to feel that they can make a difference and (2) demands on parents' time often make it difficult for them to participate in schools. Hoover-Dempsey and Sandler (1997) also stressed the decision-making process of the parents who are actively making choices regarding their involvement in school. They argued that parents' decisions are influenced by their own beliefs as well as the action of the schools. They also argued that parents' involvement can have positive outcomes for children based on how parents model their own positive involvement with the schools; thus the children also show positive behaviors toward school. Schools can assist in this decision-making process by giving parents many options to be involved and to understand that parents have various demands on their time and may need to have many options available to participate in school.

These two models stress the fact that both parents and schools are important components in producing good parental involvement. There are many challenges that exist between the parents and the schools that can make the relation between these two important contexts very weak. What both models emphasize is that negative beliefs about involvement from either the parent or the school are critical challenges for creating a positive achievement, social, and emotional environment for children. In the next section, we introduce a model that examines parents' and schools' involvement in a child's life from a developmental perspective and outline how important it is for both groups to focus on how they can work together to promote healthy and positive academic, social, and emotional development for children.

THE CONCEPT OF FAMILY MANAGEMENT

A child's life revolves around several contexts and significant individuals as well as institutions. Our model tries to incorporate these different contexts, with the focus on the management (or mismanagement) of children's academic, social, and emotional development. Children learn by interacting with their physical and social environments. They take in and process information, leading to cognitive and social learning. This flow of information occurs in a rich set of social contexts. There are multiple avenues through which information and resources reach each child. Early in life this information is generally managed by parents or parental figures through both their daily practices and the decisions they make concerning the types of information and resources the child receives. This type of management has been termed *family management* (Eccles, 1992; Furstenberg, Cook, Eccles, Elder, & Sameroff, 1999). As the child matures, teachers, religious figures, relatives, peer groups, coaches, and other significant people come to influence the informational flow and resources available to the child. Thus, over the years of childhood and adolescence, multiple significant others are engaged in managing the information and resources available to both inform children about their world and shape their growing knowledge, social and emotional development, and skill repertoire.

DEVELOPMENTAL CONSIDERATIONS

It is easy to imagine how a parent can do this coordination and management on behalf of a child during the preschool years. But what happens when the child goes to school? Parents who have had their children in day care or preschool programs have experienced major developmental transitions that are negotiated between the two contexts, school and home. Perhaps the best example of this is toilet training, where the parents and the teachers have to work closely together to coordinate this learning experience so that it is not a failure. Most preschools encourage lots of conversations with families regarding sleeping and eating habits or other major home issues that may influence the children's day at school.

It is difficult to think of similar examples for the elementary school, where this kind of regular communication does not take place unless there is substantial behavior or academic problems. In preschools, parents walk their children into the centers; in elementary school they are encouraged to drop them off and go about their day. Are elementary-age children less likely to be affected by home issues or developmental changes than preschool children? Sleeping and eating habits are still critical to the day of an elementary student as well as to their more general cognitive, social, and emotional development. So are other family issues and problems. The only difference is that opportunities for communication between parents and teachers have diminished and are often relegated to parent–teacher conferences once or twice a year.

What is becoming more apparent in the developmental literature is that dealing with social and emotional issues early is very important to the long-term outcomes for children. Research suggests that social and emotional outcomes and problem behaviors begin to stabilize and become predictive of later mental health and behaviors around 8 years of age (Huesmann & Guerra, 1997). Work on the transition to secondary school also suggests that this is a critical juncture for good parent–school communication as children move into a new structure, and those with lower social skills and self-esteem and higher anxiety begin to flounder (Eccles et al., 1993). Thus it is very important for parents and teachers to coordinate and manage their knowledge on developmental changes with each other, especially early in elementary school and as children transition into secondary school, in order for intervention to take place either in the home or at school.

The next section extends this concept of family management and builds on the concepts that have been more fully developed in work on the executive processing management in the brain. We believe this framework provides an integrative perspective on how information and resources are funneled through significant individuals in a child's environment and how these individuals need to work together in order to successfully raise a child into a competent individual able to manage his or her own emotions and problem-solving behaviors.

EXECUTIVE FUNCTION: AN ANALOGY

As discussed earlier, effective child-rearing depends on close coordination of actors in various institutions. This coordination and interaction is analogous to the smooth functioning of the multiple subsystems in the human body and brain. Work in cognitive psychology has documented the importance of executive function in the brain to allow such coordination and management to occur (Borkowski & Burke, 1996). Executive functioning is the management of information and resources coming into and being distributed within the cognitive system; it constantly evaluates and monitors the performance of various subsystems in order to make necessary adjustments for required tasks to occur (Borkowski & Burke, 1996). Effective executive functioning is critical for managing both the flow of information coming into the cognitive system from the physical and mental world and the flow of information out of the brain to the relevant subsystems throughout the body.

We are most interested in the developmental nature of effective and efficient executive functioning. Initially, efficient executive control of children's interactions with their external world needs to be scaffolded by more mature individuals for two reasons: (1) to protect the child and make sure the child has the resources necessary for survival and growth, and (2) to help the child learn both the content-specific skills and the executive functioning skills necessary to manage his or her own survival.

A good example of how this management is important for child development is in the literature on social-emotional learning (Elias et al., 1997). Here the goal is to help a child develop competence in regulating, managing, and expressing emotions in order to successfully negotiate the demands of family, school, and relationships. These processes involve parents, teachers, and other adults (e.g., coaches, religious leaders) helping to interpret and scaffold for children the different emotions and social situations that occur during development. If successful, the child begins to do this regulation independent of the significant adults and is able to successfully manage other social interactions and life tasks. If these significant adults are not successful or do not coordinate these efforts in helping the child manage emotions and appropriate social responses, then it is likely that this child will not develop the essential competences for social and emotional learning.

We believe that the idea of executive functioning provides a useful metaphor for conceptualizing individuals' interactions with their social worlds. Understanding the academic, social, and emotional developmental processes underlying the ontogeny of an efficient executive function system is critical to understanding successful social and cognitive development. Borkowski and Burke (1996) discuss the development of executive functioning skills in terms of repetition or multiple redundancies. Once a child has acquired a skill, he or she then uses it in different contexts, thus expanding skill and knowledge. Using this strategy allows the child to learn to apply the skill correctly in multiple situations. As the child continues to develop, she begins to select strategies, independent of adult assistance, from her own growing repertoire and implement them appropriately for the context. This sequence is considered to be the basis of adaptive learning and the beginning of self-regulated behavior (Borkowski & Burke, 1996).

DETAILS OF THE SOCIAL EXECUTIVE FUNCTIONING MODEL

In the previous sections we reviewed the various components that created the conceptualization and framework for the social executive functioning model (Davis-Kean & Eccles, 2000). In this section we present the details of this model and describe how it can be used to promote parent–teacher partnerships.

The contextual influences and number of significant individuals in a child's life are vast and difficult to fully capture in any model. Nonetheless, Figure 3.1 is an attempt to schematize some of this complexity. The model, derived in part from Bronfrenbrenner and Morris's (1998) nested context model, is conceptually similar to the multisystem framework presented in the Introduction. The Executive Functioning Model highlights the role of significant individuals and social institutions as the managers (executive functionaries) of children's daily lives. What isn't depicted is the critical importance of these various individuals and institutions working closely together so that all actors have sufficient information and feedback mechanisms to allow each actor to be an effective executive functioner for each child. Consequently, we assume that all elements interact and affect each other. Further, the distance of any particular element from the individual child is not indicative of the magnitude of influence—effective functioning is a dialectical property of the entire system. For example, the impact of community context will vary across communities and will depend on the effectiveness of the family's and the school's executive functioning on behalf of the child. Similarly, the demands on the family's executive functioning will depend on the characteristics of all of the other social contexts experienced by the child. Thus, the elements of this model dynamically interact, with each individual child having his own configuration of how much he is influenced by the different spheres.

The Child

At the heart of the framework is the individual child who is receiving resources and processing information from her world. This child brings to the situation her own cognitive processing ability, temperament, gender, and other constitutional factors. This is where the connection is made between the socially organized executive functioners and the child's

FIGURE 3.1. The Social Executive Functioning Model

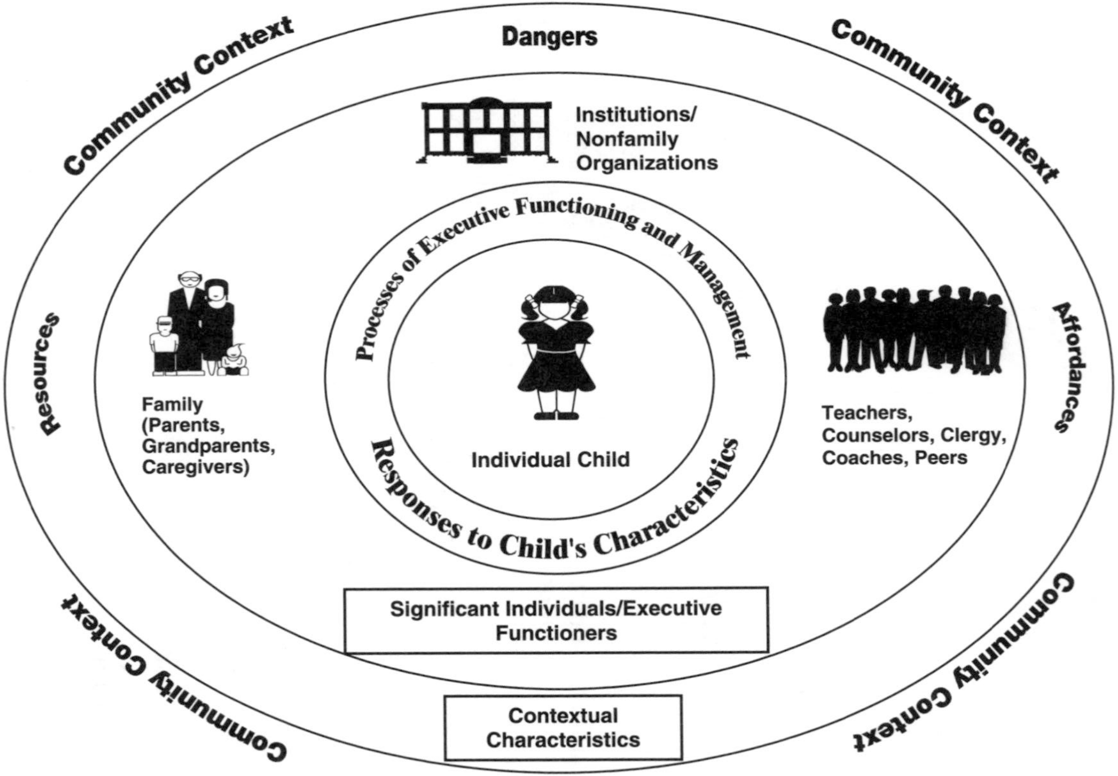

Community Context

Dangers

Community Context

Institutions/
Nonfamily
Organizations

Processes of Executive Functioning and Management

Resources

Family
(Parents,
Grandparents,
Caregivers)

Individual Child

Responses to Child's Characteristics

Teachers,
Counselors, Clergy,
Coaches, Peers

Affordances

Significant Individuals/Executive
Functioners

Contextual
Characteristics

Community Context

Community Context

internal executive functioning. The information and resources from the outside spheres present themselves to the child, who must then process and incorporate the information into his or her own cognitive processes. In order for a child to have good social skills and learn to be self-regulated, for example, a parent or significant adult in his or her life would need to provide accurate and effective information on what skills are needed for positive interactions and self-control.

Significant Individuals and Organizations

The next sphere represents the groups most likely to interact directly with the child. They are the significant people and other influences in the child's life. These groups manage information coming in from the other spheres and adjust them in response to the child characteristics. In the early years of life, the management and resources generally flow through the parents or primary caregivers. These caregivers provide information, transportation, food, values, safety, teaching, health care, social support, discipline, social and emotional learning, motivation, and other important physical and psychological resources to the child. The day care or school institution, for example, begins providing for information, safety, transportation, food, values, and other physical and psychological resources. Similarly, the teacher or caregiver provides for social support, discipline, attention, social-emotional learning, motivation, and teaching as well as the psychological and physical resources. However, for these new caregivers to be effective executive functionaries for the child, they must coordinate their activities with the activities and management of the family-level caregivers. Without this connection, neither set of social executive functionaries has the information it needs to be an effective executive functionary for the child. Recent calls for greater collaboration between school and families are an example of the growing awareness of the need for coordination among the child's various social executive functioners. But for this collaboration to occur, the two systems must have access to the same sources of information and must have shared goals against which each system is evaluating progress and potential problems.

As the child gets older, the number of potential individuals (e.g., peers, counselors, religious leaders, coaches) who can act as social executive functioners and the number of institutional settings (e.g., community programs, Head Start, juvenile court) in which executive functioning needs to occur increase. In addition, the number of opportunities and dangers that need to be managed by both the child and her social executive functionaries increases and, in many cases, become increasing more risky and difficult to manage. Thus, the demands on the child's executive functioning and on her social executive functionaries increases, leading to a need for even stronger coordination across the various executive functionaries. Unfortunately, in our society today, such coordination usually becomes increasingly difficult due to the marked independence of the various contextual systems through which our children and adolescents must navigate.

Community Context

The outermost sphere represents the community contexts that indirectly influence the child through the executive functionaries. Community contexts include the neighborhood and the larger community characteristics as well as the shared cultural context (see Chapters 4 and 5 for a detailed discussion of the issues). These contexts interact with the primary caregivers' own demographics and put certain constraints on what is proximally available to the significant people in a child's life to manage. Community contexts vary in the extent to which their potential impact on the child needs to be, or can be, managed by the child's primary caregivers. Resource-rich and safe communities with consistent norms and values can be used by the primary caregivers to provide their children with many positive opportunities. To the extent that the primary caregivers both agree with the norms and values of the community and trust their neighbors or friends to help them raise their children, the primary caregivers' executive functioning for the child is more easily distributed and shared with other members of the community. At the other extreme, primary caregivers may have to exert considerable energy as executive functioners to protect their children from the risk and dangers in their neighborhood.

When the resources and information do not exist, then the ability to reach these goals is diminished. In the case of the executive functionaries, if a primary caregiver, institution, or significant other does not have access to needed resources, then the likelihood that the child will learn to effectively self-regulate is reduced. Sometimes, however, resources not available in one social context (such as the family) may be provided in other contexts. As with the concept of family management, other significant individuals or institutions may compensate for lost or missing resources. Schools, for example, may provide resources not available in the home or community. Free or reduced-price lunch programs are one example of such resources; expert knowledge is another, and regulation of emotions and good social skills is yet another domain for which school can provide resources that may not be available in the home. Individuals at school can also connect families and children with other organizations or programs that may provide this assistance. Community outreach programs can provide assistance to families that are not able to provide certain resources (e.g., school supplies, clothes, medical care) to their children. Conversely, if the schools are not doing an adequate job of teaching a child, parents can supplement this teaching by providing it themselves or seeking out tutoring from other institutions. Hence, there is a certain amount of compensation that can and should occur between the executive functionaries in order to aid in the academic, social, and emotional development of a developing child. Within the framework of executive functioning, such compensation is likely to work best when the various systems are working together on behalf of the child.

RECOMMENDATIONS

This chapter has outlined a model of how parents, teachers, and schools can work as partners to promote the academic, social, and emotional development of children. The main recommendation stemming from this work is to create programs and curriculum that stress the importance and need for communication between parents, teachers, and schools (see Figure 3.2). This is not a new recommendation, but communication still remains one of the weakest links between parents and teachers in our schools today. Including this emphasis in curriculum in the primary and secondary schools begins with changes in the education schools where teachers receive their degrees. As Chavkin discusses in Chapter 9, there should be a focus in teacher training to better prepare teachers to work with families more effectively, especially on how teaching is not just about what happens in a classroom but is only one element of a system where all members need to be informed of issues as they relate to achievement and social and emotional development. Children spend many hours outside the home in educational settings; hence, teachers, coaches, and schools are just as responsible for the positive academic, social, and emotional development of a child as the parents. In these settings, they develop not only their academic skills but also their social skills of dealing with their peers and with adults. Such skills are valuable talents to foster and are predictive of success in adulthood. For that, it is important that avenues are created where communication, management of information, and coordination are the tools to foster both the social and academic talents of children. Highlighting a broad array of skills in educational institutions that teach future educational professionals is one way to intervene and to provide an avenue for these skills to reach the classroom.

It is also important that avenues be established where parents can bring the school into the home. Possible avenues for this involvement include monitoring homework, assisting in projects, teaching computer

FIGURE 3.2. Recommendations for Coordination Between Executive Functionaries

1. Cultivate Schools and Classrooms That Welcome and Integrate Families
   - Provide a space inside the school for parents to gather with each other, meet with school personnel, and access information and resources.
   - Make parents feel that they are a valued constituency, and welcome visits at the school.

2. Establish Effective Systems of Communication Between Home and School
   - School administrators and teachers should ensure that they have communication systems that facilitate the flow of information between parents and schools.
   - Have updated telephone and computer systems, translators, and personnel who can welcome and orient parents to the school and its practices.
   - Provide necessary infrastructure for the effective coordination and transmission of information.

3. Teach Executive Functioning Skills
   - Teachers can explicitly teach children executive functioning skills and provide parents with suggestions and tips about how to help children plan, organize, and complete homework assignments.

4. Help Families Navigate School-Related Tasks and Activities
   - Teachers and schools should be aware that some parents may not feel comfortable helping with homework assignments, while others may not know what courses are required for college attendance.
   - Schools and teachers may need to provide some parents with tools that will enable them to guide and support their children through the educational system.

skills, and promoting educational opportunities that are offered by the school such as band, sports, and journalism. Parents also are strong role models for their children and can influence the importance that children place on schooling. Thus it is important for parents to have a positive view of the schools. Communication with the schools and exchange of information about a child are good ways to promote a positive view. Teacher and parents can also meet early in each school year to jointly create developmental and educational plans for their children that target a range of activities that include academic, social, and emotional skills and those that promote both these skills. In this way, teachers and parents establish an avenue where they can discuss the issues of development and also become partners in this endeavor (Eccles & Harold, 1993).

Finally, the administration of the school is critical in creating an environment where both parents and teachers feel they can meet and discuss issues related to children. There should be parent resource rooms where information on academic, social, and emotional development and curriculum is available, as well as various ways to be involved at both the classroom and school level. There should be avenues for parents of different cultures and language ability to obtain information on the expectations of schools and the policies that govern the completion of each grade level as well as completion of schooling overall. Once this coordination and information management between the schools, teachers, and parents is established, then we should see healthier academic, social, and emotional outcomes for children.

CONCLUSION

Even though there are various models in the literature regarding parent–school involvement, we believe that conceptualizing it from a developmental perspective, with each partner having an important role in the positive academic, social, and emotional development of the child, is an important step in changing the way parents and teachers conceptualize their roles in this process. The advantage of this integrated perspective is that it focuses our attention on the interconnections among the various social contexts that influence children's lives. Understanding what facilitates efficient coordination among the various contexts in which children develop is critical to understanding how parents, teachers, schools, and other significant contexts often do a less than optimal job at this social task. There are often challenges that can lead to inadequate coordination of executive functioning across individuals and organization, such as lack of coordinated problem solving and goal-setting and lack of access to important school information.

The research we have carried out over the years has indicated that coordination between parents and teachers is critical to the social and emotional development of children, especially as they make their transitions into and out of elementary school. When social and emotional development is ignored or not managed well, then a detriment is seen to not only the mental health of the children but also to their academic achievement (Zins, Weissberg, Wang, & Walberg, 2004). Recently, the relationship between social and emotional development and academic outcomes in schools has become the focus of new curriculum and teacher training on social and emotional learning in the schools (Elias et al., 1997; Zins et al., 2004). This approach holds strong promise of pulling parents and teachers together to aid in the development of good social and emotional skills in children.

REFERENCES

Borkowski, J. G., & Burke, J. E. (1996). Theories, models, and measurements of executive functioning: An information processing perspective. In G. R. Lyon & N. A. Krasnegor (Eds.), *Attention, memory, and executive function* (pp. 235–261). Baltimore: Brooks.

Bronfenbrenner, U., & Morris, P. A. (1998). The ecology of developmental process. In R. M. Lerner (Ed.), *Handbook of child psychology, Vol. 1 (5th ed.)* (pp. 993–1028). New York: Wiley.

Comer, J. P. (1980). *School power: Implications of an intervention project.* New York: The Free Press.

Comer, J. P., & Haynes, N. M. (1991). Parent involvement in schools: An ecological approach. *The Elementary School Journal, 91*(3), 271–277.

Dauber, S. L., & Epstein, J. L. (1989). Parents' attitudes and practices of involvement in inner-city elementary and middle schools (CREMS Report 33). Baltimore, MD: Johns Hopkins University, Center for Research on Elementary and Middle Schools.

Davis-Kean, P. E., & Eccles, J. S. (2000). *It takes a village to raise a child: An executive function and community management perspective.* Unpublished manuscript, University of Michigan, Institute for Social Research.

Dornbusch, S. M. (1994). *Off the track.* Presidential address to the Society for Research on Adolescence, San Diego, CA.

Eccles, J. S. (1992). School and family effects on the ontogeny of children's interests, self-perceptions, and activity choice. In J. Jacobs (Ed.), *Nebraska Symposium on Motivation, 1992: Developmental perspectives on motivation* (pp. 145–208). Lincoln, NE: University of Nebraska Press.

Eccles, J. S., & Harold, R. D. (1993). Parent–school involvement during the early adolescent years. *Teachers College Record, 94*(3), 560–587.

Eccles, J. S., Midgley, C., Buchanan, C. M., Wigfield, A., Reuman, D., Flanagan, C., et al. (1993). Development during adolescence: The impact of stage/environment fit. *American Psychologist, 48*(2), 90–101.

Elias, M. J., Zins, J. E., Weissberg, R. P., Frey, K. S., Greenberg, M. T., Haynes, N. M., et al. (1997). *Promoting social and emotional learning: Guidelines for educators.* Alexandria, VA: Association for Supervision and Curriculum Development.

Epstein, J. L. (1986). Parents' reaction to teacher practices of parent involvement. *The Elementary School Journal, 86,* 277–294.

Epstein, J. L. (1990). Single parents and the schools: The effects of marital status on parent and teacher evaluations. In M. Hallinan (Ed.), *Change in societal institutions* (pp. 91–121). New York: Plenum.

Furstenberg, F. F. Jr., Cook, T. D., Eccles, J., Elder, G. H. Jr., & Sameroff, A. (1999). *Managing to make it: Urban families and adolescent success*. Chicago: The University of Chicago Press.

García-Coll, C., Akiba, D., Palacios, N., Bailey, B., Silver, R., DiMartino, L., et al. (2002). Parental involvement in children's education: Lessons from three immigrant groups. *Parenting: Science and Practice, 2*(3), 303–324

Henderson, A., & Mapp, K. (2002). *A new wave of evidence: The impact of school, family, and community connections on student achievement*. Austin, TX: Southwest Educational Development Laboratory.

Hoover-Dempsey, K. V., & Sandler, H. M. (1997). Why do parents become involved in their children's education? *Review of Educational Research, 67*(1), 3–42.

Huesmann, L. R., & Guerra, N. G. (1997). Children's normative beliefs about aggression and aggressive behavior. *Journal of Personality and Social Psychology, 72*(2), 408–419.

Romo, H. D. (1999). *Reaching out: Best practices for educating Mexican-Origin children and youth*. ERIC Clearinghouse on Rural Education and Small Schools, Charlestown, WV.

Zins, J. E., Weissberg, R. P., Wang, M. C., & Walberg, H. J. (Eds.). (2004). *Building school success on social and emotional learning: What does the research say*. New York: Teachers College Press.

PART II

CULTURAL AND EMPIRICAL PERSPECTIVES

4

Intercultural Considerations in School–Family Partnerships

Luis M. Laosa

The United States has increasingly become a multiethnic, multiracial society, a mosaic of diverse groups seeking to fulfill the American dream of equality of access to educational and economic opportunities while maintaining some measure of their own cultural traditions and ethnic identities. Positive and productive interpersonal interactions across all groups are essential for the successful functioning of a safe and free democracy. As previous chapters have emphasized, schools, families, and communities share major responsibilities for children, and when working collaboratively, they can play important roles in fostering cross-cultural understanding and intergroup cooperation. Scientific research increasingly shows that learning involves not only cognition but also social and emotional processes, and that culture plays a key role in how individuals interact and learn (Greenfield & Cocking, 1994; Laosa, 1999). Culture also provides individuals with a sense of identity and a frame of reference that helps them understand their worlds. Cultural discontinuities—that is, culturally determined discrepancies between the home and the school—can affect the child's success in school. Schools will likely be most successful in their educational mission when they integrate efforts to promote children's academic, social, and emotional learning in the context of each child's culture. As has been discussed in previous chapters of this book, key characteristics, such as recognizing and managing emotions and caring about others, need to be developed for children to be successful not only in school but also in life (Elias et al., 1997; Zins, Bloodworth, Weissberg, & Walberg, 2004).

 ISBN 0-8077-4600-2 (paper), ISBN 0-8077-4601-0 (cloth).

The focus of this chapter is on (a) the experiences that children and their families face in culture-contact situations—that is, situations involving interactions with a culture other than one's own primary culture—particularly when they occur at the intersection between family and school; and (b) the changes that this contact between cultures may bring about in individuals, families, and schools. One aim of the chapter is to examine current, mutually relevant social and educational issues that are intricately connected to these themes. The chapter also intends to increase awareness and understanding of the psychological, social, cultural, and educational processes, and their outcomes, that underlie intercultural change, particularly in the context of continuities and discontinuities between the family and formal schooling.

Such knowledge and understanding can be applied to foster positive relations among the nation's diverse cultural groups. It can also inform the design, implementation, and evaluation of effective partnerships among schools, families, communities, and businesses—partnerships to prepare children to function harmoniously as productive and caring adults in an increasingly diverse and complex society.

In the face of the demographic trends that are rapidly transforming the ethnic and racial composition of the U.S. population, difficult and complex questions arise concerning the relationship between family and formal schooling. An increasingly wide range of diverse cultural and linguistic groups in this country hold varied and sometimes contradictory or conflicting beliefs about appropriate family and school practices, and about the proper roles of family and school as agents of socialization and learning. For many children and their families, the school is the first—and sometimes the only—point of direct experience with a mainstream formal societal institution designed to socialize the individual. While the ethnic and racial composition of the student population grows more diverse, the nation's teaching force—the individuals who must interact with those students every day—remains relatively homogeneous, as do, as a group, the policy framers and decision makers—the individuals with responsibility and authority to formulate the guidelines or principles of action for education. In many parts of the country, a widening divide between school and family is compounded by an increasing range of cultural and linguistic variation among students.

Current policy debates concerning cultural or linguistic diversity in education typically center on some aspect of one question: How wide should be the scope of acceptance of cultural or linguistic differences? For example, in statewide referenda conducted in recent years in Massachusetts, Colorado, Arizona, and California, the electorate voted on whether schools should be permitted to teach pupils whose home language is other than English in

English only or bilingually in English and the native language. Typically, recommendations for addressing issues of cultural (or linguistic) diversity in education focus on (a) the training of educators to recognize and respect different cultures in classroom practices and to use the students' cultural or linguistic strengths to improve instruction, and (b) the inclusion of representatives from the various cultural groups in the process of devising and implementing educational practices (e.g., National Association of State Boards of Education, n.d.). Although these recommendations identify extremely important goals and ought, therefore, to be implemented, analyses of these issues must reach deeper. The analyses must encompass an inquiry into the processes and dynamics that occur when different cultures come in contact with one another. During the months following the terrorist attacks of September 11, 2001, there was a rise in this country in incidents of public expressions of ethnic prejudice and violent intergroup aggression by Americans against Americans (e.g., Goodstein & Lewin, 2001; Tolerance.org, 2002). This phenomenon made it painfully evident that many possible and unexpected factors, including foreign affairs and perceived national interest and national security, can shape public perceptions and attitudes and the way in which cultural group differences are understood, and hence the way in which diverse groups will interrelate. Nevertheless, education policies and practices, including those pertaining to the relationship of formal schooling to family, are often designed, implemented, and evaluated without a clear understanding of the cognitive, social, and emotional processes and dynamics that underlie culture-contact situations.

LIFE CHANGES AND TRANSITIONS

In societies throughout the world, individuals are required or expected to undergo some major life changes, or transitions, that significantly and permanently alter the individual's role or relationship to the society. These life changes, or *normative* transitions, which are often—but not always—accompanied by a biological change of life (e.g., school entry, graduation, marriage, parenthood, workforce entry, retirement), are predictable and publicly recognized as the major changes they are in the individual's development. In addition to these normative transitions, there exist other major life changes that are less predictable, less prevalent, and less visible and usually lack prescribed guidelines and supports. Transitions such as intercultural changes fall under this category. Compared with the large amount of published research on normative transitions, little attention has been focused on intercultural transitions generally, and particularly on their bearing on human development and education (Laosa, 1999).

The terms *intercultural transition* or *intercultural change* refer to "a process of change by means of which an individual or a social system may bridge two different cultures" (Laosa, 1999, p. 356; see also Laosa, 1997). An intercultural transition can be *subtractive* (i.e., an element of one culture is lost when an element of the other culture is acquired), *additive* (i.e., no loss occurs when an element of the other culture is acquired), or *transcending* (i.e., the culture-contact situation prompts the development of some characteristic typical of neither culture). A transition of any of these types can be *unilateral* (i.e., a change occurs in only one of the parties in the culture-contact situation) or *bilateral* or *multilateral* (i.e., both parties change). Finally, a transition can be *adjustive* or *maladjustive*; it is adjustive if it makes it possible to meet the individual's needs and expectations and most of the demands placed on the individual by the environment (Laosa, 1999).

INTERCULTURAL CHANGE AS INDIVIDUAL EXPERIENCE

A child or adult may experience considerable stress or other types of psychological distress when the demands or expectations for cultural change occur in contexts that lack sensitivity to these experiences. A heavy psychological burden can weigh down on individuals in culture-contact situations who face decisions or demands concerning intercultural change. For example, there may be issues of loyalty to the primary group, intergenerational conflict, threat to one's identity, the sheer difficulty of the task, fear of failure, fear of rejection by the primary group, or ambivalence or doubt concerning the relative worth of the outcome.

Educational research, theory, and practice have often focused on cognitive variables and have tended to neglect the psychological burden that can accompany intercultural transitions. A conflict between a real or perceived pressure to acculturate (e.g., to become "Americanized") and the individual's need to preserve his or her primary cultural identity generates psychological tension in the individual. This tension can be considerable insofar as cultural identity is an integral element of one's self. To the extent that the individual's cultural identity is closely bound to the ego, the conflict represents a threat to the ego's integrity. If the individual cannot successfully solve the conflict (e.g., by making a self-alteration resulting in an adjustive cultural transition), then the tension will find expression in one of two ways, both maladjustive: The conflict will be either internalized, becoming a debilitating inner conflict (e.g., manifested as depressive symptoms), or externalized (e.g., acted out as conduct problems).

GOALS, TRAJECTORIES, AND OPTIONS FOR INTERCULTURAL TRANSITIONS

When different cultures initially come in contact, individuals and groups face many choices regarding how they will relate to one another. In the early phases of this encounter, a wide range of variation and variableness may occur in the choices that are made by individuals from each group. Later, a fairly stable pattern emerges, becoming difficult, but not impossible, to modify.

The goals, motivation, trajectories, options, and opportunities for intercultural change differ among cultural groups and among individuals within each group. For example, two individuals from one group may differ concerning which outgroup characteristic they wish to emulate. At the same time, a person's goal for cultural transition and the likelihood of achieving it will inevitably depend on the outgroup's openness and willingness to accept members from his or her group.

Some individuals may not wish to maintain any part of their primary cultural identity (or any part of their primary culture), and rather seek to function fully in the other culture; that is, they wish to attain *full assimilation* (i.e., subtractive transition). If this trajectory is freely chosen, the resulting model is a "melting pot"; if forced by a dominant group, the model has been called a "pressure cooker" (Berry, 1990, p. 244). In contrast, other individuals place a high value on maintaining their primary cultural identity and avoid interaction with the outgroup. Their goal is *voluntary separation*, or *voluntary segregation*. When the separation is required by the outgroup or the person is otherwise kept at a distance from it, the societal model is *forced segregation*. Another category consists of individuals who wish both to maintain their primary culture and to function fully in the other culture; their goal is *integration* (i.e., additive transition). Berry (1990, 1997) distinguished integration from assimilation as follows (a distinction that is observed in the present chapter), although many writers use the two terms synonymously: maintenance of the primary culture is sought in integration, whereas little or no interest in such maintenance is sought in assimilation.

A person's goal for intercultural change may vary across domains. For example, a person may seek economic assimilation (in work) and linguistic integration (bilingualism) but ingroup marriage (Berry, 1990, p. 245).

INTERPRETING THE WE–THEY DISTINCTIONS

In the United States, many individuals tend to view with suspicion those from cultural or language groups different from their own. This unfortunate tendency, which is not unique to this country, was likely of considerable adaptive value during early phases of the evolution of the human species. In the present stage of human evolution, however, this tendency is nearly always maladaptive, a source of intergroup prejudice and antagonism. Human arrangements are universally characterized by differentiations into in groups and out groups. This is the we–they distinction that defines boundaries of group identification and loyalty among individuals (Brewer, 1997; LeVine & Campbell, 1972). Individuals tend to view all others from the perspective of the in group, shaping attitudes and values. Although this tendency is certainly normal, it can easily turn into its harmful extreme: *ethnocentrism*, an individual's tendency to use his or her own group as the frame of reference against which all other groups are judged negatively (LeVine & Campbell, 1972; Wolman, 1989).

An expression of ethnocentrism is the tendency, widespread in the United States, to view cultural or linguistic characteristics of "non-mainstream" groups as deficits or as somehow pathological. This deficit/pathology model continues to dominate much of the field of education and allied professions, suffusing many educational and assessment policies and practices. The presence of deficit/pathology thinking stigmatizes individuals who are different, ignores or denies their strengths and potential, and adds to the psychological burden in culture-contact situations.

As a point of view, *cultural relativism* opposes ethnocentrism. Cultural relativism means "not assuming, wittingly or unwittingly, the superiority of one's own society and culture" (Bourguignon, 1979, p. 13). It also means that one must expect persons from cultures different from one's own "will be full of surprises, of attitudes and behaviors not to be readily predicted or anticipated" (p. 13). In addition, it means that any behavior (verbal, nonverbal, or paraverbal) that the members of a given culture generally view positively, including skills they regard as social competencies, may be ineffectual, neutral, offensive, or even harmful in the context of a different culture.

Misunderstandings between cultural groups often occur because many significant cultural characteristics are so subtle that they are difficult to identify or articulate, even by native members of the culture (Laosa, 1999). This problem adds to the complexity of becoming socially competent in a culture different from one's own.

Openness to another culture means more than simply respecting that culture. It means opening up to the possibility of change and transformation of one's self. Rather than affirming one's own culture as the superior culture, being open to the other culture has the effect of undermining the belief that one's culture, one's way of viewing the world, is superior or absolute. Openness to the other culture challenges the claim that one's way of viewing the world is sufficient to comprehend all other cultures and that it is the only true way, the universal way, the way that reflects the true or essential nature of humanity. Such openness recognizes that human development is open-ended, that one's present way of thinking is not the ultimate, final, or highest possible point of development.

CULTURAL DISCONTINUITIES AND CONTINUITIES

A cultural discontinuity between the home and the school is said to occur whenever some culturally determined discrepancy exists between the home and the school, a discrepancy with the potential to affect the child's success in school. Research and theory have linked such discontinuities to the occurrence of cultural group differences in scholastic achievement. A practical question motivates inquiry into this issue: How can children, families, and educators successfully navigate a largely uncharted, often bewildering gulf between the cultures of home and school?

The literature offers several conceptual approaches to issues of cultural discontinuity between home and school (Laosa, 1999). Important differences among these approaches hinge on how to interpret these discontinuities. As noted before, some approaches interpret the cultural discontinuities as *deficiencies* in the child or in the home. From such a perspective, the fact that some ethnocultural groups score, on average, below national norms on academic achievement tests is blamed on child socialization practices of those groups. According to that perspective, the solution is to modify or suppress those cultural socialization practices.

In contrast, other approaches interpret the cultural discontinuities as *differences* between the culture of the home and that of the school. In this view, a solution to a cultural group's low scholastic achievement is to modify school practices in order to make them compatible with those of the home.

A variation on these approaches recognizes that in addition to home–school discontinuities, which clearly occur for some cultural groups, subtle home–school commonalities may also exist for each group (Goldenberg & Gallimore, 1995). From this viewpoint, a search for both discontinuities and commonalities may uncover more possibilities for practical application than a focus on only one.

In addition to discontinuities between the home and the school, discontinuities between the generations within a family can also lead to scholastic problems. In culture-contact situations, conflicts often arise between parents and their children. Tension within the home can explode when the child experiments with or adopts a cultural out group's attitudes, values, expectations, or behaviors that diverge from the family's ancestral culture; such a situation can trigger a series of events and processes with serious academic, social, and emotional consequences for the student.

Educational interventions based on cultural discontinuity approaches aim to reduce discontinuities by modifying the home, the school, or both. For example, some programs educate parents about the expectations of the school. Other interventions educate teachers about cultural influences on student behavior and encourage them to adopt teaching strategies that are indigenous to the student's culture. In the following section, various approaches to curriculum interventions will be discussed.

CURRICULUM INTERVENTIONS

Curriculum interventions can be designed to reduce cultural discontinuities, facilitate intercultural transitions, improve ethnic and racial perceptions and attitudes, and foster interethnic friendships. Such interventions include culturally sensitive instruction, funds-of-knowledge programs, course content that focuses on ethnicity or race, two-way bilingual education, and cooperative learning groups.

Culturally Sensitive Instruction

The term *culturally sensitive instruction* refers to pedagogical systems or techniques designed to be congruent with cultural characteristics of students and thus reduce cultural discontinuities between the home and the school. The most frequently cited example of culturally sensitive instruction is the Kamehameha Early Education Project (KEEP), designed to increase the school success of Hawaiian children of Polynesian ancestry. Through years of sustained research and experimentation, KEEP staff developed a successful program for elementary school classrooms (Au & Jordan, 1981; Tharp & Gallimore, 1988/1991). The origins of the program lie in earlier studies conducted by social scientists in Polynesian-ancestry communities in Hawaii—studies that described the population's culture and styles of interaction, thus generating hypotheses about the effects of home–school discontinuities in learning styles on scholastic performance. KEEP's effectiveness is generally credited to a type of reading lesson designed to resemble *talk-story*, a unique speech activity characteristic of Polynesian-Hawaiian culture. Talk-story consists of overlapping speech and cooperative production of narrative by several speakers. In the reading lesson, the teacher allows the children to discuss text ideas using rules for speaking and turn-taking similar to those in talk-story (Au & Jordan, 1981; Tharp & Gallimore, 1988/1991).

Funds of Knowledge

Some critics have argued against all educational programs derived from cultural discontinuity approaches, even against those that regard the discontinuities as differences rather than deficits. They have contended that by focusing on home–school discontinuities, one risks fostering the interpretation that differences between families and schools represent deficits in either the families or the schools, and that such interpretations "may provoke mutual blaming between parents and teachers while overlooking the resources for student learning that are available in each setting" (Azmitia et al., 1994, p. 4). Although this risk is real, seeking to avoid it by rejecting all programs that are based on cultural discontinuity approaches is tantamount to throwing out the proverbial baby with the bathwater. An alternative, constructive solution would incorporate into the design of these programs the safeguards needed to avoid deficit interpretations and to foster mutual respect between educators and families.

For example, a program called *funds of knowledge* (e.g., Moll, Amanti, Neff, & Gonzalez, 1992) trains teachers to interview their students' parents to discover areas of knowledge that the families possess. By thus becoming learners in a sociocultural setting different from their own, teachers may establish a "more symmetrical relationship" with the families, a relationship that may form the basis for a continuing exchange of knowledge about family and school, thereby "reducing the insularity of classrooms" (p. 139). Rather than focusing explicitly on culture in its anthropological sense, this program seeks to incorporate strategically into classrooms specific knowledge essential to the household functioning of particular families. The teachers participate in after-school study groups with local university researchers to reflect on the information gained from the home visits and to determine how to incorporate this knowledge into classroom academic content and lessons (Moll et al., 1992).

Multicultural Education

Curriculum interventions focusing on ethnicity or race, including teaching materials, course content, and curricular units, can be designed to help students acquire more positive perceptions of and less stereotypic attitudes toward specific ethnic or racial groups. For instance, the effects of reading stories about American Indians on the racial attitudes of fifth graders were examined in an early study (Fisher, 1965, cited in Banks, 1995). Students in the study participated in one of three conditions: Some students read the stories, others read and discussed the stories, and a control group had no exposure to the stories. Compared to the control group, the students in the other two conditions improved their attitudes toward American Indians. Moreover, reading and discussing the stories resulted in more attitude change than only reading them. Other curriculum interventions such as plays, folk dances, music, and role-playing can also have positive effects on the ethnic and racial attitudes of students. More generally, a multicultural curriculum that includes representations of diverse cultural groups in realistic and complex ways can help to equalize the status of all groups within the classroom or school and hence promote positive interethnic interaction (August & Hakuta, 1997; Banks, 1995).

Two-Way Bilingual Education

Two-way bilingual education programs integrate students from two different language groups in a classroom for instruction in and through both languages. Two-way programs are intended to result not only in high academic achievement and second-language learning for both groups, but also in improved cross-cultural understanding as a benefit of interaction in the classroom. These programs can provide an effective means of fostering positive intergroup relations and additive intercultural transitions.

In two-way bilingual education programs in the United States, typically some percentage (e.g., 50%) of the students are native speakers of English; the remainder are students learning English as a second language who share in common the same native language. An example is a study of 300 students who were enrolled in the Amigos two-way elementary school bilingual program in Cambridge, Massachusetts (Lambert & Cazabon, 1994, cited in August & Hakuta, 1997). Half of the students were native speakers of Spanish and half were native speakers of English. Each language was used as the medium of instruction for half of the school day. The study found that the students in the program formed close friendships with members of both their own and the other group.

Cooperative Learning Groups

An impressive body of research demonstrates that cooperative learning methods can enhance intergroup relations (August & Hakuta, 1997; Slavin, 1995). In these educational interventions, students in an ethnically mixed classroom are assigned to work in learning groups of four or five students of different ethnicities, genders, and levels of achievement, each group reflecting the composition of the class as a whole on these three characteristics. Each learning group receives rewards, recognition, and evaluation based on the extent to which the group can increase the academic performance of each student in the group. This approach stands in sharp contrast to the student competition for grades and teacher approval characteristic of traditional classrooms. Cooperation between students is emphasized both by the classroom rewards and tasks and by the teacher, who communicates an "all for one, one for all" attitude (Slavin, 1995, p. 629). Studies of cooperative learning methods indicate that when students work in ethnically mixed cooperative learning groups, they gain in cross-ethnic friendships (Slavin, 1995).

RECOMMENDATIONS

Principles for Teaching and Learning in a Multicultural Society

Based on a review of research and experience in education and diversity, an interdisciplinary panel of scholars (Banks et al., 2001) offered the following principles to help educational practitioners assist students to develop the knowledge, attitudes, and skills needed to interact positively and productively with people from diverse cultural backgrounds and thus participate effectively in a unified, pluralistic society.

1. Professional development programs should help teachers understand the complex characteristics of ethnic groups within U.S. society and the ways in which race, ethnicity, language, and social class interact to influence student behavior.
2. Teachers should help students acquire the social skills needed to interact effectively with students from other racial, ethnic, cultural, and language groups.
3. Schools should provide all students with opportunities to participate in extra- and co-curricular activities designed to foster positive intergroup relationships.
4. Schools should provide opportunities for students from different racial, ethnic, cultural, and language groups to interact socially under conditions designed to reduce fear and anxiety.
5. Schools should create or make salient "superordinate" cross-cutting group memberships in order to improve intergroup relations.
6. Students should learn about stereotyping and other related biases that have negative effects on racial and ethnic relations.

Building School–Family Partnerships

By forming partnerships with families and communities, schools can strengthen their ability to address successfully the cultural, social, emotional, and academic issues that students and educators may face in culture-contact situations. In this sense, the term *partnership* refers broadly to efforts to work collaboratively for the benefit of children. Based on a review of 64 research studies, Boethel (2003) offered the following guidance on what schools can do to lay the groundwork for building partnerships with families of diverse cultural backgrounds.

1. Adopt formal school- and district-level policies that promote parent involvement, including an explicit focus on engaging families who reflect the full diversity of the student population. Policies may address the following:

    - communicating often with families, both formally and informally;
    - adapting materials and activities to accommodate the needs of families of all cultural and language backgrounds and circumstances;
    - emphasizing family and community outreach;
    - involving families in school planning and decision-making processes;
    - training teachers to work effectively with families; and
    - offering to help families build their own capacities to support their children's schooling.

2. Demonstrate active and ongoing support from the school principal.
3. Honor families' hopes and concerns for their children. Most families, regardless of background, care about their children's future and will do what they can to support them. Let families know that you recognize and value their efforts. In orienting school staff and community organizations to family involvement strategies, include information on varied ways in which families support children's learning, as well as the varied perspectives on what family involvement should address.
4. Acknowledge both commonalities and differences among students and families. Generally, there are more similarities than differences in families' hopes and concerns for their children, regardless of their background. Differences do exist, however, in families' experiences, cultural values and practices, and worldviews. Seeking common ground while acknowledging and respecting differences is a challenging but essential process. It is important that schools and community organizations not ignore issues related to diversity and that they not act as if differences do not exist. Ignoring such issues or differences can push parents and schools further apart.

5. Strengthen school staff's capacity to work well with families. There is a need for teacher preparation programs to include instruction on how to collaborate with parents and communities. Teachers need help to build both understanding and practical strategies for engaging effectively with families, particularly when those families' backgrounds and life circumstances are very different from their own.
6. Provide supports to help families understand how schools work and what is expected of both families and students.
7. Make outreach a priority; take any extra steps necessary to make it possible for families to get involved at school as well as at home. Important elements in any outreach plan include:

    - ensuring that families and staff communicate in the language with which family members are most comfortable;
    - making it as easy as possible for families to participate in school activities;
    - finding multiple ways for families to get involved;
    - getting out into the community, particularly including home visits and participation in community activities such as social or civic events.

8. Recognize that it takes time to build trust. Do not be offended or deterred by skepticism, suspicion, or criticism, or by low initial participation.

CONCLUSION

In order to prepare every child to function as a competent, productive, caring, responsible, and peaceful adult in the diverse, complex, and changing society that is the United States, formal education must encompass not only the cognitive domain but also the cultural, social, and emotional domains, including the concerns brought forward in this chapter. Attaining this vision will require coordinated and comprehensive approaches that involve schools, families, and communities cooperating in sustained, long-term efforts.

REFERENCES

Au, K. H., & Jordan, C. (1981). Teaching reading to Hawaiian children: Finding a culturally appropriate solution. In H. T. Trueba, G. P. Guthrie, & K. H. Au (Eds.), *Culture and the bilingual classroom: Studies in classroom ethnography* (pp. 139–152). Rowley, MA: Newbury House.

August, D., & Hakuta, K. (Eds.). (1997). *Improving schooling for language-minority children: A research agenda.* Washington, DC: National Academy Press.

Azmitia, M., Cooper, C. R., Garcia, E. E., Ittel, A., Johanson, B., Lopez, E. M., Martinez-Chavez, R., & Rivera, L. (1994). *Links between home and school among low-income Mexican-American and European-American families* (Educational Practices Rep. No. 9). Washington, DC: National Center for Research on Cultural Diversity and Second Language Learning, Center for Applied Linguistics.

Banks, J. A. (1995). Multicultural education: Its effects on students' racial and gender role attitudes. In J. A. Banks & C. A. McGee Banks (Eds.), *Handbook of research on multicultural education* (pp. 617–627). New York: Macmillan.

Banks, J. A., Cookson, P., Gay, G., Hawley, W. D., Irvine, J., Nieto, S., Schofield, J. W., & Stephan, W. G. (2001). *Diversity within unity: Essential principles for teaching and learning in a multicultural society.* Seattle, WA: Center for Multicultural Education, College of Education, University of Washington. Retrieved from http://www.educ.washington.edu/coetestwebsite/pdf/DiversityUnity.pdf.

Berry, J. W. (1990). Psychology of acculturation: Understanding individuals moving between cultures. In R. W. Brislin (Ed.), *Applied cross-cultural psychology* (pp. 232–253). Newbury Park, CA: Sage.

Berry, J. W. (1997). Immigration, acculturation, and adaptation. *Applied Psychology: An International Review, 46*(1), 5–34.

Boethel, M. (2003). *Diversity: School, family, and community connections.* Austin, TX: Southwest Educational Development Laboratory, National Center for Family and Community Connections with Schools. Retrieved from http://www.sedl.org/connections.

Bourguignon, E. (1979). *Psychological anthropology: An introduction to human nature and cultural differences.* New York: Holt, Rinehart and Winston.

Brewer, M. B. (1997). The social psychology of intergroup relations: Can research inform practice? *Journal of Social Issues, 53,* 197–211.

Elias, M. J., Zins, J. E., Weissberg, R. P., Frey, K. S., Greenberg, M. T., Haynes, N. M., Kessler, R., Schwab-Stone, M. E., & Shriver, T. P. (1997). *Promoting social and emotional learning: Guidelines for educators.* Alexandria, VA: Association for Supervision and Curriculum Development.

Goldenberg, C., & Gallimore, R. (1995). Immigrant Latino parents' values and beliefs about their children's education: Continuities and discontinuities across cultures and generations. In P. R. Pintrich & M. Maehr (Eds.), *Advances in motivation and achievement: Vol. 9. Culture, ethnicity, and motivation* (pp. 183–228). Greenwich, CT: JAI Press.

Goodstein, L., & Lewin, T. (2001, September 19). Victims of mistaken identity, Sikhs pay a price for turbans. *New York Times on the Web.* Retrieved from http://www.nytimes.com.

Greenfield, P. M., & Cocking, R. R. (Eds.). (1994). *Cross-cultural roots of minority child development.* Hillsdale/Mahwah, NJ: Erlbaum.

Laosa, L. M. (1997). Research perspectives on constructs of change: Intercultural migration and developmental transitions. In A. Booth, A. C. Crouter, & N. Landale (Eds.), *Immigration and the family: Research and policy on U.S. immigrants* (pp. 133–148). Mahwah, NJ: Erlbaum.

Laosa, L. M. (1999). Intercultural transitions in human development and education. *Journal of Applied Developmental Psychology, 20*(3), 355–406.

LeVine, R. A., & Campbell, D. T. (1972). *Ethnocentrism: Theories of conflict, ethnic attitudes, and group behavior*. New York: Wiley.

Moll, L. C., Amanti, C., Neff, D., & Gonzalez, N. (1992). Funds of knowledge for teaching: Using a qualitative approach to connect homes and classrooms. *Theory Into Practice, 31*(2), 132–141.

National Association of State Boards of Education. (n.d.). *The American tapestry: Educating a nation. A guide to infusing multiculturalism into American education.* Alexandria, VA: Author.

Slavin, R. E. (1995). Cooperative learning and intergroup relations. In J. A. Banks & C. A. McGee Banks (Eds.), *Handbook of research on multicultural education* (pp. 628–634). New York: Macmillan.

Tharp, R. G., & Gallimore, R. (1988/1991). *Rousing minds to life: Teaching, learning, and schooling in social context* (1st paperback ed.). Cambridge, England: Cambridge University Press.

Tolerance.org. (2002). *Violence against Arab and Muslim Americans.* Updated June 27, 2002. Retrieved from http://www.tolerance.org/news/article_hate.jsp?id=278, http://www.tolerance.org/news/article_hate.jsp?id=412.

Wolman, B. B. (1989). *Dictionary of behavioral science* (2nd ed.). San Diego: Academic Press.

Zins, J. E., Bloodworth, M. R., Weissberg, R. P., & Walberg, H. J. (2004). The scientific base linking social and emotional learning to school success. In J. E. Zins, R. P. Weissberg, M. C. Wang, & H. J. Walberg (Eds.), *Building academic success on social and emotional learning: What does the research say?* New York, NY: Teachers College Press.

5

School and Family Influences on the Social and Emotional Well-Being of African-American Students

Ronald D. Taylor

In recent years renewed attention has been devoted to the school achievement of African-American children. The underachievement of ethnic minority students is a serious problem in this country. Findings have shown that on a variety of indicators (e.g., grades, SAT scores, college attendance and completion) African-American youngsters fare poorly compared to White children (Gandara & Maxwell-Jolly, 1999; James, Jurich, & Estes, 2001). Minority youngsters represent an increasing proportion of the school-age population and thus the nation's future workforce. The competitiveness of the United States in world markets will depend on the skills of its workforce. Jencks and Phillips (1998) have also argued that reducing the achievement gap would likely help reduce racial differences in crime, health, and family structure and thus improve the standard of living for substantial numbers of African-Americans.

Current educational and demographic trends suggest strong reasons for concern. For example, although African-American and White youngsters complete high school at similar rates, African-Americans youngsters have significantly lower grades and SAT scores. Indeed, findings show that almost 75% of African-American youngsters score below 500, the theoretical mean on the SAT. African-American students make up 12% of the population but only 9% of the students enrolled in four-year colleges (U.S. Census Bureau, 1997). Finally, African-American and Latino students take longer to graduate and leave college without completing their degrees sig-

 ISBN 0-8077-4600-2 (paper), ISBN 0-8077-4601-0 (cloth).

nificantly more often than White or Asian students (National Center for Education Statistics,1997).

Some of the explanations for the achievement gap include genetic differences between African Americans and Whites, bias in testing favoring Whites over African Americans, and racial differences in parents' socialization practices (Jencks & Phillips, 1998). Although innate differences are periodically favored as an explanation of the achievement gap, no genetic evidence has been found showing that African Americans are innately inferior to Whites (Jencks & Phillips, 1998; Lerner, 1992). The matter of testing bias is complex, and evidence suggests that while the content of tests appears to be fair measures of the abilities of African-American and White students, bias exists in the ways in which tests are used to label and select African Americans in educational settings (Jencks & Phillips, 1998). The area of parenting practices and the links to student achievement holds promise for identifying key variables linked to student achievement. Indeed, while evidence of the genetic superiority of either African Americans or Whites has neither been found nor linked to achievement differences, important differences in parenting and family processes do exist (Wilson, 1989). The present chapter will explore correlates of achievement in adolescents' family relations and processes. Findings suggest that changes in parenting practices may play an important role in reducing the achievement gap because of the links between family functioning and youngsters' schooling (Jencks & Phillips, 1998). Also, factors that may influence family functioning, including families' economic resources, the community in which they live, and the social networks in which they participate will also be examined. Indeed, findings suggest that families' interactions and experiences in one context may have implications for their behavior in the other social contexts they inhabit. Thus, factors that impact children's social and emotional well-being in the home are likely to be reflected in youngsters' behavior in school.

FACTORS INFLUENCING PARENTS' AND CHILDREN'S SOCIAL AND EMOTIONAL WELL-BEING

In order to better understand home–school links, it is important to have a clearer picture of how families are influenced by socioeconomic conditions, and how those conditions affect parent behaviors and children's well-being. Therefore, following is a presentation of demographic information on African-American families' financial resources and living patterns.

Economic Resources

African Americans in the United States number 31.4 million persons, or 13% of the nation's population (U.S. Census Bureau, 1994). Of African-American children under 18 years, 36% live with both parents. A substantial portion of the African-American population (33%) is poor, particularly families with children under age 18 (46%). The absence of a parent in the home has a significant influence on African-American children's exposure to poverty. Of African-American families maintained by a woman with children, 60% are poor. For those households headed by African-American men with children, 32% are poor. Slightly over one-half of African Americans live in the South, and approximately the same number live in central cities. Whether they live in the North or South or in the cities or suburbs, most African Americans live in segregated communities with other African Americans (Massey & Denton, 1993).

Research has revealed clear links between families' economic resources and children and adolescents' behavior and well-being (McLoyd, 1997). Numerous studies of children have shown a link between lower socioeconomic status (SES) and social and emotional problems. As many as 15% of children seen in primary care facilities presenting with at least one social and emotional problem are low-SES. For children in early and middle childhood, lower SES is associated with lower self-esteem and self-confidence, problems in peer relations, social maladaption, and psychological distress. For adolescents, low income and economic hardship have been linked to depression, problems in school, diminished self-esteem, and delinquency. In recent work findings have shown that among a sample of poor African-American families, the more that mothers reported that the families' financial resources were inadequate, the more that children reported depressive symptomatology (Taylor, Seaton, Dominguez, & Rodriguez, in press). Findings have also shown that problems associated with low income increase as children grow in age.

Finally, research on the nature and effects of poverty has grown in sophistication and has begun to distinguish the effects of current compared to long-term poverty. Results have shown that children who are occasionally and persistently poor have more problems than children who are never poor. Also, children's problems appear to grow the longer they are in poverty. Findings have also shown that for African-American children poverty tends to be of long duration (Waggoner, 1998).

Neighborhoods

Also linked to families' economic resources is their place of residence or neighborhood. African-American families, particularly those with fewer resources, have more limited options and typically reside in the most impoverished neighborhoods in cities. Findings on the links between families' neighborhood and children's social and emotional competencies have revealed that the impact of neighborhoods on children increases as children grow older and move beyond the home. As children enter school, a greater number of neighborhood factors appear associated with their cognitive and social functioning. For example, having affluent neighbors is linked to children's intellectual functioning, and African-American children appear to benefit from increasing the level of ethnic diversity in their neighborhoods. Also, African-American children are more likely to display signs of depression and misbehavior when there is greater male joblessness in their neighborhoods (e.g., Chase-Landsdale, Gordon, Brooks-Gunn, & Klebanov, 1997).

For adolescents, a greater number of neighborhood factors are linked to their social and emotional well-being. Findings have shown that teenage pregnancy, school dropout rate, delinquency, and depression are linked to neighborhood variables, including the proportion of high-status workers in the community, male joblessness, and neighborhood impoverishment (abandoned houses, population loss, etc.). Recent research (Taylor, 2000) has shown that mothers' report of the presence of important social resources in their neighborhood (e.g., banks, health clinics, markets) is associated with youngsters' social competencies, including higher self-esteem and self-reliance and less problem behavior and psychological distress in African-American adolescents.

Social Support

Also linked to families' economic prospects may be their contact with important social networks. Research has shown that for African-American adults the absence of economic resources is associated with contact and interaction with individuals' social networks. In times of need, adult males and females are likely to turn to family members, who in turn provide material and social support. For example, research has shown that economic strain is associated with economic support from kin (Dressler, 1985). Research (e.g., Taylor, Casten, & Flickinger, 1993) has shown that the more that mothers report that family resources are not sufficient, the more support they report from extended family members, including financial help and social and emotional support. Findings have also shown that social support from kin is positively related to parents' and children's social emotional well-being. For youngsters, results suggest that in the presence of social support they are more likely to display a host of social and emotional competencies at home and in school.

THE ROLE OF PARENTS

The focus of the present chapter is on family mediators, and the presentation is guided by the conceptual model shown in Figure 5.1. In the conceptual model it is proposed that families' financial resources, neighborhood conditions, and social networks are linked to children and adolescents' social skills and competencies through the association with parents' childrearing practices and parents' psychological functioning. Support for the model has shown that economic hardship in a variety of forms is associated with psychological distress in parents (e.g., McLoyd, 1990, 1997). For example, difficulty paying bills and meeting families' material needs is positively related to mothers' and fathers' depression (Conger et al., 1992). Recent research results have shown that the more that mothers report that their financial resources are inadequate, the more negative is their outlook on the future (Taylor et al., in press). Also, unemployment is linked to a variety of psychological and emotional problems, including somatic complaints and depressive symptomatology. In addition, economic hardship has been linked to marital problems and discord.

These findings raise the question of whether families' economic difficulties also affect parents' capacities for fulfilling the parenting role. Re-

FIGURE 5.1. A Conceptual Model of the Economic and Social Influences on Children and Families

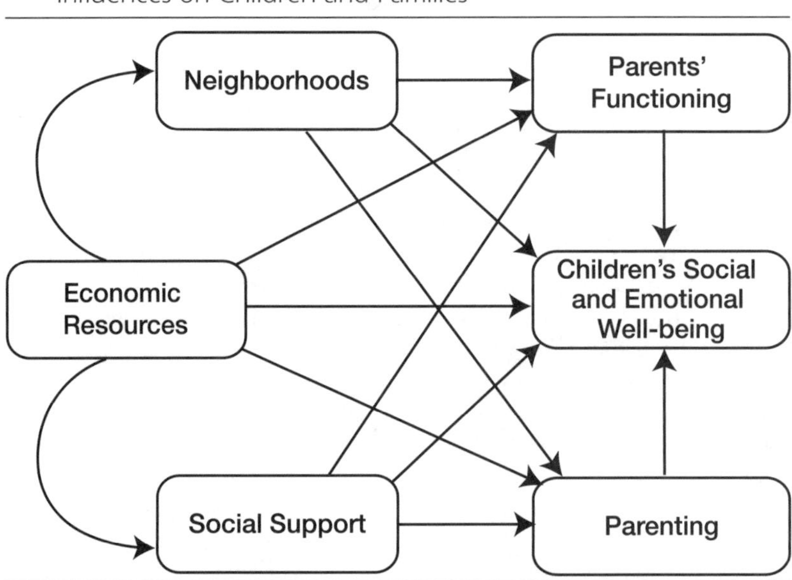

search indeed suggests that parents' emotional and psychological states and child-rearing practices are negatively affected by economic problems. Findings have shown that financial problems are associated with depressed mood and inconsistency in parenting (Ge, Conger, Lorenz, & Simons, 1994). Also, economic hardship has been linked to diminished parental emotional support and more punitive parenting. Conger, McMarty, Yang, Lahey, and Kropp (1984) have shown that mothers experiencing economic problems may spend reduced time in supportive, positive interactions with their children and increased time in harsh and punitive interactions. McLoyd (1997) has noted that low-SES parents are more likely to use harsh discipline techniques, are more likely to issue commands without explanation, and are less likely to reward children when they have behaved in desirable ways. Findings have also shown that mothers experiencing economic problems are less likely to maintain family routines and schedules (Taylor et al., in press).

Neighborhoods and Parenting

In addition to the effects of economic factors on parents' well-being and child-rearing, neighborhood conditions are associated with parents' child-rearing practices. It has been argued (e.g., Baumrind, 1991; Ogbu, 1985) that a key correlate of where families live is the parenting that children and adolescents experience. Parents adopt practices to ensure that their children are safe and can take advantage of the resources in their living environment. Research based on Baumrind's work on parenting style has produced a volume of information on the outcomes seen in children and adolescents (for reviews see Baumrind, 1989, 1991). Darling and Steinberg (1993) have argued that parenting styles are a reflection of parents' goals and values regarding social behaviors and skills desired for their children, and of their view of the important environments in which their children interact. They suggest that parenting style is best conceived as a contextual variable representing parents' attitudes toward their children and the type of emotional climate they believe is best suited for developing crucial attributes. Baumrind (1991) has suggested that as a consequence of living in neighborhoods that are often unsafe, African-American parents may adopt parenting styles that would be restrictive and authoritarian by middle-class standards. Likewise, Ogbu (1985) has argued that African-American parents living in the inner city use a harsher, more restrictive style of parenting in order to encourage the development of behaviors and attitudes valuable for growing up in a harsh, exploitive urban environment.

Findings have revealed that the more mothers report their neighborhood as safe and that social resources are accessible, the more they are emotionally supportive of their youngsters. Also, parents' report higher levels of control and monitoring of their youngsters' behavior when their neighborhood is more physically deteriorated. Research has also shown that neighborhood ethnic diversity is negatively associated with maternal warmth and support (Taylor, 1997). Also, mothers in neighborhoods at high risk of child maltreatment are more likely to rate their neighborhood as less desirable and stable than mothers in neighborhoods at low risk of child maltreatment. Additionally, neighborhoods with the highest rates of maltreatment tend to be those with conditions including poverty, unemployment, female-headed households, abandoned housing, and population loss (e.g., Coulton, Korbin, Su, & Chow, 1995).

Social Support and Parenting Behaviors

For parents, economic and social support from kin in times of need has been linked to lower depression. Also, men with social support are less depressed by stressful life events than those lacking a social support network. For mothers, those with access to a social support network during pregnancy report lower levels of stress, anxiety, and depression and are more positive about their pregnancies. In addition, for single-parent African-American mothers experiencing employment problems, social support is associated with lower depression. For economically disadvantaged African-American mothers, kin support is positively associated with self-esteem.

Like the findings for economic resources and neighborhoods, social support is linked to parents' psychological functioning, and parents' functioning is linked to their parenting practices. Findings have shown that kinship social support is positively associated with mothers' emotional well-being, and mothers' functioning is in turn associated with mothers' emotional support and acceptance of their children. Other research has simply shown that kinship support is positively associated with parents' child-rearing practices. For instance, kinship support is associated with mothers' emotional support and monitoring and supervision of their children. Social support is also positively related to the establishment and maintenance of organization and structure in the home (for a review of this research see Taylor, 1997).

Other work has shown that social support may suppress negative parenting behaviors. The availability of social support is negatively associated with severe forms of maternal punishment and with mothers' negative perceptions of the maternal role. Also, to the degree that mothers have support they are less likely to use the withholding of emotional support as a form of punishment. Finally, research has shown that the availability of social networks is negatively associated with abusive behavior by parents (e.g., Taylor, 1997).

Parents' Emotional Well-Being

In this chapter's conceptual model parents' social and emotional functioning and parenting practices are linked to children's social competencies and skills and school achievement and engagement. Research supporting these links has shown that mothers' optimistic orientation to the future is associated with lower depression and sadness in children. Poor and working class African-American children report fewer symptoms of depression when their mothers report being optimistic about the future. Other findings with working-class African-American families have shown that mothers' psychological distress is positively related to children's problem behavior, and negatively associated with self-reliance and independence. Relevant findings with European-American families have shown that parental depression is positively associated with adolescents' depression.

Child-Rearing Practices

Parents' child-rearing practices have been shown to impact children's academic, social, and emotional well-being. For example, providing emotional support and acceptance are positively linked to youngsters' self-reliance and negatively related to youngsters' problem behavior. Acceptance and support from the family is also associated with higher grades and higher scores on achievement tests (Levitt, Guacci-Franco, & Levitt, 1994). For teenage mothers, family support is a key factor in preventing school dropout. Also, mothers' control, supervision, and monitoring of their children are associated with lower dependency and lower delinquent behavior.

Families' management practices are also linked to adolescents'social and emotional functioning. Clark (1983), in an ethnographic study, examined the parenting practices distinguishing families of high- and low-achieving low-income African-American adolescents. Parents of high-achieving youngsters compared to those low in achievement were more likely to have family routines and schedules. Teenagers in such families had regular meal and bedtimes and times for doing chores and homework. Parents were also involved with students' schoolwork through help with homework and attendance at school functions. Findings consistent with the positive association of family organization with adolescents' functioning have also been obtained. Such findings revealed that family organization is positively related to adolescents' school engagement and grades and negatively associated with problem behavior and psychological distress (e.g., Taylor, 1996).

Parenting Styles

Parenting styles represent the attitudes and values parents have regarding the role of their actions and practices in promoting youngsters' social development (e.g., Darling & Steinberg, 1993). Adolescents with authoritative parents live in homes that are run democratically in which they have a voice in personal and family decisions. Youngsters in such homes tend to be the most socially competent and are least likely to engage in problem behavior or display sadness or depression (Lamborn, Mounts, Steinberg, & Dornbusch, 1991). In comparison, youngsters of authoritarian and permissive parents, respectively, are more likely to have problems with self-image or engage in a variety of problematic behaviors. Other research (e.g., Steinberg, Mounts, Lamborn, & Dornbusch, 1991) has shown that adolescents whose parents are authoritative in the home are more self-reliant, less anxious and depressed, less likely to engage in delinquency, and performed better in school. Additional findings have shown that among mother-headed working-class families, authoritative parenting is associated with self-reliance, independence, and problem behavior (Taylor et al., 1993). Adolescents of authoritative parents also are more likely to associate with peers who endorse positive values and norms (e.g., Durbin, Darling, Steinberg, & Brown, 1993).

A substantial amount of research has been devoted to examining the link between authoritative parenting and school performance (Glasgow, Dornbusch, Troyer, Steinberg, & Ritter, 1997; Steinberg et al., 1991; Weiss & Schwartz, 1996). Research has revealed a positive link between school achievement and authoritative parenting (Dornbusch, Ritter, Leiderman, Roberts, & Fraleigh, 1987; Steinberg et al., 1991). Findings have also shown that an important factor in the achievement of youngsters of authoritative parents is parental involvement in schooling (e.g., Bogenschneider, 1997; Herman, Dornbusch, Herron, & Hereting, 1997).

It is important to note that findings regarding the impact of authoritative parenting on the achievement of African-American youngsters is unclear. First, Dornbusch et al. (1987) found that when the impact of important demographic variables was controlled (e.g., family structure, parental education, adolescent gender), authoritative parenting was not associated with the achievement of African-American adolescents. Dornbusch et al. (1987) have noted that the typology of parent styles was designed for study with middle-class White parents. Therefore, the classifications are more likely to be associated with the school performance of White rather than ethnic minority adolescents. The explanation by Dornbusch et al. (1987) is similar to Baumrind's (1972, 1991) suggestion that because of their residence in harsh neighborhoods, African-

American youngsters may be exposed to and benefit from a more restrictive, authoritarian form of parenting. It is important to note, however, that Dornbusch et al. (1987) did not find that authoritarian parenting was linked to the achievement of African-American adolescents. Indeed, Steinberg, Lamborn, Dornbusch, and Darling (1992) have suggested that the lack of a link between parenting styles and the achievement of African-American youngsters may be indicative of the greater influence of peers than parents on children's school performance. It is possible, however, that for African-American parents, practices sensitive to the social contexts of family life and aimed at promoting the healthy functioning of adolescents may be linked to adolescents' school achievement. Relevant findings (e.g., Mason, Cauce, & Gonzales, 1997) have shown that African-American parents alter their parenting in part based on the peers available to their youngsters. In an environment in which positive peers are mostly available, parents may allow their youngsters greater freedom. In comparison, when they perceive that the peers available to their youngsters are problematic, parents may exert greater control. Thus, it may be difficult to discover a prototypical parenting style that may universally predict the behavior of African-American youngsters. African-American families' living arrangements are transitory and ever evolving. For example, increasing numbers of African-American families are middle-class, and the number of middle-class African-American neighborhoods is increasing. Yet substantial numbers of African-American families still live in poverty-stricken urban neighborhoods, and the impact of conditions in these neighborhoods on families is still not well understood.

Parents' Racial Socialization

An additional aspect of parenting that may be associated with the social and emotional competencies of African-American children is parents' racial socialization. Racial socialization refers to the degree to which parents teach and discuss matters of race (e.g., culture, heritage, history) with their youngsters. Peters (1985) has argued that African-American parents prepare their children for the experience of racism and racial discrimination by enhancing adolescents' self-perceptions and self-esteem regarding their race. By bolstering children's racial self-regard, parents buffer their children from the hostility and negative experiences they expect their children to see or encounter in schools, the community, or the media. Research directly assessing the impact of parents' racial socialization on youngsters' adjustment is scarce. However, research has indeed shown that African-American adolescents report incidents in which they have either individually or collectively experienced acts of racism (Seaton, 2002). Thus, parents may have the opportunity to discuss racial discrimination with their youngsters if adolescents talk about their experiences at home. Evidence of the correlation of racial socialization with youngsters' adjustment has shown that parents' emphasis upon their racial culture or heritage is positively associated with children's self-esteem (Stevenson, 1997). Also, parents' discussions of potential racial barriers to success and the effort needed to overcome them are positively associated with adolescents' grades and feelings of efficacy (Bowman & Howard, 1985) and negatively related to depression (Stevenson, Reed, Bodison, & Bishop, 1997).

SUMMARY

Research has shown that there is a gap in the school achievement of African-American and European-American youngsters, with White youngsters outperforming African-American children on virtually all measures. Research on the links among major social forces and social relations and African-American youngsters' well-being highlights the importance of attention to children's social and emotional skills, competencies, and well-being in closing the achievement gap. Findings have shown that poverty is a common experience for substantial numbers of African-American youngsters, and being poor is associated with negative outcomes for children's social and emotional skills competencies. Also, families that lack economic resources are more likely to live in dangerous, risky neighborhoods, and living in such conditions is detrimental to children's well-being. However, having the support of concerned adults, particularly extended relatives, may offset some of the negative effects of poverty or risky neighborhoods. Social support from kin is positively associated with children's social and emotional well-being. Social and economic forces and their influence on children may operate through their impact on parents. Parents lacking economic resources and living in at-risk communities are more likely to display symptoms of poor adjustment and tend to engage in less adequate forms of parenting. In comparison, parents who have supportive social networks tend to be better adjusted and are more likely to provide higher-quality parenting. Higher-quality parenting, in turn, is positively linked to youngsters' social and emotional well-being. Finally, African-American parents who prepare their children for experiences with racism and discrimination have children with greater social and emotional competencies, which benefits their schoolwork.

RECOMMENDATIONS

The importance of school and family processes in the school achievement and social and emotional adjustment of youngsters is clear. Practices in the home and school that provide youngsters with acceptance and emotional support are likely to promote social competence and school achievement and diminish the likelihood of problem behavior (see Figure 5.2). Also, work to provide structure and organization to African-American children's social environments is important. Youngsters are likely to display greater social skills and interpersonal competencies, higher grades, and less problem behavior when their social environments (home, school, neighborhood) are stable places with order and regularity. Schools can play a vital role in the creation and promotion of healthy social environments for youngsters. Joint lobbying efforts by schools and parents for safe and secure environments have met with success around the country. Because of the collaboration of schools and parents, youngsters in many urban areas are able to walk to and from school and participate in recreational activities in the community in relative safety.

Further, efforts to promote the social networks of African-American families and their youngsters are likely to pay dividends in the development of children's social and emotional competencies and their school performance. Parents who have supportive networks tend to provide their children with more adequate parenting, including emotional support and an organized home environment. Also, parents who are more actively engaged in social networks report greater involvement in their children's schooling, including help with homework, attendance at parent–teacher conferences and extracurricular activities, and participation in school–family partnership efforts. Indeed, recent work has shown that parents who are warm and accepting and who provide structured and organized home environments seek similar conditions for their children in the school environment. In some parts of the country schools serve as centers for the administration of community educational, social, and recreational services. Schools serving such functions seem highly likely to promote the development of supportive social networks and are likely to have the support and goodwill of the community.

CONCLUSION

The findings reviewed and discussed suggest important links among the social and emotional adjustment of African-American adolescents and the

FIGURE 5.2. Overview of Steps Schools May Take to Enhance the Social and Emotional Well-Being of Economically Disadvantaged Children and Families

Support from schools in comprehensive efforts to promote child development may enhance students' adjustment and families' well-being.	• *Schools are uniquely placed* to aid in the implementation of strategic, wholesale efforts to combat ills impacting children and families in at-risk communities (e.g., gang-prevention programs, after-school programs). • *Schools that serve as centers* for the administration of educational, social, and recreational services are likely to have the support and goodwill of the community.
Support from schools may offset some of the deficits families experience associated with neighborhood poverty and economic disadvantage.	• *Adult literacy programs* offered in neighborhood schools may enhance family literacy. • *Active home and school or parent–teacher associations* may help create and sustain supportive social networks that enhance the well-being of adults and children. • *Schools that take an active role* in addressing social problems affecting students and families (e.g., teenage pregnancy, family violence, drug abuse) may increase the level resilience evident in the community.
Support from and active involvement of schools in community civic affairs may have positive indirect effects on children and adults.	• *Interaction and communication* between schools and other organizations serving the community (e.g., police and fire departments, banks and businesses) will enhance the services families receive. • *Cooperative lobbying efforts* by schools and parents aimed at state and local governments can effect changes in school and neighborhood resources (e.g., school and neighborhood safety and security resources, for instance, crossing guards and police patrols).

processes that may underlie those relations. Patterns in these findings reveal that families' economic resources, the conditions of their neighborhood, and their social network are linked to adolescents' social and emotional functioning because parents' well-being and parenting practices are linked to these variables. The adequacy of family resources, the safety and social resources of neighborhoods, and the support of relatives affect parents and, through parents, their children.

From the standpoint of policy and practice, the economic resources of African-American families remains an important issue. Large segments of the African-American population, especially families with children, live in poverty-stricken communities with limited access to jobs and important resources. Improving the financial well-being of African-American families with children is highly important. Important steps in advancing the circumstances of African-American families include increasing opportunities for employment and improving the skills of parents to match the increasingly advanced skills needed for higher-paying jobs. Also, it is important to address the need to improve the quality of the neighborhoods in which African-American families live. Changes in policing practices and the introduction of incentives for businesses to locate in poor and working-class communities may be necessary in order to enhance the state of communities in which African-American families live.

Also, building the social networks of African-American families is important. Research has clearly shown that social support is positively associated with the social and emotional well-being of parents and youngsters and may buffer them from the negative effects of stressors in their environment. Schools can play an important role in creating opportunities for families to build and maintain families' social networks. Experimental programs in some urban schools have sought to build community social networks by locating important social services (e.g., clinics, child care, adult education) on school grounds. Finally, findings have shown that emotional support from parents and their establishment of routines and practices aimed at increasing the structure and organization of the home are positively associated with school achievement and engagement. School outreach aimed at enhancing African-American students' emotional and social support and students' sense of the school environment as structured and organized may reinforce some families' practices that have positive effects on students' academic, social, and emotional learning.

REFERENCES

Baumrind, D. (1972). An exploratory study of socialization effect on black children: Some black–white comparisons. *Child Development, 43,* 261–267.

Baumrind, D. (1989). The permanence of change and the impermanence of stability. *Human Development, 32,* 187–195.

Baumrind, D. (1991). Parenting styles and adolescent adjustment. In J. Brooks-Gunn, R. Lerner, & A. C. Peterson (Eds.), *The encyclopedia on adolescence* (pp. 746–758). New York: Garland.

Bogenschneider, K. (1997). Parental involvement in adolescent schooling: A proximal process with transcontextual validity. *Journal of Marriage and the Family, 59,* 718–733.

Bowman, P., & Howard, C. (1985). Race-related socialization, motivation, and academic achievement: A study of black youth in three-generation families. *Journal of the American Academy of Child Psychiatry, 24,* 134–141.

Chase-Lansdale, P. L., Gordon, R. A., Brooks-Gunn, J., & Klebanov, P. K. (1997). Neighborhood and familial influences on the intellectual and behavioral competence of preschool and early schoolage children. In J. Brooks-Gunn, G. J. Duncan & J. L. Aber (Eds.), *Neighborhood poverty: Context and consequences for development* (Volume 1, Chapter 4, pp. 79–118). New York: Russell Sage Foundation.

Clark, R. (1983). *Family life and school achievement: Why poor black children succeed or fail.* Chicago: University of Chicago Press.

Conger, R., McMarty, J., Yang, R., Lahey, B., & Kropp, J. (1984). Perception of child, child-rearing values, and emotional distress as mediating links between environmental stressors and observed maternal behavior. *Child Development, 55,* 2234–2247.

Conger, R. D., Conger, K. J., Elder, G. H., Lorenz, F. O., Simons, R. L., & Whitbeck, L. B. (1992). A family process model of economic hardship and adjustment of early adolescent boys. *Child Development, 63,* 526–541.

Coulton, C. J., Korbin, J. E., Su, M., & Chow, J. (1995). Community level factors and child maltreatment rates. *Child Development, 66,* 1262–1276.

Darling, N., & Steinberg, L. (1993). Parenting style as context: An integrative model. *Psychological Bulletin, 113,* 487–496.

Dornbusch, S. M., Ritter, P. L., Leiderman, P. H., Roberts, D. F., & Fraleigh, M. J. (1987). The relation of parenting style to adolescent school performance. *Child Development, 58,* 1244–1257.

Dressler, W. (1985). Extended family relationships, social support, and mental health in a southern black community. *Journal of Health and Social Behavior, 26,* 39–48.

Durbin, D. L., Darling, N., Steinberg, L., & Brown, B. (1993). Parenting style and peer group membership among European-American adolescents. *Journal of Research on Adolescence, 3,* 87–100.

Gandara, P., & Maxwell-Jolly, J. (1999). *Priming the pump: Strategies for increasing the achievement of underrepresented minority undergraduates.* New York: College Board Publications.

Ge, X., Conger, R. D., Lorenz, F., & Simons, R. D. (1994). Parents' stress and adolescent depressive symptoms: Mediating processes. *Journal of Health and Social Behavior, 35,* 28–44.

Glasgow, K. L., Dornbusch, S. M., Troyer, L., Steinberg, L., & Ritter, P. L. (1997). Parenting styles, adolescent attributions, and educational outcomes in nine heterogeneous high schools. *Child Development, 68,* 507–529.

Herman, M. R., Dornbusch, S. M., Herron, M. C., & Hereting, J. R. (1997). The influence of family regulation, connection and psychological autonomy on six measures of adolescent functioning. *Journal of Adolescent Research, 12*, 34–67.

James, D. W., Jurich, S., & Estes, S. (2001). *Raising minority acaChase-demic achievement: A compendium of education programs and practices*. Washington, DC: American Youth Forum.

Jencks, C., & Phillips, M. (1998). The black–white test score gap: An introduction. In C. Jencks & M. Phillips (Eds.), *The black–white test score gap* (pp. 1–54). Washington, DC: The Brookings Institution.

Lamborn, S., Mounts, N., Steinberg, L., & Dornbusch, S. (1991). Patterns of competence and adjustment among adolescents from authoritative, authoritarian, indulgent, and neglectful families. *Child Development, 62*, 1049–1065.

Lerner, R. M. (1992). *Final solutions: Biology, prejudice, and genocide*. University Park: Penn State Press.

Levitt, M. J., Guacci-Franco, N., & Levitt, J. L. (1994). Social support and achievement in childhood and early adolescence: A multicultural study. *Journal of Applied Developmental Psychology, 15*, 207–222.

Mason, C. A., Cauce, A. M., & Gonzales, N. (1997). Parents and peers in the lives of African-American adolescents: An interactive approach to the study of problem behavior. In R. Taylor & M. Wang (Eds.), *Social and emotional adjustment and family relations in ethnic minority families* (pp. 35–52). Hillsdale, NJ: Lawrence Erlbaum Associates.

Massey, D. S., & Denton, N. A. (1993). *American apartheid: Segregation and the making of the underclass*. Cambridge, MA: Harvard University Press.

McLoyd, V. C. (1990). The impact of economic hardship on black families and children: Psychological distress, parenting, and socioemotional development. *Child Development, 61*, 311–346.

McLoyd, V. C. (1997). The impact of poverty and low socioeconomic status on the socioemotional functioning of African-American children and adolescents: Mediating effects. In R. Taylor & M. Wang (Eds.), *Social and emotional adjustment and family relations in ethnic minority families*. Hillsdale, NJ: Erlbaum.

National Center for Education Statistics. (1997). *The condition of education 1997*. Washington, DC: U.S. Department of Education, U.S. Government Printing Office.

Ogbu, J. L. (1985). A cultural ecology of competence among inner-city Blacks. In M. B. Spencer, G. K. Brookins, & W. R. Allen (Eds), *Beginnings: The social and affective development of Black children* (pp. 45–66). Hillsdale, NJ: Erlbaum.

Peters, M. F. (1985). Racial socialization of young black children. In H. McAdoo & J. Mc Adoo (Eds.), *Black children: Social, educational, and parental environments* (pp. 159–173). Newbury Park, CA: Sage.

Seaton, E. K. (2002). Ethnic idenity as a buffer for perceived "race" related stress among African-American adolescents. Unpublished manuscript, Temple University, Philadelphia, PA.

Steinberg, L. D., Lamborn, S. D., Dornbusch, S. M., & Darling, N. (1992). Impact of parenting practices on adolescent achievement: Authoritative parenting, school involvement, and encouragement to succeed. *Child Development, 63,* 1266–1281.

Steinberg, L., Mounts, N. S., Lamborn, S. D., & Dornbusch, S. M. (1991). Authoritative parenting and adolescent adjustment across varied ecological niches. *Journal of Research on Adolescence, 1,* 19–36.

Stevenson, H. C. (1997). Managing anger: Protective, proactive, or adaptive racial socialization identity profiles and manhood development. *Journal of Prevention and Intervention in the Community, 16,* 35–61.

Stevenson, H. C., Reed, J., Bodison, P., & Bishop, A. (1997). Racism stress management: Racial socialization beliefs and the experience of depression and anger in African-American youth. *Youth & Society, 29,* 197–222.

Taylor, R. D. (1996). Kinship support, family management, and adolescent adjustment and competence in African-American families. *Developmental Psychology, 32,* 687–695.

Taylor, R. D. (1997). The effects of economic and social stressors on parenting and adolescent adjustment in African-American families. In R. Taylor & M. Wang (Eds.), *Social and emotional adjustment and family relations in ethnic minority families.* Hillsdale, NJ: Erlbaum.

Taylor, R. D. (2000). An examination of the association of African-American mothers' perceptions of their neighborhood with their parenting and adolescent adjustment. *Journal of Black Psychology, 26,* 267–287.

Taylor, R. D., Casten, R., & Flickinger, S. (1993). The influence of kinship social support on the parenting experiences and psychosocial adjustment of African-American adolescents. *Developmental Psychology, 29,* 382–388.

Taylor, R. D., Seaton, E., Dominguez, A., & Rodriguez, A. U. (In press). The association of financial resources with parenting and adolescent adjustment in African-American families. *Journal of Adolescent Research.*

U.S. Census Bureau. (1994). *Current Population Survey: 1994.* Washington, DC: U.S. Government Printing Office.

U.S. Census Bureau. (1997). *Population characteristics: 1996* (Current Population Survey, Series P20). Washington, DC: U.S. Government Printing Office.

Waggoner, D. (1988, September). New study finds that poverty is transient for many. *Numbers and Needs: Ethnic and Linguistic Minorities in the U.S., 8*(5). Retrieved from http:/www.asu.edu/educ.cber/.

Weiss, L. H., & Schwarz, J. C. (1996). The relationship between parenting types and older adolescents' personality, academic achievement, adjustment and substance use. *Child Development, 67,* 2101–2114.

Wilson, M. N. (1989). Child development in the context of the Black extended family. *American Psychologist, 44,* 380–385.

6

Parental Involvement and Children's School Success

Arthur J. Reynolds & Melissa Clements

Parental involvement is widely regarded as a fundamental contributor to children's school success. Although psychological theory and conventional wisdom have always regarded the family as essential in shaping children's development, empirical validation of the impact of specific types of parent involvement in children's education has emerged only recently. Consequently, enhancing parental involvement is integral to many educational programs and policies, which will be discussed in Chapters 7 and 8, and are illustrated by the following trends.

- Parental involvement is a major element of schoolwide reforms such as Schools of the 21st Century, the School Development Program and the charter school movement, as well as new governance arrangements that give parents greater input in decision making and thereby facilitate school–family partnerships.
- Since the beginning of Head Start in 1965, parental involvement has been an essential component of early childhood programs for disadvantaged children. The emphasis on providing comprehensive family services and strengthening family–community partnerships has expanded to Title I, IDEA, and state-run early childhood programs, which now total more than $20 billion in government funding annually.
- Stand-alone family support interventions have increased in popularity in schools and communities. These include Even Start, Early Head Start, Family Resource Centers, Parents as Teachers and other home visitation programs, and parenting education programs.

 ISBN 0-8077-4600-2 (paper), ISBN 0-8077-4601-0 (cloth).

WHY PARENTAL INVOLVEMENT?

Why is parental involvement a focus of so many programs and policies to promote child and youth outcomes? First and foremost is the amount of time children spend with their families. Children spend more time with parents and family members, especially in the first decade of life, than in any other social context. About one-quarter of children's time is spent in school and three-quarters at home during most of the formative years. Thus, changing parental involvement by just a small amount can have a larger cumulative effect than behaviors that occur less frequently. A second reason why parental involvement is a target of program and policy formulation is that it is open to influence by educators. This is due to (a) the common interest of teachers and parents in educating and socializing children effectively, and (b) the many avenues that exist due to the multidimensional nature of parental involvement by which to encourage parental attitudes and behaviors, be it reading to children or participating in school. Third, encouraging parent participation and engagement in children's education may in and of itself provide important sources of social support and personal empowerment that promote the positive school and community climate so important for learning. Thus, family involvement sets the conditions upon which other educational and personal experiences impact children's outcomes (Bronfenbrenner, 1975).

In this chapter, we summarize findings for three categories of evidence about how parental involvement impacts children's school achievement and success.

1. Interventions with a family support component positively affect children's outcomes.
2. Parental involvement is a mechanism through which the long-term effects of intervention are achieved.
3. Indicators of parental involvement are associated with significantly higher levels of school performance and success.

Evidence in these domains is based on a longitudinal study of urban children (Chicago Longitudinal Study, 1999) and on other studies. We define parent involvement within the context of school–family partnerships to include behavior with or on behalf of children at home or in school, attitudes and beliefs about parenting or education, and expectations for children's future. Common indicators include home support for learning, parenting practices, child–parent interactions, participation in school activities, involvement in school associations, involvement in school governance or in community activities, and expectations for children's success or educational attainment. This description reflects the multidimensional nature of school–family partnerships, represents the complexity and breadth of parental influence, and is consistent with human capital perspectives of involvement as investments in children (Haveman & Wolfe, 1994). After describing findings from recent studies, we will consider some limits to the knowledge base, and discuss several implications for strengthening parent involvement.

CHICAGO LONGITUDINAL STUDY

A major advance in knowledge over the past decade is that through longitudinal studies, measures of parent involvement show long-term effects on children's learning and development. We illustrate how parent involvement contributes to children's school success primarily using data from the Chicago Longitudinal Study (Reynolds, 1999). This ongoing investigation follows the progress of 1,539 low-income children (93% African American) who participated in the Child-Parent Center Program beginning in 1983–84 and a matched comparison group that enrolled in an alternative intervention in kindergarten. Three major goals of the Chicago study are to (a) evaluate the effects of the Child-Parent Center Program over time, (b) identify the paths through which the effects of participation are manifested, and (c) investigate the contribution of a variety of family and school factors on children's adjustment. Over the 17 years of the study, extensive information on parent involvement as measured by teachers and parents has been collected, along with other child and family experiences, and children's school success. The contribution to children's long-term outcomes of an established intervention with intensive parent involvement is also shown.

Chicago Child-Parent Centers

The Chicago Child-Parent Center Program (CPC; Sullivan, 1971) is a center-based early intervention that provides comprehensive educational and family support services to economically disadvantaged children and their parents from preschool to early elementary school. The CPC program began in 1967 through funding from the Elementary and Secondary Education Act of 1965. Title I of the Act provided grants to local public school districts serving high concentrations of low-income children. As the nation's second-oldest federally funded preschool program, the CPCs operate in 23 centers across the city.

As shown in Figure 6.1, the centers provide comprehensive services under the direction of a Head Teacher and in collaboration with the

Figure 6.1. *Child–Parent Center Program*

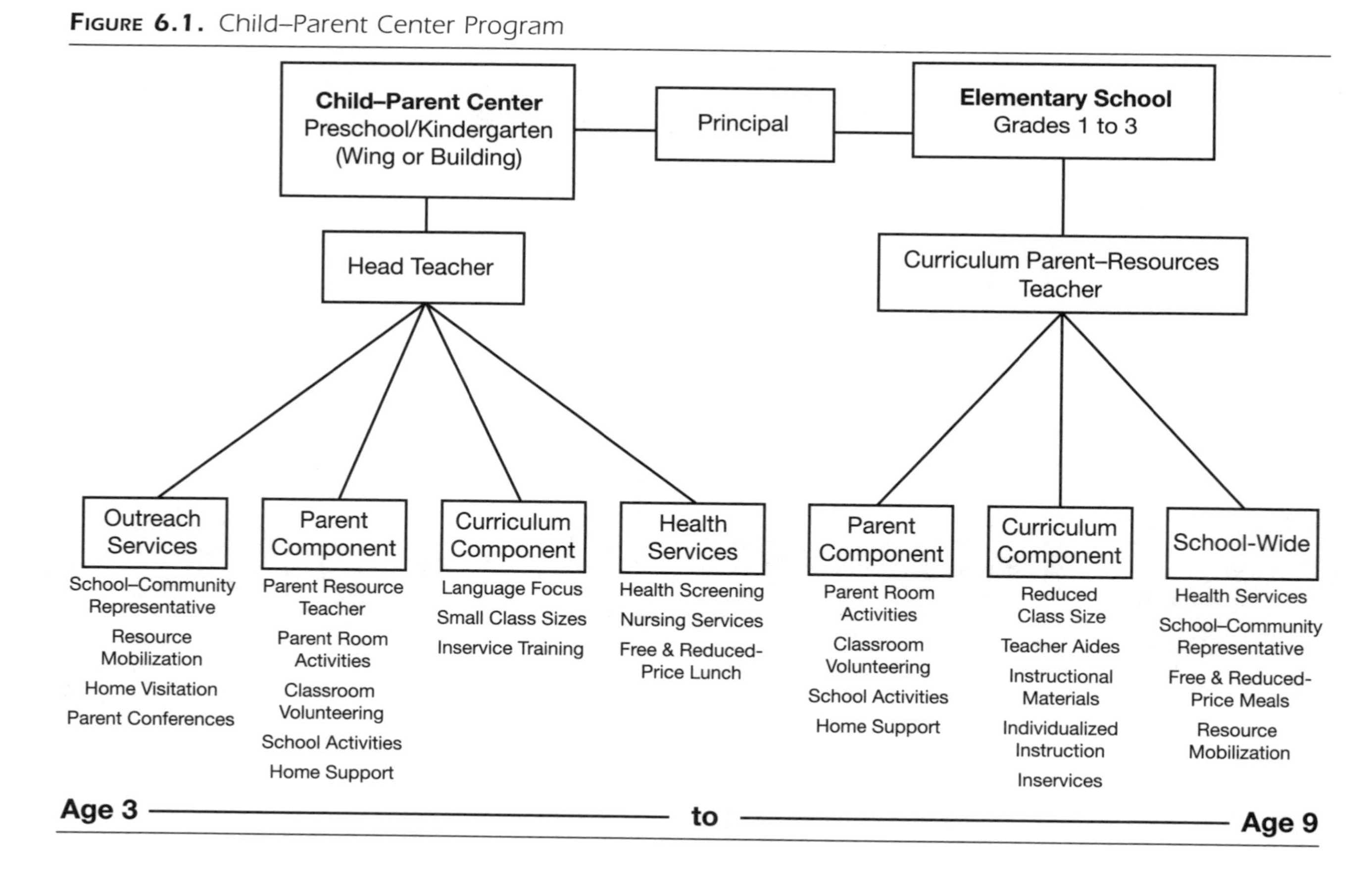

elementary school principal. Other primary staff in each center are the parent resource teacher, the school–community representative, classroom teachers and aides, nurses, speech therapists, and school psychologists. The major rationale of the program is that the foundation for school success is facilitated by the presence of a stable and enriched learning environment during the entire early childhood period (ages 3 to 9) and when parents are active participants in their children's education. Five program features are emphasized: early intervention, parent involvement, a structured language/basic skills learning approach, health and social services, and program continuity between the preschool and early school-age years.

Sullivan (1971) described the philosophy of the Child-Parent Centers as a way to enhance the family–school relationship: "In a success-oriented environment in which young children can see themselves as important, they are 'turned on' for learning. Attitudes toward themselves and others, interest in learning, increased activity, conversation, and enthusiasm are all evidences of the change. Parents are increasingly aware of the role of the home in preparing children for school and have renewed hope that education will develop the full potential of their children" (p. 70). To accomplish this, the centers offer a structured program of parent involvement and language enrichment.

Parent Involvement Component. We briefly describe the parent component of the program. As the program's title indicates, direct parent involvement in the program is expected to enhance parent–child interactions, parent and child attachment to school, family partnerships with schools, and social support among parents, and consequently to promote children's school readiness and social adjustment. Unlike most other programs, the centers make substantial efforts to involve parents in the education of their children, so much so that it is hard to overestimate. At least one half-day per week of parent involvement in the program is required, though this can take a variety of forms. The unique feature of the parent program is the parent resource room, which is physically located in the center adjacent to the classrooms. The full-time parent-resource teacher organizes the parent room in order to implement parent educational activities, initiate interactions among parents, and foster parent–child interactions. With funds for materials, supplies, and speakers, areas of training include consumer education, nutrition, personal development, health and safety, and homemaking arts. Parents may also attend GED classes at the centers as well as serve on the School Advisory Council. To enhance family–school partnerships, a wide range of activities are encouraged in the program, including parent room activities (e.g., arts & crafts projects), classroom volunteering, participation in school activities, class field trips, helping to prepare meals, and engaging in education and training activities. The diversity of activities is designed to accommodate parents' daily schedules.

EARLY CHILDHOOD INTERVENTIONS WITH A PARENT INVOLVEMENT COMPONENT INFLUENCE CHILDREN'S SUCCESS

Given the extensive family support services offered in the program, we consider the impact of the intervention on children's outcomes as the first category of evidence on parent involvement. CPC participation beginning in preschool has been consistently associated with better educational performance and social adjustment (Reynolds, Temple, Robertson, & Mann, 2001). Table 6.1 summarizes major findings in the Chicago Longitudinal Study, the largest study ever of the CPC program. Preschool participation at ages 3 or 4 is associated with educational and social outcomes spanning ages 5 to 22, up to 18 years after the end of intervention. As expected, the program had the largest impact immediately, as about one-half the program group scored at or above national norms on the Iowa Tests of Basic Skills scholastic readiness composite at school entry, compared to only one-quarter of the comparison group. This represented an 86% improvement over the comparison group.

Substantial differences were detected through the school-age years, as participation in the CPC program was associated with improvements ranging from 21% to 51% over the comparison group. The large reduction in child abuse and neglect is particularly important given that other early childhood programs have not reported these effects. Major reasons why the CPC program reduces child maltreatment may be the focus of the parent program in increasing social support and parenting skills, and reducing social isolation among families, which reduces family conflict and punitive parenting practices, which can enhance family cohesion and positive parenting practices (Reynolds & Robertson, 2003).

Reductions in special education placement and grade retention are consistent with much previous literature (Barnett, 1995; Reynolds, 2000). The significant program impacts on delinquency and high school completion are rare for a large-scale program and are particularly notable give their economic and social impact. High school completion is a basic requirement for nearly all career endeavors and for access to college. Besides contributing to a more productive life, avoidance of crime leads to major government savings on justice system treatment.

These findings indicate that interventions with strong family support components can impact many domains of performance. CPC intervention

Table 6.1. Proportion of the Chicago Child–Parent Center Program Preschool and Comparison-Group Children Achieving School and Social Competence

Child Outcome	*Age*	*Program Group*	*Comparison Group*	*Difference*	*Change*[a] *(%)*
At/above national norm on scholastic readiness	5	46.7	25.1	21.6	86
Child maltreatment	4–17	5.0	10.3	5.3	51
Repeated a grade	6–15	23.0	38.4	15.4	40
Special education	6–18	14.4	24.6	10.2	41
Juvenile arrest	10–18	16.9	25.1	8.2	33
Arrest for violent offense	10–18	9.0	15.3	6.3	41
Completed high school	18–22	65.7	54.5	11.2	21

Note: All differences were statistically significant. Rates are adjusted for group differences in sex of child, race/ethnicity, participation in school-age intervention, and family risk status.

[a] Percentage of change over the comparison group.

that continues into the elementary grades also contributes significantly to children's later success. Compared to children who participated for 1 to 4 years, children with 4 to 6 years of intervention had higher reading and math achievement in the elementary grades, lower rates of special education (13.5% vs. 20.7%) and grade retention (21.9% vs. 32.3%), and lower rates of child maltreatment (3.6% vs. 6.9%). These findings are aligned with recommendations by the Collaborative for Academic, Social, and Emotional Learning (CASEL), indicating that the most effective programs promote academic as well as social and emotional learning through coordinated and comprehensive services that include school–family partnerships as a key component (Denham & Weissberg, in press; Greenberg et al., 2003; Zins, Bloodworth, Weissberg, & Walberg, in press).

The pattern of findings from other intervention research is that early childhood programs with family support components are more likely to provide long-term benefits for children than programs that do not have such components (Olds et al., 1997; Schweinhart, Barnes, & Weikart, 1993). Providing intensive child education and family support, the High/Scope Perry

Preschool Project and the CPC program have shown the broadest impact on child and family outcomes. In the High/Scope Perry Preschool Project (Barnett, Young, & Schweinhart, 1998), the mother's participation in their child's education significantly influenced school success. Other frequently cited effective early childhood programs with family support components include the Syracuse Family Development Program, Early Training Project, Houston Parent-Child Development Center, and Yale Child Welfare Research Program (see Consortium for Longitudinal Studies, 1983). In a study of Head Start children, long-term effects on cognitive, social, and personal abilities were enhanced when a stronger and longer-lasting parental involvement component was included (Zeece & Wang, 1998). In their review of early interventions, Clarke and Clarke (1989) found that a variety of programs had enduring effects only when parents continued to be involved with the program after it had officially ended. Fantuzzo, Davis, and Ginsburg (1995) examined the effects of an intervention partly designed to enhance parental involvement with the child's education and found that the parental involvement component led to higher childhood scholastic and behavioral adjustment.

PARENTAL INVOLVEMENT IS A MECHANISM OF LONG-TERM EFFECTS OF INTERVENTION

What is the specific contribution of the family component of the CPC program to children's outcomes? Although it is difficult to separate the effects of parent involvement from those of child education and other services, one approach is to investigate the extent to which the main effects of program participation are explained by parent involvement and other factors. Parent involvement as a pathway or mechanism of effects of intervention indicates that it has an indirect influence on children's outcomes. Because indirect effects are more subtle than direct effects, they are largely unrecognized in the intervention literature. Very few studies have investigated parent involvement as a mechanism of intervention effects.

Figure 6.2 shows five hypotheses as identified in the accumulated literature that explain the link between early intervention and competence outcomes. In addition to the family support hypothesis, intervention effects may be explained by cognitive advantage, social adjustment, motivational advantage, and school support hypotheses. To identify the precise contribution of family support behavior to the long-term effects of program participation, measures of parent involvement must be considered together with these alternative hypotheses to document its unique contribution, or "added value."

FIGURE 6.2. *Alternative Paths Leading to Social Competence*

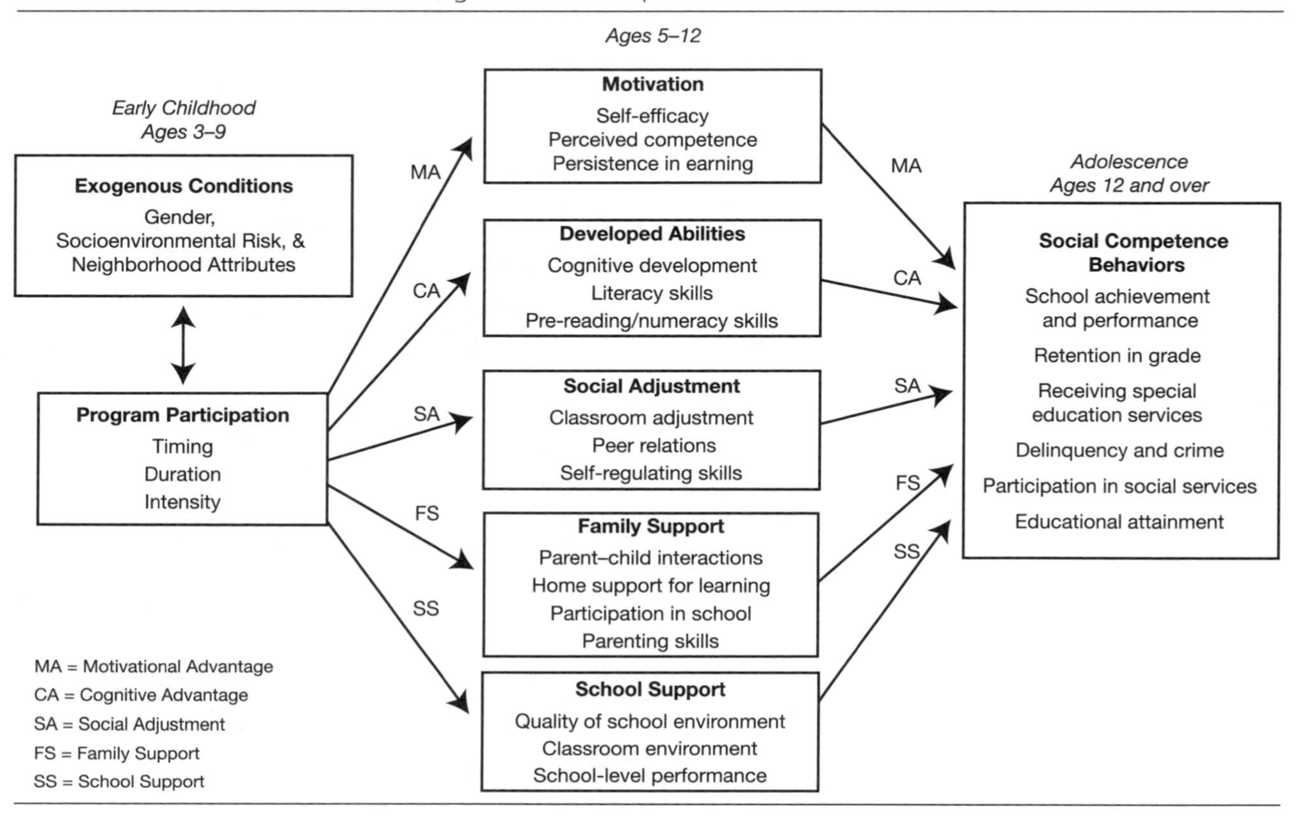

We summarize the contributions of the hypotheses in Figure 6.2 to the long-term effects of CPC participation on high school completion and juvenile arrest. The family support hypothesis was measured in two ways. The first was the frequency of teacher and parent ratings from ages 8 to 12 on the item "parents' participation in school." We used teacher and parent ratings to minimize possible reporter bias, and substantiated reports of child abuse and neglect between ages 4 and 12. The family support hypothesis accounted for 28% of the effect of preschool on high school completion and 21% of the effect of preschool on juvenile arrest. In other words, controlling for other hypotheses, program participation was directly associated with higher levels of parent involvement, and these higher levels of involvement were significantly linked to higher rates of school completion and lower rates of juvenile arrest.

Though not the focus of the current chapter, it is noteworthy that the school support and cognitive advantage hypotheses also accounted for a substantial portion of the link between CPC preschool participation and high school completion and juvenile arrest. The social adjustment and motivational advantage hypotheses made smaller contributions. Overall, these findings indicate the substantial contributions of parent involvement to long-term outcomes.

To illustrate the potential economic implications of these findings, a cost-benefit analysis of the CPC program (Reynolds, Temple, Robertson, & Mann, 2002) showed that for every dollar invested in the preschool component, seven dollars were returned to society at large through government savings on remedial education and justice system treatment, and increases in economic well-being. About two dollars of these economic benefits can be attributed to the family support program, or about $14,000 per participant in discounted 1998 dollars. This estimate is conservative to the extent that parental involvement has synergistic effects with other components of the program. Nevertheless, the cognitive benefits of participation and school support experiences in the elementary grades contribute significantly to the total effect of preschool.

Through an intensive parent program in the centers, CPC intervention encourages parent involvement in school and in children's education so that when the intervention ends, parents are more likely to continue to provide the nurturance and support necessary to maintain benefits, which makes later school attainment and prosocial behavior more likely. That parent involvement directly predicted the rate of juvenile arrest is especially significant, since this has not been documented previously. In the High/Scope Perry Preschool Study (Barnett et al., 1998), parent involvement was positively related to later educational attainment but was not related to preschool participation. Nevertheless, family support was a major component of the program through biweekly home visits. Indeed, the only early childhood programs that have shown effectiveness in preventing delinquency are interventions that have significant family support components (Zigler, Taussig, & Black, 1992). Our findings indicate that school-based parent involvement provides another avenue to enhance family support behaviors.

INDICATORS OF PARENTAL INVOLVEMENT PREDICT CHILDREN'S LEARNING AND DEVELOPMENT

In most of the research on parental involvement, natural variation in parental attitudes and behaviors is associated with children's outcomes, including school performance, achievement test scores, and educational attainment. Overall, higher parental involvement is associated with higher academic performance for young children and adolescents (Graue, Weinstein, & Walberg, 1983; Patrikakou, 1997; Reynolds & Bezruczko, 1993; Seefeldt, Denton, Galper, & Younoszai, 1998). These predictive relations usually remain after taking background factors into account, such as SES. Furthermore, variables such as maternal education, SES, child sex, and child motivation level have all been found to influence child academic performance by impacting parental involvement (Reynolds, 1989; Stevenson & Baker, 1987), further supporting the importance of the parental influence. Parent involvement has also been related to social and emotional learning, including self-regulation and self-concept.

One measure developed in the Chicago Longitudinal Study is the number of years between first and sixth grades that teachers rate parent participation in school as average or better. This provides a cumulative index of involvement over many years. As shown in Table 6.2, rates of high school completion increase as ratings of parent involvement increase. Children with 6 years of positive parent ratings had a high school completion rate of 83% by age 21. These same children had a delinquency rate of 8%. Both rates are substantially better that those of other inner-city children. The findings did not change when family demographic factors such as race/ethnicity, sex of child, and SES were taken into account (Barnard, 2001).

In further support of the impact of this measure and others, we examined its independent influence on high school completion at age 20 for 1,286 youth after taking other important factors into consideration, including school performance, family socioeconomic status, school commitment, and school quality. We found that a 1-year change in parental school involvement was associated with a 16% increase in the odds of high school completion. A change in involvement from 1 to 4 years, for example,

TABLE 6.2. Juvenile Delinquency and High School Completion Rates by Parent Involvement Rating

Parent Involvement [a]	*Juvenile Delinquency* [b] *(%)*	*High School Completion* [c] *(%)*
0	22.6	37.3
1	22.4	49.5
2	16.1	66.2
3	12.6	69.6
4	13.9	70.6
5	11.3	83.1
6	8.7	82.6

[a] Number of years of positive teacher ratings of involvement in school from grades 1–6; 0 = no ratings were "average" or better; 6 = all ratings were "average" or better.

[b] 1 = juvenile delinquent; 0 = not juvenile delinquent.

[c] 1 = completed high school; 0 = not completed high school.

increased the odds of high school completion by 48%. Parent expectations for children's educational attainment also were associated with significantly higher levels of completion. A change of 1 year in expected years of education were associated with a 9% increase in the odds of high school completion. A change of 4 years (expecting college vs. high school completion) increased the odds of high school completion by 36%. These findings are suggestive of the long-term benefits. Notably, parent educational attainment also significantly predicted high school completion above and beyond parent involvement and expectations. Its impact was similar to parent expectations.

In other Chicago studies, parent involvement predicted academic success and socioemotional adjustment in first grade (Reynolds, 1989) as well as increased reading achievement, lower grade retention, and fewer years in special education by age 14 (Miedel & Reynolds, 1999). Parent involvement also significantly predicted school achievement across two successive school years and predicted academic growth from year 1 to year 2, after controlling for demographic variables.

While CPC parent involvement at the site level was found to be the only program factor (i.e., instructional approach, size of site) associated with early academic outcomes in kindergarten, parent involvement at the individual or child level was significantly related to early and later academic achievement as well as high school completion and juvenile delinquency (Clements, Reynolds, & Hickey, 2004). This finding suggests that children's early school success can be enhanced by providing opportunities for parents to be involved in school and in children's education.

Many other studies support the positive and significant link between measures of parental involvement and children's school success (Reynolds, 1992). A meta-analysis of 25 studies by Fan and Chen (2001) found that parent expectations or aspirations had the largest impact (effect size of $z = .40$) in relation to measures of academic achievement, with parental supervision at home (effect size of $z = .09$) having the smallest effect size. The size of the impact of parent involvement on academic achievement was also larger for global (i.e., GPA) as compared to subject-specific academic achievement measures. Parental commitment and volunteer behavior have smaller positive influences (Fan, 2001). Parental contact and supervision tend to have small but negative influences on achievement outcomes. It is likely that children receiving more supervision and contact with the school were more in need of such supervision and contact, possibly because of prior problematic behavior or academic achievement.

Parent involvement also has been found to positively influence motivational outcomes such as academic self-concept, attributions for academic achievements, and self-regulation as well as high school dropout and truancy behaviors (Gonzalez-Pienda et al., 2002). In reports from the National Educational Longitudinal Study (NELS; Keith, Keith, Troutman, & Bickley, 1993; Patrikakou, 1996, 1997; Trivette & Anderson, 1995), parent expectations or aspirations for children's education were most consistently associated with eighth-grade achievement even after controlling for the influence of SES and ethnicity. Moreover, the association between parental expectations and achievement was strongest for higher-SES students. Relatively few studies have examined the relation between involvement and children's social and emotional learning, and even fewer have investigated effects across SES and ethnicity.

LIMITATIONS OF THE OVERALL KNOWLEDGE BASE

Despite the overall positive evidence described in this chapter, three limitations remain in the knowledge base on parent involvement. First, while parental involvement is associated with higher school success and can predict school performance, its link to children's outcomes should not be regarded as causal. The quality and amount of parent involvement, for example, may be key ingredients rather than involvement per se. In addition, parent involvement may be an effect of children's performance as much as a cause, and few studies have investigated this and other possibilities. Moreover, understanding of specific mechanisms linking parental involvement to different children's outcomes, a key element for establishing cause, is just beginning.

Second, the definition and measurement of parent involvement vary greatly from study to study. Findings vary as a function of children's age and circumstances, whether the source of report is parents or teachers, and whether the behaviors and attitudes concern home or school support for children's learning. The reliability and validity of these different operationalizations can be variable, and the correspondence between parent and teacher ratings relatively low. The lack of consensus about the definition and measurement of parent involvement has hindered integration of knowledge for policy decisions.

Third, there is a presumption among researchers and educators that all types of parent involvement are positively associated with children's adjustment. This is not necessarily the case. Involvement in response to child problems is not generally associated with positive outcomes in several studies. The conditions under which parent involvement yields the most and least positive links with children's outcomes need further investigation.

CONCLUSION AND RECOMMENDATIONS

Enhancing parental involvement is a goal of many educational policies and practices. Our review indicates that parental involvement, in all its forms, can contribute substantially to children's school success. In the past decade, two of the greatest advances in knowledge are that parent involvement in the education of low-income children helps explain the long-term effects of early childhood intervention and that parent involvement in the elementary grades is associated with significantly higher rates of later educational attainment. These findings have not been reported before. We found stronger evidence for some indicators of parental involvement than others. Parent expectations for children's educational attainment and participation in school activities had the most consistent influence on children's outcomes. Parental supervision had smaller impacts. This pattern of findings is consistent across studies, from the Chicago Longitudinal Study to meta-analyses of the whole field.

Early Childhood Programs

The policy implications of the findings presented in this chapter are directed to early childhood programs and to schools and families more generally (see Figure 6.3). In regard to early childhood intervention, greater investments in programs that provide child education and intensive resources for parent involvement are needed. Findings from the Child-Parent Centers and other similar programs demonstrate their positive long-term effects on a wide range of outcomes that span the ages of 5 to 22. To expand effective programs such as the CPCs, one recommendation is to increase the proportion of expenditures on Title I programs beyond the 5% that currently goes to preschool programs.

Are large payoffs from early childhood education inevitable? Not if the programs are low in quality or poorly coordinated. Judging from the accumulated evidence, four elements are critical to the success of early education programs. First, a coordinated system of early care and education should

FIGURE 6.3. Recommendations for Enhancing Parent–School Partnerships in Early Educational Settings

Provide greater investments in programs that provide child education and intensive resources for parent involvement.

- Increase the proportion of expenditures on Title I programs that goes to preschool programs.

Provide intensive and comprehensive family services and parent involvement activities, especially for children who have special needs or reside in low-income families.

- School communities should take an active role in providing a variety of ways for parents to get involved that require both physical and human resources.
- The Child-Parent Centers have a staffed parent room in each preschool and elementary school site to provide a comprehensive set of activities tailored to the needs of families.

Provide teachers with greater opportunities for training and professional development in working with families.

- Areas of training include: overcoming barriers to involving parents, alternative ways of involving parents in children's education and schooling, promoting effective communication with families, and resolving and reducing conflicts with parents and other family members.

Emphasize high parent expectations and values toward education in addition to high parental involvement. School–family partnerships that provide many ways to strengthen involvement are the most likely to impact children's school success.

be in place that spans at least the first five years of a child's life. Public schools, where over 90% of children end up, are in the best position to take a leadership role in partnership with the community. Second, preschool teachers should be trained and compensated well. It is no coincidence that the Perry and Chicago programs were run by staff with at least bachelor's degrees and certification in their specialties. Third, educational content should be responsive to all of children's learning needs, but special emphasis on literacy skills is needed. Fourth and most relevant, family services and parent involvement activities should be available that are intensive and comprehensive in scope, especially for children who have special needs or reside in low-income families. One advantage of the school-based model of the Child-Parent Centers is that they help forge school–family partnerships.

Schools

With regard to enhancing parent involvement more generally, we have three recommendations. First, school communities must actively provide a variety of ways for parents to get involved. As noted in previous chapters, more than lip service is required; physical and human resources are needed. Instead of reacting to children's difficulties, proactive strategies are required, often with appropriate resources. To increase involvement, the Child-Parent Centers have a staffed parent room in each preschool and elementary school site to provide a comprehensive set of activities tailored to the needs of families. The provision of physical and staff resources for a parent program is a primary reason that the centers maintained an 80% rate of regular participation, compared to less than 50% in many other programs. This approach warrants expansion in other settings.

Second, teachers would benefit from greater opportunities for training and professional development in working with families. Existing teacher training usually deemphasizes or ignores parent–teacher collaboration (see Chapter 9 for a detailed discussion of teacher preparation). Among the areas needing greater attention are: overcoming barriers to involving parents, alternative ways of involving parents in children's education and schooling, promoting effective communication with families, and resolving and reducing conflicts with parents and other family members.

Finally, based on the accumulated evidence that parent expectations for children education are a key factor in school success, greater emphasis on promoting high expectations and values toward education is needed. This indicates that strategies to increase behavioral involvement are not the only or even most desirable approaches for enhancing children's school success. School–family partnerships that provide many ways to strengthen involvement are the most likely to impact children's academic, social, and emotional learning, and lead to school success.

REFERENCES

Barnard, W. M. (2001). *Early intervention participation, parent involvement in schooling, and long-term school success*. Unpublished doctoral dissertation, University of Wisconsin—Madison.

Barnett, W. S. (1995). Long-term effects of early childhood programs on cognitive and school outcomes. *The Future of Children, 5*(3), 25–50.

Barnett, W. S., Young, J. W., & Schweinhart, L. J. (1998). How preschool education influences long-term cognitive development and school success: A causal model. In W. S. Barnett & S. S. Boocock (Eds.), *Early care and education for children in poverty: Promises, programs, and long-term results* (pp. 167–184). Albany, NY: State University of New York Press.

Bronfenbrenner, U. (1975). Is early intervention effective? In M. Guttentag & E. Struening (Eds.), *Handbook of evaluation research* (Vol. 2, pp. 519–603). Beverly Hills, CA: Sage.

Chicago Longitudinal Study. (1999). *Chicago Longitudinal Study: A study of children in the Chicago public schools*. [User's guide, version 6]. Madison: University of Wisconsin, Waisman Center.

Clarke, A. M., & Clarke, A. D. B. (1989). The later cognitive effects of early intervention. *Intelligence, 13*, 289–297.

Clements, M., Reynolds, A. J., & Hickey, E. (2004). Site-level predictors of children's school and social competence in the Chicago Child-Parent Centers. *Early Childhood Research Quarterly, 19*, 273–296.

Consortium for Longitudinal Studies. (1983). *As the twig is bent . . . lasting effects of preschool programs*. Hillsdale, NJ: Erlbaum.

Denham, S., & Weissberg, R. (in press). Collaborative for Academic, Social, and Emotional Learning (CASEL): Social-emotional learning in early childhood: What we know and where to go from here. In M. Bloom & T. P. Gullotta (Eds.), *A blueprint for the promotion of prosocial behavior in early childhood* (pp. 13–50). New York: Kluwer/Academic Publishers.

Fan, X. (2001). Parental involvement and students' achievement: A growth modeling analysis. *Journal of Experimental Education, 70*(1), 27–61.

Fan, X. T., & Chen, M. (2001). Parental involvement and students' academic achievement: A meta-analysis. *Educational Psychology Review, 13*, 1–22.

Fantuzzo, J. W., Davis, G. Y., & Ginsburg, M. D. (1995). Effects of parent involvement in isolation or in combination with peer tutoring on student self-concept and mathematics achievement. *Journal of Educational Psychology, 57*(2), 272–281.

Gonzalez-Pienda, J. A., Nunez, J. C., Gonzalez-Pumariega, S., Alvarez, L., Roces, C., & Garcia, M. (2002). A structural equation model of parental involvement, motivational and aptitudinal characteristics and academic achievement. *Journal of Experimental Education, 70*(3), 257–287.

Graue, M. E., Weinstein, T., & Walberg, H. J. (1983). School-based home instruction and learning: A quantitative synthesis. *Journal of Educational Research, 76*(6), 251–260.

Greenberg, M. T., Weissberg, R. P., Utne O'Brian, M., Zins, J. E., Fredricks, L., Resnik, H., et al. (2003). Enhancing school-based prevention and youth development through coordinated social, emotional and academic learning. *American Psychologist, 58*, 466–474.

Haveman, R., & Wolfe, B. (1994). *On the effects of investments in children*. New York: Russell Sage Foundation.

Keith, T. Z., Keith, P. B., Troutman, G. C., & Bickley, P. G. (1993). Does parental involvement affect eighth-grade achievement? Structural analysis of national data. *School Psychology Review, 22*(3), 474–496.

Miedel, W. T., & Reynolds, A. J. (1999). Parent involvement in early intervention for disadvantaged children: Does it matter? *Journal of School Psychology, 37*, 379–402.

Olds, D. L., Eckenrode, J., Henderson, C. R., et al. (1997). Long-term effects of home visitation on maternal life course and child abuse and neglect: Fifteen-year follow-up of a randomized trial. *Journal of the American Medical Association, 278*, 637–643.

Patrikakou, E. N. (1996). Investigating the academic achievement of adolescents with learning disabilities: A structural modeling approach. *Journal of Educational Psychology, 88*, 435–450.

Patrikakou, E. N. (1997). A model of parental attitudes and the academic achievement of adolescents. *Journal of Research and Development in Education, 31*, 7–26.

Reynolds, A. J. (1989). A structural model of first-grade outcomes for an urban, low socioeconomic status, minority population. *Journal of Educational Psychology, 81*, 594–603.

Reynolds, A. J. (1992). Comparing measures of parental involvement and their effects on academic achievement. *Early Childhood Research Quarterly, 7*, 441–462.

Reynolds, A. J. (1999). Educational success in high-risk settings: Contributions of the Chicago Longitudinal Study. *Journal of School Psychology, 37*, 345–354.

Reynolds, A. J. (2000). *Success in early intervention: The Chicago Child-Parent Centers*. Lincoln, NE: University of Nebraska Press.

Reynolds, A. J., & Bezruczko, N. (1993). School adjustment of children at risk through fourth grade. *Merrill-Palmer Quarterly, 39*, 457–480.

Reynolds, A. J., & Robertson, D. L. (2003). School-based early children intervention and later maltreatment in the Chicago Longitudinal Study. *Child Development, 74*, 3–26.

Reynolds, A. J., Temple, J. A., Robertson, D. L., & Mann, E. A. (2001). Long-term effects of an early childhood intervention on educational achievement and juvenile arrest: A 15-year follow-up of low-income children in public schools. *Journal of the American Medical Association, 285*(18), 2339–2346.

Reynolds, A. J., Temple, J. A., Robertson, D. L., & Mann, E. A. (2002). Age 21 cost-benefit analysis of the Title I Chicago Child-Parent Centers. *Educational Evaluation and Policy Analysis, 24,* 267–303.

Schweinhart, L. J., Barnes, H. V., & Weikart, D. P. (1993). *Significant benefits: The High-Scope Perry Preschool study through age 27.* Ypsilanti, MI: High/Scope Press.

Seefeldt, C., Denton, K., Galper, A., & Younoszai, T. (1998). Former Head Start parents' characteristics, perceptions of school climate, and involvement in their children's education. *Elementary School Journal, 98*(4), 339–349.

Stevenson, D. L., & Baker, D. P. (1987). The family–school relation and the child's school performance. *Child Development, 58,* 1348–1357.

Sullivan, L. M. (1971). *Let us not underestimate the children.* Glenview, IL: Scott, Foreman.

Trivette, P., & Anderson, E. (1995). The effects of four components of parental involvement on eighth grade student achievement: Structural analysis of NELS-88 data. *School Psychology Review, 24*(2), 299–318.

Zeece, P. D., & Wang, A. (1998). Effects of the family empowerment and transitioning program on child and family outcomes. *Child Study Journal, 28*(3), 161–177.

Zigler, E., Taussig, C., & Black, K. (1992). Early childhood intervention: A promising preventive for juvenile delinquency. *American Psychologist, 47,* 997–1006.

Zins, J. E., Bloodworth, M. R., Weissberg, R. P., & Walberg, H. J. (in press). The scientific base linking social and emotional learning to school success. In J. E. Zins, R. P. Weissberg, M. C. Wang, & H. J. Walberg (Eds.), *Building academic success on social and emotional learning: What does the research say?* New York: Teachers College Press.

PART III

POLICY ISSUES

7

School–Family Relations and Student Learning: Federal Education Initiatives

Oliver C. Moles, Jr.

Parent roles with schools can take a variety of forms, such as communicating about academics, attending parent education workshops, serving on school advisory committees, or choosing educational programs for their children. These school–family relationships create the possibility of true partnerships characterized by continuous exchange of information, mutual respect, and shared power and responsibilities for the education and development of children. The Introduction and the first three chapters of this book presented discussions of theoretical issues in school–family partnerships, whereas Chapters 4, 5, and 6 focused on the cultural and empirical evidence related to home–school relations. The focus of this chapter is to examine how much school–family partnerships are actually encouraged in federal education programs.

Besides academic achievement, schools may contribute to the development of social and emotional competencies in children. Core competencies include awareness of self and others, self-management, interpersonal skills, and responsible decision making. Social and emotional learning (SEL) programs assume that the best learning derives from supportive and challenging relationships (CASEL, 2002). While the focus of most federal education programs is on academic achievement, a few address important aspects of SEL, and some also have significant roles for parents.

The federal government greatly expanded its role in education in the 1960s, responding to social concerns about disadvantaged Americans and underachieving students. This chapter analyzes major programs funded by the U.S. Department of Education (hereafter the Department) since the Elementary and Secondary Education Act of 1965, which

 ISBN 0-8077-4600-2 (paper), ISBN 0-8077-4601-0 (cloth).

included provisions for parent involvement with schools to strengthen academic, social, and/or emotional learning of K–12 students. For present purposes, *parents* are defined broadly to include all family members who have responsibilities for the care and upbringing of children and youth. *Family involvement* and *parent involvement* are also used interchangeably. (See "Parent Involvement in Federal Education Programs," Moles, 2001, for an analysis of earlier initiatives before the NCLB legislation.)

The eight Department programs to be discussed begin with Title I, Safe and Drug-Free Schools, and Character Education. The federal Even Start program is not discussed, since it funds projects for children 0–7 outside of schools. Other Department programs not directly related to schools are also omitted from the following review. Several aspects of each program will be presented: the evolving legislation regarding parent roles, the nature and scope of required activities with parents, how these activities may affect various kinds of learning, and any recent large-scale federally funded evaluations of parent aspects of the programs. Table 7.1 notes the presence of key features of each program regarding parent involvement, student learning, and roles of different parties. State and district policies and programs will be discussed in the following chapter by Redding and Sheley.

The No Child Left Behind Act of 2001 (NCLB) (Public Law 107-110, 2002) is a centerpiece of the Department's education strategy. It addresses various aspects of academic achievement, school safety, student drug use, character education, and a wide array of other educational issues. NCLB also demands a strong research base for programs; the need for guidance from "scientifically based research" is mentioned over 100 times. The Department interprets this as studies involving random assignment to intervention or control groups. Because of its scope, any review and analysis must begin with NCLB and its Title I programs.

TITLE I OF THE NO CHILD LEFT BEHIND ACT

Title I Part A is the first section of the Elementary and Secondary Education Act of 1965 as amended by the No Child Left Behind Act of 2001 (Public Law 107-110, 2002). The overall goal of Title I Part A (hereafter Title I) is to improve the teaching and learning of children in high-poverty schools so that they can meet challenging academic content and performance standards. Title I received $12.34 billion in federal funding in 2004 and serves more than 10 million students. While not funded to the maximum authorized by NCLB, Title I funding has increased substantially since the inception of NCLB.

TABLE 7.1. Parental Involvement and Learning Requirements in Federal Education Programs

Program	*Parent Involvement (PI)*	*Academic Learning (AL)*	*Social & Emotional Learning (SEL)*	*Implementation, Monitoring, & Evaluation (IME)*
Title I	Yes	Yes	No	Yes
Safe & Drug-Free Schools	Yes	No	Yes	Yes
Character Education	No	No	Yes	Yes
After-School Programs	Yes	Yes	Yes	Yes
Partnership for Family Involvement in Education	Yes	Yes	Yes	Yes
Parent Information and Resource Centers	Yes	Yes	Yes	Yes
English Language Acquisition	Yes	Yes	No	Yes
Special Education	Yes	Yes	Yes	Yes

Originally there were no parent involvement requirements in Title I. In 1971 states and localities were required to consult with parents on the development and operation of programs and to establish parent advisory councils, but in the 1981 amendments the parent advisory functions were greatly reduced. With the 1988 amendments local education agencies were again required to involve parents in program planning and implementation, to give parents information in their own language, and to evaluate parent programs. Districts and schools were encouraged to develop parent advisory councils, resource centers, parent liaison staff, and resources for home learning (D'Agostino, Hedges, Wong, & Borman, 2001).

The Improving America's Schools Act of 1994 added several new Title I provisions for parent involvement in Section 1118 of the Act. School districts that received more than $500,000 per year in Title I funds had to reserve at least 1% for parent involvement activities. Each Title I school was also required to develop jointly with parents a school–parent compact. These compacts describe the school's responsibilities to provide high-quality curriculum and instruction, the parents' responsibilities to support their children's learning at home, and the ongoing school–home communication necessary for achieving high standards. Parents and key school staff are urged to sign the compacts. The 1994 Act also required that schools develop with parents a written parent involvement plan that included shared responsibility for high student performance and make the plan available to parents of participating children. Schools must also convene parents annually to inform them of the school's parent involvement program.

NCLB continues and strengthens these requirements in its Section 1118. It also defines parental involvement for the first time. Such involvement is to include "regular, two-way, and meaningful communication" on student learning and school activities, ensuring that parents play an "integral role" in assisting their child's learning and act as full partners in their child's education. NCLB requires materials and training for parents on working with their children, and training for educators on the value of parent contributions and how to reach and work with them as equal partners. Parents may help develop this training. (See Chapter 9 on preparing educators for partnerships.)

These are all potentially powerful ways of communicating, sharing power and responsibilities, and building mutual respect between schools and families—the essence of partnerships. However, the Department's dissemination and enforcement efforts have been focused on a new aspect of NCLB—the choices parents can make if their child is in a failing school; that is, one lacking state-defined adequate yearly progress for two years. Beginning in 2002–2003, annual report cards for all schools must show overall student performance by subgroups in each school and district on state assessments of basic subjects in grades 3–8, graduation and retention rates for secondary schools, and teacher qualifications. If the school is failing, parents must be given an explanation of the situation, what the school is doing to address the problem of low achievement, and how parents can become involved in working on the academic issues. Parents may then choose to transfer their child to another public school with transportation paid or to have the child given supplemental educational services (tutoring) outside of school hours. Districts are to conduct, with parent input, annual evaluations of the content and effectiveness of parent involvement policies. States are to review district parent involvement policies and practices.

Building comprehensive programs to involve parents faces serious obstacles. Representatives of the National Coalition of Title I/Chapter 1 Parents have reported that school personnel and parents across the country are unfamiliar with the parent involvement provisions of Title I and do not know about the resources that Title I and other programs provide to support parent involvement. While national organizations are attempting to inform parents and community organizations on the provisions of NCLB, it is unclear how much information is reaching parents. Whether states will develop guidelines to support schools and districts and will provide technical assistance to schools in implementing the NCLB Title I parent involvement provisions, as most did in the past, remains to be seen.

States have been slow to implement the NCLB, which demands quick changes. Written guidance from the Department was issued in 2002 for education choice options, but guidance for other aspects of parent involvement has lagged. Assistance and oversight regarding these Section 1118 requirements seem minimal.

The new choices for parents have not yet been systematically studied. On the other hand, there is considerable large-scale research evidence on some parent involvement provisions of the earlier Title I programs that warrants a separate discussion.

RESEARCH ON PARENT ROLES IN TITLE I

The Prospects Study of the national Title I program (then called Chapter 1) in the early 1990s included considerable information on parent involvement. Prospects found no connection between student achievement and school-based parent involvement such as volunteering or serving on school governance boards, or between student achievement and Title I school activities to involve parents. But parents reported more home-based involvement when schools offered home learning materials, open communication with schools, learning compacts, and other aspects of a comprehensive program. Since stronger home-based learning is linked to Title I student reading achievement, the authors concluded that Title I may influence achievement indirectly by helping parents to become better home educators and that comprehensive programs are needed if parents are to be more involved in their children's home learning (D'Agostino et al., 2001).

A study of 71 Title I schools in 1996–99 found evidence that teacher outreach to parents of low-performing students helped improve student achievement in both reading and math (Westat & PSA, 2001). It followed the progress of students as they moved from third to fifth grade. Outreach was measured in terms of face-to-face meetings, sending learning materials home, and routine phone calls. These personal learning-oriented contacts are examples of how teachers can help parents become better home educators.

Under NCLB an annual state evaluation must identify barriers to greater parent participation, especially for those with economic disadvantage, disabilities, minority background, or limited English proficiency. A number of such barriers have been reported by principals in Title I schools, according to a report to Congress (U.S. Department of Education, 1997b). Lack of time by parents and school staff head the list, followed closely by lack of staff training to work with parents and lack of parent education to help with homework. In the highest-poverty Title I schools, principals reported less parent attendance at regular parent–teacher conferences and special school events than in the lowest-poverty Title I schools.

In conjunction with the report to Congress, 20 Title I schools with successful approaches to involving families were studied. These approaches included providing information and training to parents and school staff, restructuring schools to support families better, and bridging school–family differences. This information has been disseminated widely in an Idea Book (U.S. Department of Education, 1997a) and in workshops held in many locations.

By 1998 compacts were used in 75% of Title I schools, and most schools with them said that parent involvement was enhanced by the compacts (D'Agostino et al., 2001). The development and implementation of new and promising compacts were analyzed in five Title I schools. All of these schools viewed the written compact as part of their larger parent involvement efforts. Still, stakeholders said it was difficult to maintain active support for the compacts after their initial introduction. Compact-related discussions and activities became less visible over time, making the compact's impact sometimes questionable (Funkhouser, Stief, & Allen, 1998). On the other hand, in a national longitudinal study of students in 300 schools, those schools with compacts produced higher student achievement. This held true even after accounting for family background and initial achievement (Borman, D'Agostino, Wong, & Hedges, 1998). Overall, serious challenges appear to limit full use of Title I services by parents. There is much yet to learn about the effects of the NCLB Title I provisions for parent involvement in their children's education.

SAFE AND DRUG-FREE SCHOOLS

In response to the increasing awareness of alcohol and other drug abuse among youth, Congress enacted the Drug-Free Schools and Communities Act in 1987 to expand and strengthen related education and prevention programs across the nation. As safety became a more pressing concern—reflected in the National Education Goal for the year 2000 to make all schools safe, disciplined, and drug-free—the law added safety in 1994 to the 1987 Act. Promoting a sense of individual responsibility for one's behavior is a prominent authorized activity under this program.

The Safe and Drug-Free Schools and Communities Act was reauthorized as Title IV Part A of NCLB, and $674 million were appropriated for fiscal year 2004. School districts must consult with parents in developing and operating their programs, promote the involvement of parents in their activities, and evaluate whether they have conducted effective parent involvement and training programs. In addition to these state and local programs, national programs include the development and demonstration of innovative prevention activities and strategies for training school personnel, parents, and community members. An evaluation of these national programs that includes parent involvement and training is required every 2 years.

Drug use and violence prevention programs clearly are social and emotional learning programs. Studies show that academic failure, drug use, and violence are associated with similar risk factors (CASEL, 2002). Many who study these problem behaviors delineate risk and protective factors in the school, community, family, and peer group that influence the individual child. Such factors that have been proven through longitudinal research or strong theory are recommended for use in this NCLB program.

The evidence for the overall effectiveness of the program has not been encouraging. A 1997 study of school-based drug prevention programs in selected school districts determined that they often suffered from inconsistent delivery, weak effects, and disregard for the research base (Silvia & Thorne, 1997). This led the Department in 1998 to establish principles of program effectiveness such as basing programs on research-proven strategies, setting measurable objectives, and evaluating local programs. A broader national evaluation from that period concluded that local drug and violence prevention efforts have wide scope but questionable quality. Based on 1997–98 school year data, the report suggests that schools need to improve their needs assessment, planning, use of research-based approaches, and monitoring of implementation (Crosse, Burr, Cantor, Hagen, & Hantman, 2002). Whether the 1998 principles of effectiveness, now incorporated into NCLB, will actually spur higher-quality local programs remains to be seen.

CHARACTER EDUCATION

A program of character education grants to local education agencies and collaborating organizations was initiated in the 1994 Improving America's

Schools Act and continued in NCLB. The purpose of the grants is to promote elements of character such as caring, civic virtues and citizenship, justice and fairness, respect, and responsibility. These are all social and emotional learning competencies, although SEL also addresses drug use, violence, social relationships, service learning, and other areas.

In this program character education is required to be integrated into the classroom instruction and to be consistent with state academic content standards such as for health or civics education. Student, parent, and community involvement is expected in the design, implementation, and evaluation of each local program. State and local grant recipients must conduct a comprehensive evaluation of the program, including its impact on students, teachers, administrators, parents, and others. Effects on parent involvement, discipline issues, and academic achievement are factors that may be considered in evaluating programs according to NCLB. Character education may also be a component of drug and violence prevention programs. From 1994 to 2002 projects operated in 45 states. Under NCLB annual funding has increased from about $8 million to over $24 million annually.

A number of character education programs have been developed across the country, and a few appear to have strong evidence of effectiveness (CASEL, 2002). One such program that combines school and home learning is the Child Development Project. It seeks to create supportive social relationships and a commitment to pro-social values such as caring and showing concern for others. This project combines cooperative learning, literature with ethical implications, student-centered classroom organization and management, schoolwide programs to build connections among students, and parents doing home activities with students that are linked to classroom activities. This comprehensive program shows positive results on indicators of student social skills and problem behaviors (Solomon, Battistich, Watson, Schaps, & Lewis, 2000).

AFTER-SCHOOL PROGRAMS

Federal funding of an after-school program called 21st Century Community Learning Centers was begun in 1997. It has grown rapidly to over $990 million in fiscal years 2003 and 2004. The President's budget request for only $800 million in 2004 left many projects uncertain of their future, although eventually the previous funding level was retained. Local projects were originally funded through schools, although under NCLB community-based organizations can also run centers. They are designed to provide academic enrichment, including tutoring for students in schools with many low-income families, to help students meet state and local academic standards in core subjects. Additional purposes of the Centers reflect SEL program activities. These include youth development activities, drug and violence prevention programs, character education, counseling, and other activities designed to augment the regular academic program of students. Families of students served are to be offered programs for family literacy and parent involvement, but no more specifics are presented in NCLB. About 6,800 rural and inner-city public schools and other organizations in over 1,400 communities now participate in the Centers program. Principles of effectiveness similar to those for the Safe and Drug-Free Schools program are enunciated here. As with other programs, state education agencies are charged to evaluate the effectiveness of their programs and disseminate their findings.

First-year findings from a national evaluation show that 21st Century after-school centers changed how students spent time after school and increased parent involvement. Middle school parents were more involved at their children's school as volunteers, and attending open houses and parent–teacher organization meetings. Parents of elementary school children were more likely to help their children with homework and ask about their classes. However, the programs had limited impact on academics, did not increase students' sense of safety after school, and had negligible effects on ability to plan, goal-setting, and teamwork (Dynarski et al., 2003). Critics note the preliminary nature of the findings and point out that this program was not originally designed to focus on academic achievement.

THE PARTNERSHIP FOR FAMILY INVOLVEMENT IN EDUCATION

The U.S. Department of Education started a Partnership for Family Involvement in Education in 1994 that expanded to more than 7,000 family and education, community, business, and religious organizations. The Partnership aimed to increase opportunities for family involvement in their children's education at home and at school, and to further children's learning and achievement. It held forums and conferences, provided partnership-building tools, and used local best practices and research findings to mobilize partnership interest and activity (U.S. Department of Education, 2001a). The author was actively involved in the development of the Partnership and several related publications while he was a member of the Department.

An extensive white paper called *Strong Families, Strong Schools* (U.S. Department of Education, 1994) presented the rationale for the Partnership. It reviewed research and promising practices to show how families, schools, and other organizations can collaborate and contribute to children's learning. The Partnership developed activities such as kits for back-to-school time, materials for preparing teachers to work with families, suggestions for employers, and a report on improving the participation of fathers in children's learning.

The Partnership is no longer identifiable as such. The department now focuses on faith-based and community initiatives rather than working with the many organizations from its previous network. However, the department's Educational Partnerships and Family Involvement Unit continues to develop publications and information on involvement opportunities for parents. It has produced research-based materials for parents on early learning, helping children read and do homework, and succeeding in school. It also publishes booklets for parents on the academic, social, and emotional development of children from preschool to adolescence, responsible citizenship, and keeping children drug-free. This Unit also disseminates many Department publications through faith-based and community organizations.

PARENT INFORMATION AND RESOURCE CENTERS

In 1994 the Goals 2000: Educate America Act (Public Law 103-227, 1994) established a program of state-based Parent Information and Resource Centers. The Centers were reauthorized in the NCLB Act of 2001. Their broad purpose is to increase parents' knowledge of child-rearing activities, to strengthen partnerships between parents and education professionals, and to promote the development and academic achievement of assisted children. Training and support is given to parents of children from birth through high school and to individuals and organizations that work with them. Projects must have parent input on their governing boards. Annual reports from grantees must address the effectiveness of centers in improving home–school communication, student academic achievement, and parent involvement in planning, review, and improvements. At least half of each project's funds must go to areas with many low-income families. Thirty percent or more of project funds must also be used to operate the Home Instruction Program for Preschool Youngsters (HIPPY), the Parents as Teachers program, or other early childhood parent education program.

For parents of 0–3-year-olds, the Parents as Teachers (PAT) program provides regular home visits by certified parent educators, group meetings, developmental screenings, and links with other community services. For 4–5-year-olds, HIPPY provides resource materials and other activities on a regular personal basis to help parents improve learning at home. Studies of both programs have shown positive changes in parent and child behaviors (Moles, 2001; see also a discussion of state and local initiatives in Chapter 8).

In 2004 more than 70 Parent Information and Resource Centers operated in almost all states, with a $42 million appropriation. (The President had requested no funds in 2004 for this program, leading to much concern as to whether projects would be continued.) Since the centers began in 1995 at least 750,000 families and 60,000 school personnel have received direct services. In order to increase the flow of information, the centers use strategies from disseminating flyers and newsletters to presentations in care centers, Head Start, and local school classes. The most frequently offered services to increase parents' knowledge of educational issues are (in order) brochures and pamphlets, parent and school personnel workshops, referral services, Web sites, audio/videotapes, and parent support groups (McFarland, 2002). Centers are now required to provide information to parents on key features of Title I of NCLB. No systematic study of center effects has been performed, although much information on individual program activities has been provided by the centers. A discussion of the activities of the Illinois center can be found in the next chapter.

ENGLISH LANGUAGE ACQUISITION

Title III of NCLB authorizes local programs that serve children with limited English proficiency (LEP). These programs are administered by a new Office of English Language Acquisition that replaces the longstanding Office of Bilingual Education and Minority Language Affairs. While the origin of this program is the 1968 Bilingual Education Act, the present NCLB emphasis is on development of the English language skills and academic achievement of LEP children. For fiscal year 2004, $681 million were appropriated for this Office.

Under the law, parents must be consulted in developing local plans, and local programs must institute culturally and linguistically appropriate family education programs. They must also promote parent and community participation in local programs for limited English proficient children. Funds may be used for parent outreach and training activities to help parents become more actively involved in their children's education. School districts must inform parents about the program and why their children were selected, and obtain their consent before placing children in the program. Parents may decline to enroll their children in the program, and if the program fails to help the child make progress parents are to be informed immediately. As with other NCLB programs, all communication with parents must be in an understandable format and, if practicable, in the parent's own language. There were about 700 local projects in 2002. The NCLB requires local evaluations done biennially as well as national evaluations of practical use to educators and parents. No national data exist on the parent involvement aspects of these bilingual education programs, and the role of parents in design and conduct of these evaluations is not mentioned.

SPECIAL EDUCATION

Special education is a form of instruction designed for the unique needs of children with disabilities. Special education can be conducted in the classroom, the home, an institution, or other settings, but a majority of services are in regular classroom settings. During the 1999–2000 school year over six million children and young people 3 to 21 years of age received special education and related services in the United States (U.S. Department of Education, 2001b). In fiscal year 2004 this program received $11.2 billion.

The number of children with disabilities receiving special education and related services has grown steadily since the Education for All Handicapped Children Act was passed in 1975. According to its successor, the Individuals with Disabilities Education Act (IDEA) of 1990, for special education services to be provided the disability must affect the child's educational performance. The most common types of disabilities among students age 6 to 21 are specific learning disabilities (51%), speech and language impairments (19%), mental retardation (11%), and emotional disturbance (8%) (U.S. Department of Education, 2001b).

Many children with disabilities that are diagnosed in infancy are provided early intervention services. Once a child is school-age and parents think their child has a disability, they can ask the school for an evaluation. They can also contact a nearby Parent Training and Information Center to learn more about special education, their rights and responsibilities, and the law. In addition, Community Parent Resource Centers help meet the needs of racially and ethnically diverse communities.

For eligible children, an Individualized Education Program (IEP) is drawn up to set learning goals and outcomes and to identify the services to be provided for the child. Parents have the right to be involved in developing their child's IEP and to consent to or refuse decisions about their child's initial placement. A recent national survey shows that about 90% of parents of special education students attend such conferences, although a third of them say that school staff primarily develop the IEP goals and they would like to be more involved in the process. These tend more often to be minorities and low-income families (U.S. Department of Education, 2001b).

The IDEA amendments of 1997 expanded opportunities for parents and school staff to work together at the state and local level. A new federal continuous monitoring process reviews state activities and solicits input from steering committees of parents, agencies, and other stakeholders in each state. The monitoring process has found that some states provide joint training that includes parents and educators. States must now also offer a mediation system to resolve conflicts with parents (U.S. Department of Education, 2001b). Pending reauthorization of the IDEA would give parents more information and options such as new flexibility in IEP procedures and focusing parent training centers more on the needs of low-income, minority, and limited English proficient parents (Committee on Education and the Workforce, 2004).

SUMMARY

The legislation for some large programs of the U.S. Department of Education describes in considerable detail the supporting role mandated for parents. This often takes the form of involving them in the planning, implementation, and review of programs. Family involvement also appears in provisions for outreach and training to help families strengthen the home learning environment, and provisions to coordinate their efforts better with schools and other child development programs. Such provisions need to be extended to all Department programs so that the views parents bring are considered respectfully and the resources they bring fully utilized.

The school–family partnership theme is most fully developed in provisions of the Title I program to improve learning in high-poverty and low-achieving schools. In Title I parents are involved in a variety of ways. Both parents and educators are to be trained for working collaboratively, and a number of strategies for outreach to parents are identified. The Special Education and English Language Acquisition programs also make specific provisions for establishing true partnerships. Provisions in other programs are more general. Fortunately, whenever parental involvement is mentioned, as with 21st Century Community Learning Centers, the partnership-oriented definition in NCLB can be invoked.

Social and emotional learning is the focus of several Department programs. A range of SEL behaviors are addressed, including avoidance of drugs and violence, respect and responsibility, caring for others, and habits of citizenship. Except for the 21st Century after-school program, Department programs seem to emphasize either academic achievement or social and emotional learning rather than all these kinds of learning. The success of the Safe and Drug-Free Schools program is unclear, although some character education programs appear to be effective.

Local projects are required to evaluate their activities on a regular basis in some programs reviewed, and most national programs have overall program statistics. However, in most current programs strong evaluations on their benefits are not available. Strong evaluations of program implementation and outcomes are essential to provide convincing evidence on which to base program improvements.

RECOMMENDATIONS

This review suggests a number of recommendations for action. Some bear on the roles of federal and state policymakers, some on what schools and school districts can do, and others on the actions parents can take in the education of children. The main recommendations are described in Figure 7.1. As background, stable federal funding and presidential leadership are prerequisites to the vitality of all these programs.

While Title I includes many specific provisions for parent involvement, the Department has emphasized the new NCLB provisions for parent choice. More written guidance, technical assistance, and oversight from the Department and the states are needed for all the parent involvement requirements of Title I and the other parts of NCLB to assure that these requirements are fully met. Truly shared power and responsibilities are needed in all programs to build commitment and knowledge among parents to work with schools as real partners. Finally, the Department should conduct rigorous large-scale evaluations of all its programs, for only in this way will their real value by known.

Schools and school districts have responsibilities under some NCLB programs to develop, implement, and review policies with parent participation. These requirements should be carried out fully to gain the benefits of school–family partnerships, including their translation into policy guidance and assistance for administrators and teachers on required and permitted activities with parents. Regular communication with parents is important to advise them on children's progress, school programs, and ways of working together. Personal outreach and a welcoming school environment will do much to create the mutual respect needed for effective partnerships.

Despite the fact that federal education initiatives are devised and promulgated from afar, parents can still play an important role in how these programs are enacted in the community. For example, Henderson (2003) had offered a guide for parents on the requirements of NCLB. Building on

FIGURE 7.1. Recommendations for Action Regarding Parents

Federal and state policymakers	1. Provide written guidance on all parent involvement aspects of NCLB, and consider amending the law to encourage more sharing of power and responsibility with parents. 2. Furnish technical assistance and oversight to see that the law is well-implemented. 3. Request reports on program compliance and evaluations of performance from states, school districts, and schools. Make these available to the public. 4. Conduct rigorous large-scale evaluations on the effectiveness of programs.
Schools and school districts	1. Develop, implement, and review policies for parent involvement with strong parent participation. Evaluate programs on a regular basis. 2. Provide ongoing guidance and assistance to teachers and parents on required and permitted activities with parents. 3. Communicate regularly with parents about their children's progress, the school's programs, and parent involvement activities. 4. Reach out and welcome parents as partners in their children's education.
Parents	1. Obtain a copy of the school or district policy on parent involvement. Check that it covers all points in the law and is carried out in practice. 2. Study the annual school and district report card, and ask how schools plan to address any problem areas. 3. Take part in available activities such as parent–teacher conferences, parent training, options in failing schools, and developing and implementing plans for parents, including school–parent compacts. 4. Talk regularly with teachers about children's needs and progress and how parents and schools working together can help them learn.

Henderson's recommendations, by comparing school and district policies on parent involvement to state and federal standards, parent groups can determine whether the whole law is covered by the policies. How well local policies are being enforced is a further point to check. In addition, when the annual school and district report card arrives, parents have an opportunity to ask schools how they plan to address problem areas. Parents can also take part in available activities such as parent–teacher conferences, compacts, and options in failing schools. When parents get involved in such activities, talk regularly with teachers, and help develop school plans for parent involvement, they are helping to build strong school–family partnerships and support the academic, social, and emotional learning of all children.

AUTHOR'S NOTE

For more information on these programs of the U.S. Department of Education, readers may consult the Department Web site at www.ed.gov or call 1-800-USA-LEARN.

REFERENCES

Borman, G., D'Agostino, J., Wong, K., & Hedges, L. (1998). The longitudinal achievement of Chapter 1 students: Preliminary evidence from the Prospects study. *Journal of Education for Students Placed at Risk, 3,* 363–399.

CASEL, the Collaborative for Academic, Social, and Emotional Learning. (2002). *Safe and sound: An educational leader's guide to evidence-based social and emotional learning programs.* Chicago: University of Illinois at Chicago, Department of Psychology.

Committee on Education and the Workforce. (2004). *Strengthening and renewing special education: Bill summary.* Updated November 17, 2004. Washington D.C.: U.S. House of Representatives.

Crosse, S., Burr, M., Cantor, D., Hagen, C., & Hantman, I. (2002). *Wide scope, questionable quality: Drug and violence prevention efforts in American schools.* Rockville, MD: Westat.

D'Agostino, J., Hedges, L., Wong, K., & Borman, G. (2001). Title I parent-involvement programs: Effects on parenting practices and student achievement. In G. Borman, S. Stringfield, & R. Slavin (Eds.), *Title I: Compensatory education at the crossroads* (pp. 117–136). Mahwah, NJ: Erlbaum.

Dynarski, M., Moore, M., Mullens, J., Gleason, P., James-Burdumy, S., Rosenberg, L., et al. (2003). *When schools stay open late: The national evaluation of the 21st-century community learning centers program.* Washington, DC: U.S. Department of Education, Office of the Under Secretary.

Funkhouser, J., Stief, E., & Allen, S. (1998). *Title I school-parent compacts: Supporting partnerships to improve learning.* Washington, DC: Policy Studies Associates, Inc.

Henderson, A. (2003). *No child left behind: What's in it for parents.* Fairfax, VA: Parent Leadership Associates.

McFarland & Associates, Inc. (2002). *Parental Information and Resource Centers' annual report.* Silver Spring, MD: Author.

Moles, O. (2001). Parent involvement in federal education programs. In D. Hiatt-Michael (Ed.), *Promising practices for family involvement in schools* (pp. 25–38). Greenwich, CT: Information Age Publishing.

Public Law 103-227. (1994). Goals 2000: Educate America Act. Washington, DC: Government Printing Office.

Public Law 107-110. (2002). No Child Left Behind Act of 2001. Washington, DC: Government Printing Office.

Silvia, E., & Thorne, J. (1997). *School-based drug prevention programs: A longitudinal study in selected school districts. Final report.* Research Triangle Park, NC: Research Triangle Institute.

Solomon, D., Battistich, V., Watson, M., Schaps, E., & Lewis, C. (2000). A six-district study of educational change: Direct and mediated effects of the child development project. *Social Psychology of Education, 4*(1), 3–51.

U.S. Department of Education. (1994). *Strong families, strong schools: Building community partnerships for learning.* Washington, DC: Author.

U.S. Department of Education. (1997a). *Family involvement in children's education: Successful local approaches.* Washington, DC: U.S. Department of Education, Office of Educational Research and Improvement.

U.S. Department of Education. (1997b). *Overcoming barriers to family involvement in Title I schools: Report to Congress.* Washington, DC: U.S. Department of Education, Office of Educational Research and Improvement.

U.S. Department of Education. (2001a). *The Partnership for Family Involvement in Education: Who we are and what we do.* Washington, DC: Author.

U.S. Department of Education. (2001b). *To assure the free appropriate public education of all children with disabilities: Twenty-third annual report to Congress on the implementation of the Individuals with Disabilities Education Act.* Washington, DC: Author.

Westat and Policy Studies Associates. (2001). *The longitudinal evaluation of school change and performance (LESCP) in Title I schools. Final report. Volume I: Executive summary.* Washington, DC: U.S. Department of Education, Office of the Deputy Secretary.

8

Grass Roots from the Top Down: The State's Role in Family–School Relationships

Sam Redding & Pamela Sheley

The previous chapter describes the federal government's interests in the role of parents and families in children's education. This chapter will examine state legislative and policy actions that have defined the relationship between parents and their children's schools.

Public education is primarily the responsibility of the states, and historically state actions regarding public education have impacted families by: 1) expanding the role of schooling in children's upbringing through such actions as compulsory attendance statutes and the extension of public education to include high school, kindergarten, and prekindergarten, and 2) targeting state resources to specific localities or categories of students in an attempt to redress inequities in family and community resources. Thus, states have carved out a larger portion of child-rearing as the province of the school rather than the family, and they have targeted resources to compensate for imbalances in family and community support for children's learning.

A policy search conducted by the Education Commission of the States (ECS) for this chapter reviewed legislation proposed in the 50 states over a 7-year period (1996–2003) that touched upon the role of parents. Table 8.1 demonstrates the types of legislation proposed and the success of the bills.

 ISBN 0-8077-4600-2 (paper), ISBN 0-8077-4601-0 (cloth).

TABLE 8.1. State Legislation Relevant to Families and Schools (July 1, 1996, through December 31, 2003)

Purpose of Proposed Legislation	*Number of Bills*	*Percentage of Total Bills*	*Number Passed*	*Passed (%)*
RIGHTS AND RESPONSIBILITIES				
Special Education	6	6.0	5	83.3
Time off from work for parents	8	7.9	3	37.5
Parents' rights	20	19.8	12	60.0
Schools' legal responsibilities relative to criminal and negligent behavior of students and parents	19	18.8	14	73.7
PREPARING CHILDREN FOR SCHOOL SUCCESS				
Parents as first teachers	3	2.9	3	100.0
Family Literacy	4	3.9	4	100.0
Support for parental engagement in children's education	23	22.8	15	65.2
CONNECTING FAMILIES WITH COMMUNITY RESOURCES				
Family Resource Centers	5	5.0	3	60.0
OTHER POLICIES [a]	13	12.9	8	61.5
TOTAL	101	100	67	66.3

[a] *Other policies:* The remaining legislative proposals were varied and did not fall into easily ascertainable patterns. They ranged from Tennessee's policy for an oversight committee to study issues relative to relationships among administrators, teachers, parents, and students and admission and retention of students in public schools, to Wisconsin's legislation defining "parent."

PARENTS IN SCHOOL GOVERNANCE

As school districts have grown ever-larger, encompassing more schools and wider swaths of population, the distance between a parent and that parent's civic representative—the elected school board member—has likewise grown. The American ideal of local citizens watching closely over their children's schools has been dramatically diluted by the expansion in size and reduction in number of school districts. As a consequence, some states and districts have empowered school-based councils to restore the place of parents and citizens in the decision-making processes of their children's schools. School-based councils can be elected or appointed, and some have significant legal stature and advisory roles.

Site-Based Decision Making

Illinois was an early leader in transferring control in a big-city school district from the central office to the school site. The 1988 Chicago school reform law, enacted by the state legislature, required every Chicago public school to elect a Local School Council (LSC). The majority of members of the LSC in each school were to be parents with children in school. Among other tasks, LSCs were to select the principal and develop a yearly school improvement plan. A 1998 study of the LSCs found that schools making substantial progress on reading assessments tended to have strong councils, while schools with level or declining reading achievement tended to have weak councils. The study concluded that cooperative adult efforts among the teachers, parents, community members, and administrators involved in the school was a force for improving student learning outcomes (Moore, 1998).

Charter Schools

Charter schools have been a favored vehicle for reengaging parents in the polity of their local schools. Charter schools typically operate with school-based governance, and parents choose to send their children to them. The U.S. Department of Education reports that the charter school movement took flight with state legislative action; in 1991, Minnesota passed the first charter school law. The U.S. Department of Education has provided grants to support the states' charter school efforts since 1994. According to the Center for Education Reform (2003), 2,687 charter schools were operating in 36 states and the District of Columbia as of fall 2002, and the number increased to 2,996 by the beginning of the 2003 school year, a one-year growth rate of 10%. These schools served more than 700,000 students.

School Choice

School choice places parents at the fulcrum of power in determining the best educational placement for their children. Without the necessity of prescriptive, regulatory legislation, school choice simply allows the marketplace of competing educational approaches to find best matches with parents' expectations for their children's well-being. While voucher-based school choice programs have been limited to a few districts, a June 27, 2002, decision by the U.S. Supreme Court upholding the Cleveland voucher program has encouraged proponents to aggressively seek legislation, state by state, to proliferate school choice (Gehring, 2002). Bills introducing school choice legislation will likely constitute a growing proportion of education-related actions by state legislatures, and states will look more closely at the experiences of Cleveland, Milwaukee, and other districts with choice programs.

PARENTAL RIGHTS AND RESPONSIBILITIES

Parents' rights relative to school districts have been the source of abundant legislative action in recent decades, especially in the area of special education.

Special Education

Since the Education for All Handicapped Children Act of 1975 was passed by Congress (reauthorized as the Individuals with Disabilities Act in 1997 and again in 2004), litigation and legislation have cascaded out through the states to define and enact the provisions of the statute. The language of the original act, commonly known as Public Law 94-142, specifically includes parents and guardians in asserting a set of rights and responsibilities relative to disabled children: "It is the purpose of this Act to assure that all handicapped children have available to them . . . a free appropriate public education which emphasizes special education and related services designed to meet their unique needs, to assure that the rights of handicapped children and their parents or guardians are protected, to assist States and localities to provide for the education of all handicapped children, and to assess and assure the effectiveness of efforts to educate handicapped children" (Fraser, 2001, p. 310). As the processes of special education—diagnosis, staffing, placement, delivery of services, documentation—have become mechanically procedural to meet the requirements of the law, state legislatures have continually refined legislation to account for new discoveries of inequity. Always, the role of the parent is central in these deliberations, as the parent has typically been the greatest advocate for the child with disabilities, and states have granted parents considerable leverage in negotiating appropriate services for their children.

The new state legislation proposed during the seven-year period (1996–2003) dealing exclusively with parents of special education students included the need to incorporate the parents of children who receive special education services into Parent Advisory Councils; the rights of surrogate parents and foster parents; and requirements for reporting of information by the school district to parents.

Time Off from Work for Parents

From 1996 through 2003, seven states proposed legislation (approximately 8% of total bills dealing with parents) that would encourage employers to provide time off, without pay, for parents to participate in school programs, activities, or parent–teacher conferences. Of these seven states, only three passed the bills that were proposed. The one exception to time off without pay was Hawaii, which requires employers to provide time off with pay.

Other Parental Rights and Responsibilities

Other state legislative actions regarding parental rights and responsibilities ranged from providing parents with an effective means for contacting schools regarding the location or safety of their child (Virginia), to several states' concern with issuing questionnaires or surveys to students without parental consent or notification. This category also included a parent's right to review material concerning sex education or other curricula and instructional materials.

Superintendents in Delaware can now issue subpoenas to compel the attendance of parents, guardians, or custodians and students at school parent conferences. Texas required parents' participation in parenting courses and counseling in certain suits involving children. Georgia allowed juvenile courts to fine parents for failing to attend parent–teacher conferences. Courts now have the authority to order parents to participate in programs or treatment that the court determines necessary to improve student behavior.

The Delaware law changed the school crime reporting statute to include violent felonies and sex crimes, and requires the parent or guardian to be notified if a pupil is the victim of criminal conduct, or if a pupil is found to possess a controlled substance or weapon on school property. North Carolina increased the criminal penalty for making a bomb threat; made parents civilly liable for children who do so; required suspension for 365 days; and will further study the issue of students who make or carry out threats of violence.

STATE INITIATIVES: PREPARING CHILDREN FOR SCHOOL SUCCESS

Over the past two decades, cognitive psychology and the burgeoning evidence from "brain science" have revealed the physiological and psychological bases for what educators have long assumed: a child's ability to learn in school is determined, in part, by the conditions of life outside the school, especially during the critical years from birth until schooling begins.

Birth to Five

One response to this evidence has been to extend schooling downward, reaching younger children. Another response has been to recognize the importance of parents, and to provide programs and support to improve the quality of home environments. Maine established a task force to study strategies to support parents as a child's first teacher, and in New Hampshire the policy created a "Parents as Teachers" pilot program. The research evidence is instructive here. Early childhood programs that train parents to work with their children at home have proven effective in preparing the children for school (Baker, Piotrkowski, & Brooks-Gunn, 1998; Mathematica, 2001; Starkey & Klein, 2000).

Missouri. In 1984, Missouri became the first state to make parent education and family support mandatory in every school district. This came 10 years before the 1994 Educate America Act that stipulated, "all children in America will start school ready to learn" (U.S. Department of Education, 1994). This mandate stemmed from early research in the 1960s stressing that the first three years of a child's life are critical in development and that parental involvement is the key to a child's success.

From research presented at a conference held in 1981 by the Department of Education, the idea for the Parents as Teachers (PAT) program was developed. The pilot program of PAT had seven specific goals: increased knowledge of child development, improved confidence in child-rearing, better social development, better cognitive and language development, fewer undetected hearing and vision or handicapping problems, positive feelings for the program, and positive feelings toward the school district.

Research and Training Associates of Overland Park, Kansas, evaluated the pilot program. They found that the children were "significantly more advanced than comparison group children in language development"

(PAT National Center, 1999), had better problem-solving skills and coping skills, and positive relationships with adults. These results led the Commissioner of Education to designate PAT as the model for parent education. The support of PAT goes beyond the school doors, however. Health care providers, mental health services, churches, and businesses all saw the benefits from such a program, and members of all areas of community sit on the board for PAT.

Since its inception, PAT has been evaluated and studied many times over. A longitudinal follow-up revealed that the higher achievement of PAT children at age three was maintained when they became first graders. Their parents were found to demonstrate high levels of school involvement, which they frequently initiated, and to support their children's learning in the home (Pfannenstiel, Lambson, & Yarnell, 1996).

PAT's success can also be measured in the way it has spread from its small beginnings in Missouri to being offered in school districts all over the state and in over 2,000 other sites throughout the nation and even worldwide. What has made this program so successful? The heart of PAT's message is that the home, or specifically, the parent, is the child's first learning environment and most influential teacher. They promote this concept by home visits by PAT-certified parent educators who model activities for parents and involve them in interacting with their child in ways that promote learning. They encourage parents to read to their child. The relationship the parent forms with the parent educator is the strongest indicator of the success of the program for the child. When the child reaches age three or four, PAT parents are more likely to enroll their child in preschool. In kindergarten readiness testing, PAT children scored significantly higher than their non-PAT counterparts and continued to score higher on standardized tests of reading, math, and language through the third grade.

In 1999, the National Center for PAT began its *Born to Learn* curriculum, based on studies in neuroscience from Washington University School of Medicine. PAT is continually researching new developments in the study of brain development and early education and implementing these ideas into their practices. While for many it may just "make sense" that the parents will interact and teach their child from the time he or she is born, these skills are not universal. For a child to come into school ready to learn and able to learn, the groundwork must be done well in advance. PAT developed a program that has stood the test of science, evaluation, and time to prepare the soil of the child's fertile mind to be planted and nurtured by parents and later schools. It benefits the child, the parent, the school, and the community.

School Age

The parent's role in supporting children's school learning is not confined to the preschool years, of course. Research that has articulated the "curriculum of the home" outlines factors of the home that correlate with success in school. State legislatures have recently begun to turn attention to this research and to include provisions for its application as part of efforts to improve student learning.

Beginning with the alarm call of *A Nation at Risk* (National Commission on Excellence in Education, 1983), the states have moved aggressively to establish learning standards, implement assessments, and focus sharply on student learning. With support from the U.S. Department of Education through comprehensive school reform (CSR), schools have attempted dramatic alterations in the way they do business, with state assessment scores as the benchmarks of success. Most recently (2001), the federal reauthorization of the Elementary and Secondary Education Act, titled No Child Left Behind, requires states to channel federal dollars to schools not making adequate yearly progress toward state assessment goals. With all of this attention on student learning, one might expect keen interest in family influences on school success. In fact, while research has clearly connected family behaviors with student learning, the evidence of successful school intervention to alter family behaviors in constructive ways has been lacking. Also, school reform efforts have directed their resources at the variables schools can most readily control—teacher quality, instructional practices, curriculum, and school organization.

Federal Title I funding carries with it the requirement that districts target 1% of their dollars for parental involvement. Especially in large districts, this can be a considerable pool of money. But the programs and practices spawned by this monetary set-aside have not distinguished themselves with evidence of effect. Once again, research that demonstrates the family's influence on children's learning has not been matched with the procedural knowledge necessary for schools to effectively alter family behaviors. To say that children do better in school if their parents are involved with their learning at home and at school is one thing; to increase the level of involvement by parents is quite another.

Belatedly, but increasingly, legislative attention has been directed toward the role of parents in children's learning, accompanying the movement toward standards, assessments, and accountability for school performance. Legislation that promotes the positive aspects of parental involvement included South Carolina's Parental Involvement in their Children's Education Act, which established a framework for increased parental involvement in the education of children, for parental involvement training for educators and school staff, for parental responsibilities for children's academic success, and for efforts to increase parent–teacher contacts. New Mexico proposed a Legislative Parental Involvement Act that would promote school and community partnerships to increase parental involvement in public schools; to promote social, emotional, and academic growth of children; to foster relationships and improve communication between home, school, parent, student, and community; and to provide greater accountability among parents, school personnel, and students. This proposed legislation, while providing the correct rhetoric and ideas, failed. A successful bill in Indiana allows schools to include in their school improvement plans strategies for encouraging parental involvement. The law requires the state to make available to schools models of parental involvement for use in developing school improvement plans. The Indiana law also requires a parent to enter into a written compact with the child's school, setting forth expectations for the child, the parent, the teachers, and the school. Louisiana recently passed legislation that establishes a parental involvement demonstration program to be implemented in local schools.

Four percent of the total of the proposed bills reviewed in the ECS policy search reflect the importance of family literacy. Virginia required parental involvement components in their at-risk programs for four-year-olds, including activities to promote family literacy. Maryland provided grants for family literacy as part of legislation addressing adult education. Colorado established grants for family literacy within its Department of Education; local education providers may apply for the grants to promote family literacy education, adult literacy education, or English language literacy education.

Illinois. In 2000, the Illinois state legislature established a line item in the state budget for parental involvement. The item was included in the budget of the Illinois State Board of Education (ISBE), which is the state's department of education. ISBE decided to collaborate with the Academic Development Institute (ADI), a nonprofit organization based in Illinois and the state's Parent Information and Resource Center (PIRC), as designated by the U.S. Department of Education.

ISBE and ADI agreed upon a plan to focus primarily on the state's high-priority schools, defined as those with low test scores, typically serving high-poverty communities. ADI would establish eight regional centers throughout the state, and each would provide training, materials, and technical assistance to elementary schools in their region. The grant to ADI also supported research and development projects to produce effective parent engagement strategies and materials for middle school, high school, and parents of children in special education. But the heart of the ISBE/ADI initiative, dubbed Solid Foundation, was focused on grades K–5, and sought to build the capacity of low-performing schools to engage parents in ways that would enhance children's reading habits, study habits, and responsible behavior.

A chief aim of Solid Foundation is to build within each school the structures and capacities that will enable it to continue and to expand upon the initiatives established during a two-year period in which close support is provided by ADI's regional staff. Secondly, in each region ADI is forging a network of the participating schools, and that network will continue beyond the 2-year program implementation phase. Key components of the Solid Foundation program are a site-based support team that includes parents, home visits, parent education, and reading school–home links that are aligned with state reading standards.

In the 167 schools that began implementation of Solid Foundation in 2001, 64% of the students received free or reduced-price lunch, 45% of the students were African-American, and 12% of the students were Hispanic. In the first 2 years of implementation, Solid Foundation proved that a critical mass of constructive school–home activity can be generated in a relatively short period of time in schools most in need of programming to engage parents in children's learning.

Most encouraging in the implementation of Solid Foundation was that participating schools achieved an improvement in the percentage of students who met or exceeded state expectations on the Illinois Standards Assessment Test (ISAT) of 4.5% over the 2-year period, compared with a gain of only .1% for the state as a whole and 2.5% for a control group with identical beginning scores (Redding, Langdon, Meyer, & Sheley, 2004).

In 2003, the Illinois State Board of Education revamped its system of support to schools making inadequate yearly progress under No Child Left Behind, replacing its statewide program with a regional system of support providers. Solid Foundation is now demonstrated in high-implementing schools in each region, and a Solid Foundation Academy in each region trains school-based teams of administrators, teachers, and parents to implement its components.

Connecting Families with Community Resources

Family resource centers accounted for 5% of the total legislation proposed in the ECS study of state legislative actions relative to parents and schools. The legislation proposed in Tennessee would establish family resource centers in schools or other community centers, and would provide such services as after-school care for school-age children. Connecticut legislation mandated the location of centers, using facilities that are cost-effective and easily accessible.

Kentucky. In 1990 the Kentucky Education Reform Act (KERA) was passed in response to the state's Supreme Court ruling that declared the entire state's system of education unconstitutional. The Family Resource and Youth Service Centers are a part of the effort to reform public elementary and secondary education. The mission of the centers is to "enhance students' abilities to succeed in school by developing and sustaining partnerships that promote: early learning and successful transition into school, academic achievement and well-being and graduation and transition into adult life" (Kentucky Office of Family Resources and Youth Services Centers, 2002). This broad mission statement gives the centers the freedom to interpret the needs of each community and to enact programs that best fit the people they serve. Centers are school-based, with the requirement of a school having 20% of their students receiving free or reduced-price lunches. Nine out of 10 schools qualify. Currently, 83% of all public schools have a Family Resource Center and/or a Youth Service Center. The Family Resource Centers serve elementary schools with child care, parent training, parent and child education services, or other programs that best support families with children up to age 12. Youth Centers serve secondary schools with employment, drug/alcohol, and mental health counseling and other services more suited for the older child.

Each center is funded by state education dollars and administered by the Cabinet for Human Resources. Advisory councils serve as local governance. The advisory council plays a central role in designing and delivering services. The council's members must be one-third parents and no more than one-third school district personnel. In 1999, the Kentucky Department of Education implemented consolidated planning, which required all schools to develop one coordinated plan for all its state-funded and federally funded programs. The policy included the Family Resource and Youth Services Centers so that their services would be in alignment with other school programs that are designed to improve student performances.

The legislation that defined the center's mission was broad-based: "The centers shall provide services which will enhance students' abilities to succeed in school" (Denton, 2001, p. 3). What Kentucky began to realize was the importance of meeting children's needs before they ever reached the classroom. If a child enters the school without adequate housing, without a stable or safe home environment, that child is not going to become a willing and able student. The needs of children are more widely defined than being able to read. If they do not come into the school in the frame of mind and with the proper emotional and social skills, that child will not perform well in school. With the implementation of the Family Resource Centers, the teachers began to see improvements in the social and learning behavior of their students.

The Youth Centers also serve a growing need in the state. Teachers tell of students who are "tormented by issues like abuse, STDs, anger, suicide, pregnancy, hunger, alcohol, gangs, drugs, violence, hyperactivity, families in turmoil—issues that really get in the way of education" (North Central Regional Educational Laboratory, 1994, p. 1). Teachers cannot teach if they are also being social workers. Their time is limited and they have many students. The Youth Centers keep the students' problems from being ignored or overlooked. They have a place outside the classroom to go to get the help they need. The Youth Centers are also bringing their messages into the classroom by promoting curriculum that deals with violence, conflict resolution, and stress management. One principal stated, "We are still in the very dawn of these services, but this is the first time in my career of 26 years that we have been able to speedily and efficiently address the social, emotional, and psychological needs of youngsters" (North Central Regional Educational Laboratory, 1994, p. 5).

Each center has a core of programs they must offer. However, they are also not limited to these core programs. While they are based in the schools, they are not limited to serving only the students being identified as low-income. The centers were set up to meet the needs of all students to improve their academic achievement. The very diversity that enables each of the centers to function in their communities also makes it harder to evaluate their services in a comprehensive manner. R.E.A.C.H. of Louisville, Inc., conducted an evaluation in 1999. In a study of 20 centers it was found that the centers "effectively helped families and students deal with nonacademic problems that placed them 'at risk for negative outcomes' in school" (Denton, 2001, p. 9). The Kentucky Office of Evaluation Accountability stated in its 2000 annual report, "Evidence is mounting that these [the centers'] services are making a difference" (Denton, 2001 p. 10). Policymakers are finding their constituents pleased with the contributions the centers are making to their communities. While it may not be the magic bullet, taking care of the emotional, social, and physical needs of families and children frees children from the worries of daily life and allows them to concentrate on the academics of learning.

SUMMARY AND RECOMMENDATIONS

The evidence has mounted: family behaviors at home and parental involvement with their children's schooling are strongly associated with school learning (Henderson & Mapp, 2002). Children do better in school if their parents are engaged with their learning at home and are in communication with the school. How, then, does the school influence parental behaviors? How would state and district policies and programs establish the conditions by which schools could help parents provide the support requisite to their children's school success? (See Figure 8.1 for recommendations for state policies.)

For school-age children, comprehensive efforts to engage parents at various points and in different ways seem most productive (Gordon et al., 1979; Swap, 1993). A 2002 study by Westat and Policy Studies Associates for the U.S. Department of Education considered student achievement in 71 Title I elementary schools. The study looked at the connection between the school's practices of outreach to parents and improvement in reading and math by low-performing students. Outreach was measured along three scales:

- Meeting face-to-face
- Sending materials on ways to help their child at home
- Telephoning both routinely and when their child was having problems

Outreach to parents was more strongly associated with student learning gains than any other variable considered in the study.

FIGURE 8.1. Recommendations for State Policies

1. Comprehensive, school-based processes that outreach to parents, train teachers in school–home relations, connect parent engagement to children's learning, and provide consistent contact with parents over time.
2. Line items in school budgets for parent engagement connected with children's learning.
3. Training and certification for school personnel who work with parents.
4. Parents on site-based bodies that oversee policy and practice, especially as they relate to areas where the responsibilities of the school and the home overlap.
5. School accountability for effective parent engagement.
6. Rigorous evaluation of parent engagement programs and models and dissemination of the results.

Epstein has summed up the components of successful school–home partnerships, and the need for support for these endeavors from states and districts: "We . . . found that it was the teachers' practices, not the education, marital status, or work place of parents that made the difference in whether parents were productive partners with schools in their children's education" (Epstein, 1988, p. 58). Improvement of partnership practices, Epstein asserts, requires a school-based process to assess strengths and weaknesses, set a course of action, and monitor incremental goals over a period of several years. This will happen systemically only when parental involvement receives serious support from states and districts.

Because engaging parents is often viewed as peripheral to a school's mission of educating children, personnel and budgets assigned to parental engagement are scarce, and schools are not typically held accountable to the state for the effectiveness of their outreach to parents. In January 2003, the New York City Department of Education created 1,200 parent liaison positions and budgeted $43 million to put staff in place in schools to improve communication between teachers and parents (CNN.com, 2003). In another promising action, in August 2003 the Boston Public Schools created the position of Deputy Superintendent for Family and Community Engagement. These high-profile, district-level initiatives are hopeful signs that states will take parent engagement seriously, hold schools accountable for it, and see that schools have sufficient resources of staff and money to effectively engage parents around children's learning.

As states and districts provide opportunities for experimentation with various approaches to connecting families and schools, some approaches will no doubt emerge successful, and the right blend of several approaches will be discovered to provide the best formula. The research trail linking home behaviors and conditions to children's school learning is long and convincing. Proof of successful state, district, or school interventions to alter home behaviors and conditions in ways that enhance student learning is less apparent. Large-scale programs with adequate budgets for both implementation and evaluation, and enough sticking power to allow for corrections in course, will provide the best evidence to guide further program development.

In the national clamor to boost children's learning, most attention and certainly the lion's share of the resources are being devoted to professional development for teachers, assessment and accountability systems, and school restructuring. But when results are disappointing, parents become an easy target as the root of the problem. It seems fair, then, that states and districts channel reasonable resources to help parents provide for their children the support and guidance that success in school requires. Just as we are discovering what works instructionally for children in various settings and circumstances, we will learn what programs and practices are most efficacious for engaging parents.

REFERENCES

Baker, A. J. L., Piotrkowski, C. S., & Brooks-Gunn, J. (1998). The effects of the Home Instruction Program for Preschool Youngsters (HIPPY) on children's school performance at the end of the program and one year later. *Early Childhood Research Quarterly, 13*(4), 571–588.

Center for Education Reform. (2003). Washington, DC. http://www.edreform.com.

CNN.com. (August 30, 2003). New York schools hire parents. http://ww.cnn.com/2003/US/Northeast/08/30/sprj.sch.nyc/.

Denton, D. (2001). *Helping families to help students: Kentucky's Family Resource and Youth Services Centers.* [Electronic version]. Atlanta: Southern Regional Education Board.

Epstein, J. (1988). How do we improve programs for parent involvement? *Educational Horizons, 66,* 58–59.

Evans, R. (1996). *The human side of school change: Reform, resistance, and the real-life problems of innovation.* San Francisco: Jossey-Bass.

Fraser, J. W. (2001). *The school in the United States.* New York: McGraw-Hill.

Gehring, J. (2002, April 10). Voucher battles head to state capitals. *Education Week, 21*(42), 1, 24, 25.

Gordon, I. J., et al. (1979, July). How has Follow Through promoted parent involvement? *Young Children, 34*(5), 49–53.

Henderson, A., & Mapp. K. (2002). *A new wave of evidence: The impact of school, family, and community connections on student achievement.* Austin, TX: Southwest Educational Laboratory.

Kentucky Office of Family Resources and Youth Services Centers. (2002). http://cfc.state.ky.us/frysc.

Mathematica Policy Research, Inc., and Center for Children and Families at Teachers College, Columbia University. (2001). *Building their futures: How early Head Start programs are enhancing the lives of infants and toddlers in low-income families.* Washington, DC: U.S. Department of Health and Human Services.

Moore, D. R. (1998, April). *What makes these schools stand out: Chicago elementary schools with a seven-year trend of improved reading achievement.* Available from Designs for Change, 6 North Michigan Avenue, Chicago, IL 60602.

National Commission on Excellence in Education. (1983). *A nation at risk: The imperative for educational reform.* Washington, DC: U.S. Government Printing Office.

North Central Regional Educational Laboratory. (1994). *Kentucky's Family and Youth Services Centers break new ground.* http://www.ncrel.org/sdrs/cityschl/city1l_1e.html.

Parents as Teachers National Center, Inc. (1999). *Parents as Teachers history.* Program Administration Guide Curriculum. St. Louis, MO: Author.

Pfannenstiel, J., Lambson, T., & Yarnell, V. (1996). *The Parents Teachers Program: Longitudinal follow-up to the second wave study.* Overland Park, KS: Research & Training Associates.

Redding, S., Langdon, J., Meyer, J., & Sheley, P. (2004, April). *The effects of comprehensive parent engagement on student learning outcomes.* Paper presented at the annual meeting of the American Educational Research Association, San Diego, CA.

Starkey, P., & Klein, A. (2000). Fostering parental support for children's mathematical development: An intervention with Head Start families. *Early Education and Development, 11*(5), 659–680.

Swap, S. (1993). *Developing home–school partnerships: From concepts to practice.* New York: Teachers College Press, Columbia University.

U.S. Department of Education. (1994). *Goals 2000: Educate America.* http://www.ed.gov/legislation/GOALS2000/TheAct/sec102.html

Westat and Policy Studies Associates. (2002). *The longitudinal evaluation of school change and performance in Title I schools. Volume 1: Executive Summary.* Washington, DC: U.S. Department of Education.

9

Preparing Educators for School–Family Partnerships: Challenges and Opportunities

Nancy Feyl Chavkin

Earlier chapters extensively describe and document the benefits of school–family partnerships for both the academic and social–emotional learning of students. It would seem logical that colleges of education would be preparing educators for working with school–family partnerships. Universities have a tremendous potential to improve the academic achievement and social and emotional learning of all students by preparing future educators to work with school–family partnerships; however, very few universities are actively preparing educators to do so.

To understand the reasons why universities are not preparing educators for school–family partnerships, this chapter reviews surveys of higher education institutions, examines state certification policies and national standards, describes curriculum models, and explores promising programs about preparing educators for school–family partnerships. The chapter will consider both the challenges and the opportunities ahead and will close with key recommendations for the future. The issues about preparing educators for school–family partnerships are complex and cannot be attributed to a single cause and effect; it is imperative that educators examine these compelling issues.

SURVEYS OF HIGHER EDUCATION

Recent surveys of higher education institutions continue to reveal a dismal picture about educator preparation for school–family partnerships.

 ISBN 0-8077-4600-2 (paper), ISBN 0-8077-4601-0 (cloth).

Teacher education continues to take place in colleges and universities, apart from the real world of families and communities with whom teachers are being prepared to work. Surveys of educational institutions are consistent in their findings that although universities are offering some modules about school–family partnerships, a dearth of training exists for this important educational activity.

Shartrand and colleagues at Harvard (1997) surveyed 60 teacher education programs in 22 states that mentioned family involvement in their certification requirements. Only 37% of the respondents reported that a full course on family involvement was required in their curriculum; most (83%) reported that family involvement was taught as part of a course, and the majority (63%) identified the student teaching semester as the most popular time to teach students about family involvement.

When the teacher educators were asked about the type of content in their family education curriculum, they reported that the most frequently addressed topics included: parent–teacher conferences, parent teaching child at home, parent as class volunteer, and parent as school decision maker. The definition of family involvement, teaching methods, and the mode of delivery were reported to be very traditional; little attention was given to creative practices for involving families. The curriculum did not stress how to use the strengths of families for increasing social-emotional learning.

Epstein, Sanders, and Clark (1999) at Johns Hopkins University conducted a national survey of randomly selected colleges and universities about the preparation of educators for school–family–community partnerships. More than half (59.6%) offered a full course on family involvement or on school, family, and community partnerships, and 67.5% of these courses are required (often targeted to graduate students). Only 8.7% offered more than two courses on working with families. Almost all colleges and universities reported that they discussed the topic of family involvement or partnerships in at least one course.

Early childhood programs covered the topic of family involvement most often; it has only been recently that content on family involvement has been added to the curriculum for teachers at other grade levels, administrators, and counselors. The most commonly covered topics include theories of partnerships, research, and practical activities such as the parent–teacher conference. The programs report that they less often cover topics about complex organizations or integrations of programs such as developing interactive homework assignments, planning parent workshops, or working with school-linked social service programs.

Hiatt-Michael (2001) surveyed 147 department chairs or deans of education in 50 states and asked about the number of courses, types of courses, topics, and instructional methods. Approximately 23% replied that they offered a course about family involvement but that it was not required. These courses were most often developed as special education, early childhood, or elective courses. Most respondents (93%) reported that the topic of family involvement was integrated into existing courses. Again, the most popular topic covered was the parent–teacher conference.

STATE CERTIFICATION AND NATIONAL STANDARDS

Because most educators in the United States receive teaching and administration certification from a state-approved program, Radcliffe, Malone, and Nathan (1994) conducted a review of these teacher education requirements in 50 states and the District of Columbia. Because states express their requirements in different ways (either by requiring courses or requiring skills or some combination of the two), the study looked at a range of requirements. The central finding, however, was clear no matter which requirements were examined: there were very few requirements for courses or skills about parent involvement. Only 7 states required principals or administrators to study parent involvement, and only 15 states required most of their teachers to study and develop abilities in parent involvement. The two certification areas with the most requirements were early childhood and special education.

Shartrand and colleagues (1997) found that only 22 states had specific standards relating to working with parents or families; many states did not even mention the topic. Most of the state certification standards that did refer to family involvement rarely defined the term in precise terms. The researchers concluded early in their study that family involvement was not a high priority in state certification.

There are some notable exceptions to this dismal generalization about state certification standards. Hiatt-Michael (2001) reports that California is the first and only state to mandate that teachers work in partnership with parents (California Education Code, 44261.2, 1993). Kirschenbaum (2001) reports that the State Department of New York requires that by 2003 all teachers receiving certification will have training that includes "factors in the home, school, and community" and "participating in collaborative partnerships."

Because the National Council for Accreditation of Teacher Education (NCATE) has included family–school–community partnerships in its standards, it is expected that more states will continue to add requirements and mandates in the coming years. Given that it takes states a long time to get documents approved officially, it may be that more standards are forthcoming. Hiatt-Michael (2001) reported a significant increase in the number of states that were working on changes in their credentialing documents, and many were considering statements about working with families.

FRAMEWORKS AND MODELS

Since the mid-1980s, there has been a renewed interest in developing some frameworks and models of how to prepare educators for working with families. The list below contains examples that help demonstrate the linkage between educator preparation for achieving academic, social, and emotional learning success for students.

SEDL's Ideal Model for Preparing Educators

After finding out how little preparation was being offered about how to help teachers work with parents, the Southwest Educational Development Laboratory (SEDL) developed a prototype plan for preparing educators about parent involvement in education that is still used today. Their ideal teacher preparation framework was developed by utilizing a comprehensive process of recommendations from the previous surveys of teachers, principals, parents, administrators, and teacher educators; key points from a thorough review of the literature; and the results of comments from 150 inservice directors and college and university faculty regarding what teacher preparation in parent involvement should include (Chavkin & Williams, 1988).

Four essential components for a prototype parent involvement teacher preparation program were identified and described: the personal framework, the practical framework, the conceptual framework, and the contextual framework. An overlapping of elements from the first three components is the ideal program. Figure 9.1 illustrates SEDL's suggestions for an ideal model in more detail.

It is important to give attention to the knowledge, understanding, and skills areas that are contained within each framework. For example, the personal framework focuses on teachers' knowledge about their own beliefs and values, their understanding of the school, their comprehension of the diversity within the community, and the importance of individual differences among parents. The practical framework contains information about various models of parent involvement, effective methods, interpersonal communication skills, and potential problems in developing parent involvement programs. The conceptual framework highlights the theories, the research, the history, and the developmental nature of parent involvement.

FIGURE 9.1. SEDL's Ideal Model for Effective Preparation of Educators

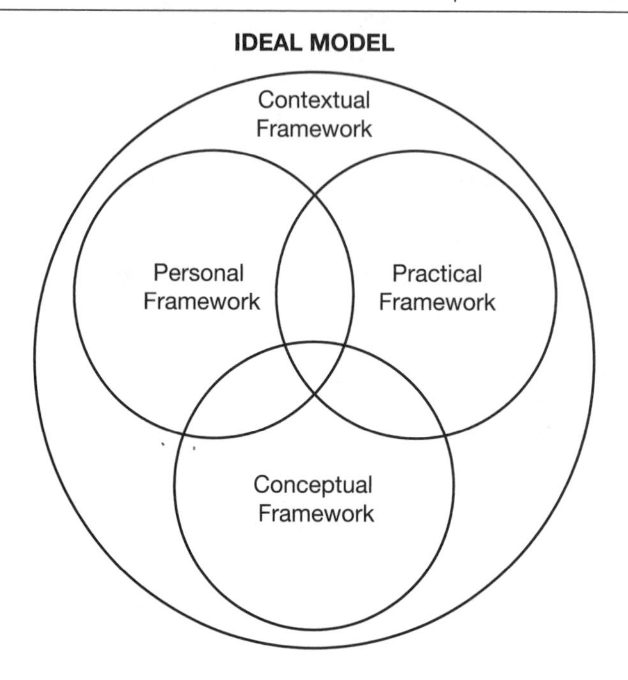

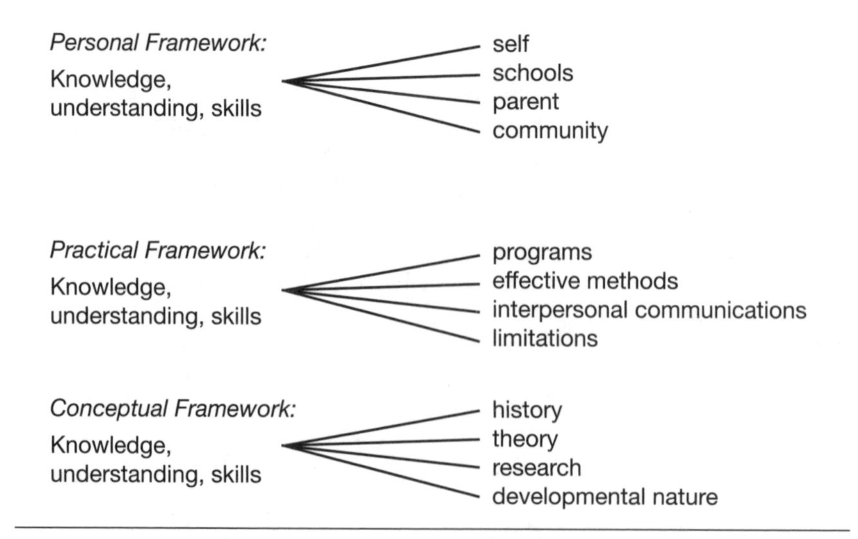

The contextual framework includes both the physical and the attitudinal environment where the training occurs.

Harvard Family Research Project: Seven Key Knowledge Areas

The Harvard Family Research Project (HFRP) has emphasized the importance of preparing teachers to work more effectively with families and communities using a meaningful framework. They have identified seven key knowledge areas about family involvement that teachers need to know, and recommend that they be included in teacher training programs (Shartrand et al., 1997). These seven areas include:

1. *General Family Involvement:* knowledge of the goals of, benefits, and barriers to parent involvement
2. *General Family Knowledge:* knowledge of different families' cultures, child-rearing, lifestyles, etc.
3. *Home–School Communication:* provision of techniques and strategies for two-way communication between school and home
4. *Family Involvement in Learning Activities:* information about how to involve parents in their children's learning at home or in the community
5. *Families Supporting Schools:* information on ways that families can help the school, both inside and outside the classroom
6. *Schools Supporting Families:* information on possible ways that schools can help support the social, educational, and social services needs of families
7. *Families as Change Agents:* information on possible roles that families can play as decision makers, researchers, and advocates in the improvement of policies, programs, and curriculum.

The group also presented four approaches for how to teach the attitudes, knowledge, and skills that are needed in these seven content areas. The approaches include: (1) a functional approach that clearly describes the roles of schools and parents; (2) a parent empowerment approach that is based on the strengths of disenfranchised families; (3) a cultural competence approach that focuses on an inclusive, respectful school where diversity is valued; and (4) a social capital approach that builds on community assets and parental investment in their children's education. Each of the four approaches can be used alone or in combination.

The functional approach is probably best represented by the work of Joyce Epstein, whose work has been discussed in the Introduction and earlier chapters. Universities would promote the attitude that all teachers should learn skills in working with parents, and they would prepare teachers to take the lead in engaging families by providing educators with knowledge about both the benefits and the barriers to family involvement. Schools of education would advocate for teachers and administrators involving diverse families of all backgrounds in their children's education. They would emphasis developing teacher knowledge and attitudes about different childrearing practices, family structures, living environments, belief systems, and lifestyles.

The parent empowerment approach has the same basic goal as the functional approach, but it places more emphasis on the role of the family. The attitude expressed by university educators using this approach is that all parents want the best for their children and are the most important teachers of their children. This approach not only has respect for the importance of the family in a child's life but also focuses on beginning with the family and the community to find the most useful knowledge about rearing and educating children. The university curriculum would focus on the strengths of parents and the knowledge of power differences among groups in our society. This approach recognizes that many families belong to disenfranchised groups, and works with the effects of this disadvantaged status on a family's relationship with the school. Shartrand and her colleagues (1997) cite the work of Moncrieff Cochran as an example of this approach.

The cultural competence approach is again similar to the other approaches, but it places the primary emphasis on the knowledge that family involvement benefits the academic achievement of minority and low-income students. Universities who use this approach would teach both knowledge and skills in using culturally appropriate curriculum and outreach methods. They would teach skills in understanding and combating stereotypes and prejudices. The authors cite the work of Luis Moll as an example of the cultural competence approach.

The social capital approach is a strengths-based approach that focuses on building on and using the skills and knowledge that parents already possess. Universities who use this approach would teach content about differences and commonalities in norms and values across cultures. They would emphasis skills in conflict negotiation and consensus-building. The authors offer James Coleman's work as an example of the social capital approach.

Kirschenbaum's Model: Knowledge, Attitudes, and Skills

Kirschenbaum (2001) presents another model that describes the components of professional preparation for family–school partnerships. Kirschenbaum uses a chart that outlines the knowledge, attitudes, and skills that are important. He provides a beginning list of what educators need to know about family involvement in education under each category and stresses that participants need to add additional items that are relevant to their own learning needs. He stresses that it is not enough for professionals to just *know* how to develop a family–school partnership, but they must also *want* to do it and to *believe* that they can do it. He states that educator preparation must include cognitive, affective, and behavioral components.

Some brief examples of the areas covered under knowledge, attitudes, and skills from Kirschenbaum's (2001) chart about what teachers should know about family involvement include:

Knowledge—Understandings and Perspective

- Theoretical understandings
- Paradigm shift about role of professional
- Types of family involvement
- Implementation models

Attitudes

- Comfort with diverse populations
- Self-knowledge
- Receptivity about working in partnership

Skills

- Two-way communication
- Conducting parent–teacher conferences
- Conduct home visit
- Communicate across language, culture, class

Kirschenbaum makes it clear that these areas of family involvement learning for teachers are not discrete categories. The areas of knowledge, attitudes, and skills overlap. For example, teachers will want to develop more skills with families if they have knowledge about the positive effects of strong family involvement. Their comfort (attitude) about working with families will be at a higher level when they have developed some skills in conducting home visits or conducting parent–teacher conferences. Knowledge, attitudes, and skills work together to produce effective family involvement in education, and university educators can help future teachers learn the most when they develop curriculum that includes content on all three areas.

Leuder's Self-Renewing Partnership

Another important model, the Self-Renewing Partnership Model, was developed by Leuder (1998) with his text *Creating Partnerships with Par-*

ents: An Educator's Guide. This model emphasizes changing the traditional parent involvement approach from single-dimensional, with parent involvement coming into the school for the sole purpose of supporting the school, to a multidimensional model that focuses on reaching and involving the "missing families."

The goal of Leuder's model is to create learning communities, and it requires a new outreach dimension. Instead of just the traditional "energy-in" components of family involvement where families are supporting the school by volunteering and giving their time, Leuder's model expands on traditional roles and also adds an "energy-out" component. The energy-out component is a series of strategies that the school uses to reach out to parents. In other words, the school uses its resources to create a collaborative relationship with families and communities. Leuder believes that it is the school's responsibility to work with the hard-to-reach families, and his model focuses on the active roles that educators can play in developing self-renewing partnerships with families.

Leuder's case study of Apple Hill Elementary School provides an excellent example of how to implement this model. After completing the Pre-Intervention Planning Phase of the Strategic Partnership Planning System, the School Improvement Council members and the faculty of Apple Hill decided that there were two family populations that they wanted to target for interventions. These two populations were hard-to-reach families who were not involved and the parents of students who were bused in. The Council decided that there was not enough good communication between the groups and the school. They decided that their first goal was to strengthen the connections between the families and the school, and the second goal was for families and the school to be communicating effectively. Because the second goal was largely dependent on meeting the first goal, the group focused all of their efforts on increasing the two-way communication flow between the school and parents.

The Council decided on objectives for each goal and then selected and implemented best practices. For example, for Goal One, they changed the warning signs at the front of the building to "WELCOME" signs and allocated some of the parking spaces at the front of the school for "PARENTS." As the connections were made and the goals were being met, the Council was able to enhance the Parent Partner Role skills and knowledge and get the families working with the school to make sure that the partnership would continue and the families were fully involved in the partnership to educate their children. Working through case examples about real schools and real families is an excellent way of teaching the Self-Renewing Partnership Model.

Epstein's Practice Model: Six Types of Involvement

Epstein (2001), who has been a leading expert in working with family–school partnerships and whose work has been discussed throughout this book, also recently developed a text, *School, Family, and Community Partnerships: Preparing Educators and Improving Schools*. The book is written specifically for educators and is divided into two parts: Part 1, understanding school, family, and community partnerships; and Part 2, applying research on school, family, and community partnerships. In Part 1, she provides a historical overview, discusses the theory, and presents the latest research. In Part 2, she presents policy implications, a framework for developing comprehensive partnership programs, practical applications for linking family involvement to student learning and productively using volunteers, and strategies for action.

The centerpiece of the book is her practice model for developing school, family, and community partnerships, which is discussed in earlier chapters. Although developed from theory, research, and current policy, the major focus of the model is good practice on the part of educators. She uses her framework of the six types of involvement as the core for helping prepare educators for family involvement practice. Each of the six types (parenting, communicating, volunteering, learning at home, decision making, and collaborating with the community) is followed by sample practices, challenges, redefinitions, and results for students, parents, and teachers.

The National Network of Partnership Schools uses Epstein's model and works with school–family–community teams to plan, coordinate, and implement partnership activities. This program provides inservice education in conjunction with a school's ongoing reform effort. Educators learn about family involvement partnerships as they are actively developing them in their own schools (Epstein et al., 1997). The text can also be used by preservice educators, particularly when students are working in the schools as part of their student teaching experience.

PROMISING CURRENT INITIATIVES

Recent program evaluations and research have confirmed the benefits of preservice preparation for educators. Of particular note are service-learning opportunities where preservice students are actively engaged in local family involvement programs. They are learning through hands-on experience, and the family–school partnership is benefiting from their volunteer hours. Katz and Bauch (2001) reported that students who had had preservice preparation were more comfortable with family involvement activities and actually reached more families in their classes than teachers who did not take a course or unit on family involvement.

The American Association of Colleges for Teacher Education (AACTE) and the Metropolitan Life Foundation have recently developed a new initiative on infusing parental engagement education in teacher education programs. They selected five national sites through a competitive grant process to become partners. The five partners are each developing and evaluating new approaches to preparing teachers to engage families. These approaches often engage students early in their teacher education career through volunteer work with local school–family involvement programs. As they progress through their teacher education curriculum, students are often involved with research about family involvement. Students sometimes interview family members and community leaders about their involvement with schools. The emphasis on infusion differs from the typical separate course on parent involvement because it adds content on families and family involvement with schools to many courses across the curriculum.

INFUSED OR SEPARATE COURSES?

There are indeed a variety of ways to offer preparation for working with families. Some universities feel strongly that content should be infused into many courses, while others believe it is best to have a separate course or a sequence of courses. Either way can work, but it seems more likely that the process of educating more beginning teachers about family involvement will take place faster if content can be infused throughout the curriculum rather than waiting for new courses to be developed. Mandating an entire course takes time and sometimes tends to isolate the content instead of integrating it into the curriculum. The reality is that in many schools there are many competing courses for time slots and there is only so much room in the degree plan. This dilemma is not unique to family involvement content; it is also true for multicultural issues, special education, character education, and a number of other important content areas.

Martha de Acosta (1996), who argues for the inclusion of family involvement content in teacher preparation curriculum, suggests that this family involvement content should be infused throughout foundation courses, and advocates for community-based learning to help teachers develop reflective family involvement practices. She proposes three themes for foundation courses: families and schools; communities and schools; and the social context of teaching. She suggests that the use of themes in foundational courses will help students think critically about family and community involvement. Students will not just be memorizing strategies but will, instead, be examining the pros and cons of alternative courses of action. The community-based learning component would give students hands-on experiences in applying what they have learned.

The Peabody Family Involvement Initiative (PFII), described by Katz and Bauch (2001), has developed a sequence of family involvement courses that focus on three major areas: (1) general knowledge, (2) skills, and (3) authentic "real life." The PFII emphasizes six key themes about the importance of respecting and building upon the strengths of all types of families. The program presents a clear message that because the family is the child's first and most important teacher, it is essential that schools work in collaboration with families. Family involvement includes activities at both home and school and is most effective when it not only strengthens the relationship between the child and the family but also addresses the teacher's needs.

The PFII program begins with a required one-semester course called "Parents and Their Developing Children." Students are taught both the traditional typologies (e.g., Epstein's six family involvement categories) and innovative strategies such as electronic voice mail and interviewing parents in their homes. The practice component of the initiative allows students through course assignments and student teaching placements to implement some of the strategies they have been discussing in class.

Kirschenbaum (2001) describes Rena Rice's required course at the Bank Street College of Education. All early childhood and elementary teacher education students must take this course to help "develop competency in working with families and communities." A central part of the course focuses on helping these prospective teachers develop positive attitudes about working with families. Some of the student assignments include: autobiographical essays, critiquing family involvement policies, analyzing parent involvement practices, and designing a work plan for collaboration.

The University of Houston—Clear Lake developed an educator preparation course specifically designed in collaboration with a school district's inservice education and offered as a graduate course. Andrea Bermúdez (1993) describes the university–school district collaborative education program that was developed to integrate knowledge about multicultural systems and family involvement in education in the inservice teacher training curriculum. The program developed a curriculum guide for helping teachers work with non-English-speaking parents. The guide had three major parts: (I) English for Survival; (II) Topics for Parent Education Programs (English); and (III) Temas Para Programas Educativos Para Padres de Familia (Spanish). Under Part II, some of the topics focused on understanding the educational system, working with teachers, going to the library, and helping students with their study habits. The research on this inservice education program showed gains for parents, teachers, and students.

Kirschenbaum (2001) discusses three major methods (classroom, home, and field) and lists several beginning suggestions in each category. For example, under classroom methods he lists lectures, case studies, role-playing, guest speakers, debates, and reports. Under assignments at home, he describes readings, reaction statements, autobiographical work, simulated class newsletters, notes, and class Web pages. In the category of field experiences, he suggests interviews with parents from diverse backgrounds, home visits, community visits, community service, observing decision-making teams, attending PTA meetings, videotaping, and parent–teacher conferences. He stresses that these are beginning methods and should serve only as catalysts for creative teacher educators. He provides some examples of guest speakers such as parents who can describe their successful and unsuccessful experiences and teachers who have developed excellent partnership programs. He suggests some useful assignments where students can practice developing interactive homework assignments, writing newsletters, designing Web sites for parents, revising a school district's parent involvement policy, or writing about their own family involvement in education experiences.

The Family Involvement Network of Educators (FINE) is a national network of over 2,000 people who are interested in promoting strong partnerships among children's educators, their families, and their communities. FINE's membership is composed of faculty in higher education, school professionals, directors and trainers of community-based and national organizations, parent leaders, and graduate students. The FINE network does not take a position about whether the curriculum should be infused or separate courses, but instead offers examples of cases for teaching about family involvement in course modules and complete syllabi.

Whether the content is in a required class, an elective class, a sequence of courses, or infused throughout several classes, there is a wide variety of methods that have been very successful in helping to prepare educators for family involvement. The important point is that we need to pay closer attention to teachers' knowledge, beliefs, emotions, and attitudes about working with parents. Research by Graue and Brown (2003) clearly supports the notion that we must provide more opportunities for teacher education students to expand their theoretical background and experiences with families in a variety of settings during preservice education or else prospective teachers develop strategies for working with families based solely on their own middle-class experience, rather than the reality of today's diverse schools which often do not mirror the background of the teacher pool.

CHALLENGES, OPPORTUNITIES, AND RECOMMENDATIONS

Even if colleges and universities do offer new courses, it is probable that few current educators would be willing to take additional courses because of the lack of time and money. Yet educators must be informed about family–school partnerships and must be provided with materials to enable them to work with families. Preparing current educators to work with school–family partnerships is a daunting task.

From the educator's perspective, the call for increased contact with parents has added to the demands traditionally associated with the teacher's role. In addition to the skills that pertain to classroom instruction, new teachers are now expected to develop skills in working with parents and to assume leadership in working with advisory groups. In many parts of the country, educational reform legislation already has added more classes, more recordkeeping, more self-evaluations, more teacher appraisal systems, and career ladders that require additional course work. Teachers are mandated to attend inservice education on many other topics.

It is understandable that many teachers are feeling short of time and overburdened with all of these additional responsibilities. It is not likely that teachers on their own will become knowledgeable about school–family partnerships in education. Certainly it is unfair to place the responsibility for family involvement on teachers without giving them some assistance. Policies will need to be changed so that teachers are given adequate time for these additional duties.

Inservice education is one way to help the large cadre of current educators who have not had the opportunity to take a course on developing school–family partnerships. Evans-Schilling's 1996 proposal for a Continuum of Family Involvement Training might be one way to address the ongoing needs of teachers, the dynamics of the field, and the personal growth process of educators.

Toni Griego Jones (2003) researched the perspectives of Hispanic parents on teacher preparation. She found that consulting Hispanic parents and bringing them into the teacher education classroom had a significant impact on the beliefs and attitudes of teacher candidates. Her work suggests that teacher education programs must do more than just add multicultural content and field experiences to the curriculum. She suggests that although influencing beliefs and attitudes may be the most difficult part of preparing teachers, it should be the focus of teacher preparation because teacher beliefs help determine the expectations that teachers have for children.

As Figure 9.2 illustrates, future work in the area of educator preparation for school–family partnerships needs to be approached from multiple

FIGURE 9.2. Recommendations for a Multilevel Approach to Increasing Educator Preparation for School–Family Partnerships

At the Preservice Level	• More content on family involvement across subject areas (horizontal integration) • A curriculum that builds upon learning from introductory classes to advanced classes (vertical integration) • More opportunities for hands-on learning and/or service learning • More linkage between preservice and inservice educators • Collaboration across disciplines, especially in the fields of health, social services, home economics, and child development • An assessment plan that examines the outcomes of school–family partnership curriculum
At the Inservice Level	• Time to work on school–family partnership plans • Access to tools and resources about working with families • Collaboration with families and community members to develop plans for involvement in schools • Assistance with incorporating evaluation into plans • Opportunities for continuing education • Time to share and collaborate best practices with other teachers
At the State and National Level	• Include competency in school–family partnerships as a credentialing requirement • Develop networks to support educator preparation for school–family partnerships • Encourage continuing education about school–family partnerships • Stress the importance of preparation for school–family partnerships at both the preservice and inservice level

levels. At the preservice level, there is a strong need for activities that help beginning teachers understand their own social and emotional learning first through the development of self-awareness and interpersonal skills. One of the recommended ways to accomplish this is through more opportunities for hands-on learning or service learning. These experiential activities increase the linkage between preservice and inservice educators and encourage collaboration across disciplines. At the inservice level, the recommendations focus on time, access, resources, and opportunities to continue learning and developing partnership plans. Time is indeed a premium for the competing demands on teachers, and teachers in poor areas will have an increasingly difficult time meeting their classroom instructional challenges as well as involving families. Both preservice and inservice education need support from state and national associations and policymakers. It is critical that competency in school–family partnerships is a credentialing requirement, that schools support continuing education about school–family partnerships for all faculty, and that schools provide supports for teachers to work with school–family partnerships.

CONCLUSION

The evidence for increasing educator preparation for school–family partnerships is clear, and the case for approaching the task by using a multilevel approach is promising. Preparing teachers to develop school–family partnerships will have strong academic and social-emotional learning benefits for children. The only question remaining is, when will we accept this challenge to unite families and schools in their common task of educating American children? I hope that we will rise to the challenge sooner rather than later for the benefit of all children.

REFERENCES

Bermúdez, A. (1993). Teaming with parents to promote educational equality for language minority students. In N. F. Chavkin (Ed.), *Families and schools in a pluralistic society.* Albany: State University of New York Press.

Chavkin, N. F., & Williams, D. L. Jr. (1988). Critical issues in teacher training for parent involvement. *Educational Horizons, 66*(2), 87–89.

de Acosta, M. (1996). A foundational approach to preparing teachers for family and community involvement in children's education. *Journal of Teacher Education, 47*(1), 9–15.

Epstein, J. L. (2001). *School, family, and community partnerships: Preparing educators and improving schools.* Boulder, CO: Westview Press.

Epstein, J. L., Coates, L., Salinas, K. C., Sanders, M., & Simon, B. (1997). *Partnership 2000 schools manual: Improving school–family–community connections.* Baltimore, MD: Johns Hopkins University.

Epstein, J. L., Sanders, M. G., & Clark, L. A. (1999). *Preparing educators for school–family–community partnerships: Results of a national survey of colleges and universities* [Report No. 34]. Baltimore, MD: Johns Hopkins University Center for Research on the Education of Students Placed at Risk.

Evans-Schilling, D. (1996). Preparing educators for family involvement: Reflections, research, and renewal. *Forum of Education, 5*(1), 35–41.

Graue, E., & Brown, C. P. (2003). Preservice teachers' notions of families and schooling. *Teaching and Teacher Education, 19,* 719–735.

Griego Jones, T. (2003). Contribution of Hispanic parents' perspectives to teacher preparation. *The School Community Journal, 13*(2), 73–97.

Hiatt-Michael, D. (2001). *Preparing teachers to work with parents.* ERIC DIGEST, EDO-SP-2001-2. (ERIC Clearinghouse on Teaching and Teacher Education)

Katz, L., & Bauch, J. P. (2001). Preparing pre-service teachers for family involvement: A model for higher education. In S. Redding & L. G. Thomas (Eds.), *The community of the school* (pp. 185–204). Lincoln, IL: The Academic Development Institute.

Kirschenbaum, H. (2001). Educating professionals for school, family and community partnerships. In D. B. Hiatt-Michael (Ed.), *Promising practices for family involvement in schools* (pp. 185–208). Greenwich, CT: Information Age Publishing.

Leuder, D. C. (1998). *Creating partnerships with parents.* Lancaster, PA: Technomic Publications.

Radcliffe, B., Malone, M., & Nathan, J. (1994). *The training for parent partnerships: Much more should be done.* Minneapolis: Center for School Change, Humphrey Institute for Public Affairs, University of Minnesota.

Shartrand, A. M., Weiss, H. B., Kreider, H. M., & Lopez, M. E. (1997). *New skills for new schools: Preparing teachers in family involvement.* Cambridge, MA: Harvard Family Research.

Conclusion

School–Family Partnerships: Dimensions and Recommendations

Evanthia N. Patrikakou, Roger P. Weissberg, Sam Redding & Herbert J. Walberg

School–family partnerships are a strong force in children's academic, social, and emotional development. Positive and planned home–school relations greatly improve student academic, social, and emotional learning, and serve as mediators of school and life success. The challenge to raise student achievement compels all of us to take advantage of all the benefits that school–family partnerships offer for establishing long-term, sustained learning and achievement.

Contributors to this book discussed the different perspectives, dimensions, and levels that constitute the complexity of school–family partnerships, provided evidence for their effectiveness, and offered recommendations to improve and benefit from them. The dimensions and common themes in recommendations stemming from this book's chapters, as well as the need for a multidisciplinary approach, were also emphasized at the National Invitational Conference in Washington, D.C., in December 2002. Conference participants agreed that (a) strengthening links between evidence-based research and partnership activities and enhancing the links between schools and families would increase the learning opportunities for all students, and (b) such partnerships should be an important factor in national education reform.

In the next section, we outline the cross-cutting themes and recommendations that were raised by all chapters and have been identified as essential ingredients for establishing and maintaining school–family partnerships.

DIMENSIONS

The conceptual, empirical, and policy-related evidence presented in this book builds a strong case about the importance and multidimensional nature of school–family partnerships. Such evidence refers to all systems involved in an individual's life, from the proximal, such as the microsystem, which includes the child's characteristics, to the more distal, such as the macrosystem, which includes cultural beliefs. As we presented in the Introduction, the dimensions that cut across all systems and constitute the multidimensional nature of school–family partnerships include: (a) child development characteristics; (b) beliefs and expectations; (c) roles that parents, teachers, and students play; (d) cultural perspectives; and (e) policies.

Some of the major points contributors made under each school–family partnerships dimension include the following.

Child Development

School–family partnerships are crucial to set a solid foundation for learning and development in early childhood. Young children whose parents actively participated in early childhood programs not only display a head start in academic, social, and emotional learning, but also engage in fewer risky and delinquent behaviors later in life.

Child characteristics can also influence the magnitude and nature of school–family partnerships. For example, as children move into adolescence they mature cognitively, emotionally, and socially; become more independent; and are faced with more complex learning tasks. During the junior high and high school years, parent involvement and teacheroutreach decline drastically. It is during these years that parents may experience less self-efficacy to become involved due to the increasingly complicated curriculum of secondary education and the increasing complexity of middle and high school structures. Parents may perceive that schools are not extending an invitation to them to be involved, or that their children want them to participate less. The same issues of complicated structure and demanding curriculum in secondary education make it harder for educators to reach out to families systematically, as they usually can in primary education.

Beliefs and Expectations

Parent and teacher beliefs and expectations are crucial to shaping school–family partnerships. Beliefs about individuals and group characteristics drive an individual's choices and behaviors, and are crucial in shaping relationships and partnerships. A positive, welcoming school climate and consistent invitations to parents with ways to become involved in their children's education at home and school positively predispose parents about the school's efforts. Parents' positive perception highly influences their children's perception of school, which, in turn, positively contributes to students' academic, social, and emotional learning.

Parent and teacher expectations regarding the academic, social, and emotional development of children have been shown to be among the best predictors of school success. Educators should make parents aware of this powerful factor, and assist them to communicate their expectations clearly and in developmentally appropriate ways to their children. Having a clear picture of what is expected and specific strategies to achieve the goals set fosters consistency and more learning opportunities.

Parent, Teacher, and Student Roles

Within a school–family partnership framework, parents, teachers, and students play a critical role in establishing and maintaining the partnership. Roles are affected by and also affect beliefs and expectations about what are the desirable outcomes of a school–family partnership, who is responsible for these outcomes, how important each member's contribution is, and in what ways each member can contribute. If parent roles are limited to chaperoning field trips and organizing bake sales, home–school relations remain on the surface level, with no true partnership occurring. Parent roles must expand in other areas such as (a) decision making at school or on other levels (e.g., the district); (b) collaborating with the community to offer additional resources and support to families and students; and (c) establishing new learning opportunities, such as interactive homework assignments and classroom presentations, to enhance involvement both at school and at home.

When deemed developmentally appropriate, students can also play an important role in school–family partnerships. Involving older students in parent–teacher conferences and intervention planning creates continuity between home and school. Students get the strong message of valuing education and collaboration, and expectations are clarified and reinforced. Receiving a unified message and clear expectations from both parents and teachers is a strong motivation for students to succeed.

Cultural Perspectives

In recent years, with the increasing cultural and linguistic diversity of families, the home and school environments may hold varied and sometimes diverging beliefs about the appropriate degree and nature of school–family partnerships. Being aware of such factors and addressing them in appropriate ways makes home–school interactions more positive and productive for the benefit of all students in a multicultural society. By creating culturally aware school–family partnerships, school systems can reduce cultural discontinuities, facilitate intercultural transitions, improve ethnic and racial perceptions and attitudes, and foster interethnic friendships. For example, providing parents with materials and activities that are adapted to accommodate the needs of families from different cultural and linguistic backgrounds will enhance parent involvement and contribute to the creation of a positive home–school climate. In this way, more learning opportunities will be created and students will be better prepared to acquire not only knowledge, but also attitudes and skills necessary to interact positively and productively with people in a pluralistic society.

Policies

For decades, federal, state, and local policies have impacted families and the roles they should play in public education. Based on research evidence indicating the positive impact on student academic, social, and emotional learning, major legislation has acknowledged the importance of school–family partnerships by mandating parent participation on various levels of policy design and implementation. The No Child Left Behind Act of 2001, which reflects the U.S. Department of Education's educational strategy, provides a very broad definition of parent involvement. Such involvement is to include ongoing home–school communication, and parents should be included as full partners in their child's education. Still, current policies could provide better research-based guidance, which practitioners need in order to better define and integrate school–family partnerships in their daily routines. Contributing to this vagueness is also the lack of systemic preservice and inservice training to prepare educators to engage students and their families in a positive, meaningful, and effective way. By preparing teachers and administrators to develop school–family partnerships, we have the potential to improve the learning opportunities for all students, so no one is left behind.

RECOMMENDATIONS

In addition to the analysis of the multiple dimensions that underlie home–school relations, as well as ways that school–family partnerships influence student learning, chapter authors offered a variety of recommendations for building and maintaining successful school–family partnerships. The common themes of those recommendations include the following (see Figure C.1):

Communication. Positive, ongoing, two-way communication is a suggestion embedded in *all* dimensions and levels involved in school–family partnerships, and it is *key* to forming and maintaining meaningful and productive home–school connections. It is imperative that communication systems be established to facilitate the flow of information between home and school. Parents must be kept informed about their children's progress, the schools' programs, and ways in which they can be involved.

FIGURE C.1. The Multidimensionality of Successful School–Family Partnerships

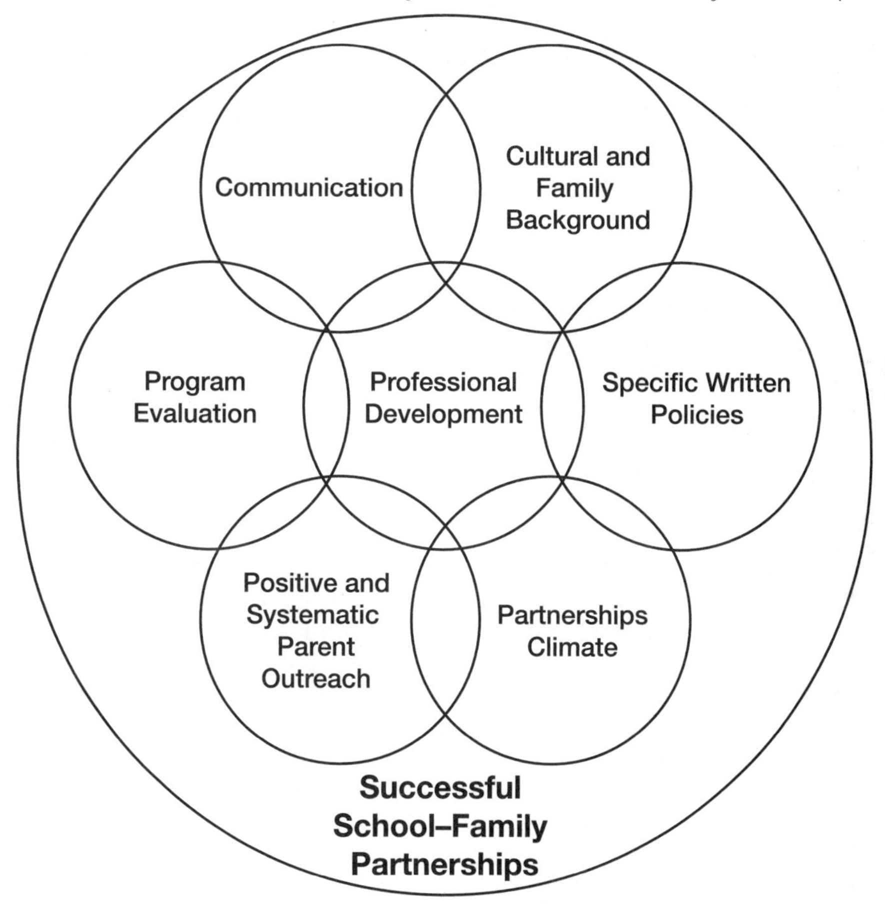

Information regarding best ways and times to contact school personnel is also crucial to facilitate a two-way pattern in communication. Expectations and roles should be clearly communicated to parents and students so that all parties involved are on the same page and work collaboratively to achieve the same goals. Interaction and communication between schools and other organizations serving the community will also enhance partnerships and enrich the services available to students and their families.

Specific, Written Policies. Establishing school–family partnerships is both a process and a goal. In order for both to be recognized and specified, there should be written declarations of (a) the importance of such a partnership, (b) the great, positive impact it exerts on the academic, social, and emotional learning of children, and (c) specific ways through which it can become reality. To that end, policymakers should provide written guidance on all parent involvement provisions of major legislation, such as those found in No Child Left Behind. Districts and schools must adopt formal policies and ensure that their mission statements explicitly state the value and describe the nature of partnerships. Specific information regarding ways in which parents can be involved, as well as the expectations of such involvement, must also be clearly stated.

Partnerships Climate. Districts and schools must cultivate a welcoming environment for parents to participate in multiple ways in the educational process. Districts and schools can establish a school–family team, which can be instrumental in designing, implementing, and monitoring parent involvement practices. Also, establishing a school environment that overtly values and respects parental input and presence will clearly convey a positive, inviting message to parents. Having a space inside the school for parents to meet each other and with school personnel further cultivates a welcoming environment.

Cultural and Family Background. Acknowledge the unique contribution of families from different backgrounds. Inviting family members to share information about their culture or family traditions provides an excellent way to include parents, acknowledges cultural differences as part of social reality, and sends the message that different families are respected and welcome. Districts and schools should provide the necessary supports for parents to understand how schools work and what the expectations are for students and parents. Such a process involves adapting materials and activities to accommodate families from different cultural and linguistic backgrounds. Parents and other community members can also be instrumental in partnering with the school to (a) contribute to professional development programs that call attention to the cultural characteristics of students and their families and (b) design welcoming practices that respect and build on family culture, traditions, and strengths.

Positive and Systematic Parent Outreach. Making parent outreach a priority is essential to facilitate home–school relations. It has been shown that reaching out to families positively and systematically can be the most effective way to motivate parents to be involved. However, it is not unusual for outreach to be initiated by schools only when a student is experiencing academic or behavioral problems. If outreach is limited to that, it usually serves as a deterrent rather than a motivator for parent involvement. Taking the extra step to contact a parent when there is good news to share will add to the positive climate and increase parent involvement.

Program Evaluation. School–family partnership programming varies tremendously in its design, implementation, and evaluation. Especially, the latter has lacked clear design and rigorous data collection. This becomes problematic on the classroom, school and district, and broader policy levels. First, classroom teachers should monitor and evaluate their parent outreach efforts in order to assess, review, and improve their practices and involve more families in a meaningful way. Second, school and district accountability for effective parent engagement must be measured by using systemic, long-term evaluation methods. Third, the effectiveness of state and federal programs must be assessed through large-scale, rigorous, longitudinal evaluations that can inform future, better-targeted policies. It will be beneficial if a more interdisciplinary approach to designing and evaluating school–family programs is followed. Researchers and practitioners from the fields of education, psychology, sociology, and public policy can collaborate to integrate theoretical approaches and research methodologies that enhance the design and evaluation of school–family partnerships.

Professional Development. Preservice and inservice training of administrators, teachers, and other school personnel is *instrumental* in facilitating *all* of the above recommendations. Incorporating competency in school–family partnerships as a certification requirement will trigger a much-needed emphasis on both the preservice and inservice level. Either through more content on family involvement across subject areas or by having separate courses, colleges of education must enhance their curricula to better prepare prospective teachers for their mission. More opportunities for hands-on experiences will further enrich the experiences of preservice educators. Teachers already in the field must also have opportunities for continuing education on school–family partnership issues and be given access to tools, resources, and practices that have been shown to be effective when working with families. It is important to note that training supporting school personnel, in addition to teachers and administrators, on ways to best work with families is critical for creating a global, positive school climate, increasing parent involvement, and fulfilling the school's stated mission.

CLOSING THOUGHTS

School–family partnerships foster child development and academic, social, and emotional learning. Effective home–school relationships take a long time to form and become established. The process involves (a) ongoing communication, (b) developing a partnership climate with positive and trusting attitudes, (c) commitment and specific, written policies, (d) acknowledging respect and embracing diversity, (e) positive and systematic parent outreach, (f) continuous staff development, and (g) rigorous planning and evaluation.

Schools of the 21st century face the challenge of preparing students for a rapidly changing world. Rapidly changing demographics, social structures, and economies demand better prepared individuals. Students must have a solid base of academic, social, and emotional skills to increase their chances of succeeding in school and later in life. Literacy and numeracy, history, and the sciences are necessary, but not sufficient subjects to be cultivated by schools. Catering to the whole child and assisting all students to acquire skills such as the ability to recognize and manage emotions, develop caring and concern for others, make responsible decisions, establish positive relationships, and handle challenging situations effectively are key to school and life success.

The important dimensions of school–family partnerships were discussed in a systematic, integrative way, and its impact on children's academic, social, and emotional learning was addressed in the book's chapters. Experts in the fields of psychology, education, sociology, and public policy examined the theoretical base and presented the empirical evidence of home–school relations and its effects on student success. The authors provided a robust case for the multidimensional nature and importance of school–family partnerships, and also extended their discussion to include recommendations through which partnerships can be enhanced and maintained.

We hope that in this book you found the information you need to make a strong case for school–family partnerships and take practical action steps to implement effective partnerships that benefit our schools, families, and children.

About the Editors and the Contributors

THE EDITORS

Evanthia N. Patrikakou is currently a professor at DePaul University's School of Education. She is also the Director of the School–Family Partnership Program and a senior research associate with the Laboratory for Student Success at Temple University, working on the dissemination of school–family partnership efforts. Prior to joining DePaul, Dr. Patrikakou was a research professor at the Psychology Department of the University of Illinois at Chicago. Dr. Patrikakou has done extensive research in parent involvement and its effects on children's academic, social, and emotional development, and she has been directing the development of school–family programming to enhance home–school relations. She has presented her work on parent involvement, school–family partnerships, and the academic achievement of individuals with and without disabilities in numerous national and international conferences. She has authored articles and chapters on parent involvement and the academic, social, and emotional development of children and adolescents. Her publications include articles in *Diagnostique*, *Education Week*, the *Journal of Educational Psychology*, the *Journal of Prevention and Intervention in the Community*, the *Journal of Research and Development in Education*, and the *School Community Journal*. She is also the lead author on a series of brochures for parents and teachers on topics such as communication and homework. Dr. Patrikakou has been systematically working to better inform practitioners, facilitate their outreach efforts, and bridge the research–practice gap.

Roger P. Weissberg is a professor of Psychology and Education at the University of Illinois at Chicago, President of the Collaborative for Academic, Social and Emotional Learning (CASEL), and a Senior Research Associate with the Mid-Atlantic Regional Educational Laboratory for Student Success. He has published about 150 articles and chapters focusing on preventive interventions with children and adolescents. Recent publications include *Prevention that Works for Children and* Youth (2003), *Safe and Sound: An Educational Leader's Guide to Evidence-Based Social and Emotional Learning Programs* (2003), and *Building Academic Success on Social and Emotional Learning: What Does the Research Say?* (2004). Dr. Weissberg has received numerous honors, including the National Mental Health Association's Lela Rowland Prevention Award, the 2000 American Psychological Association Distinguished Contribution Award for Applications of Psychology to Education and Training, and the 2004 Distinguished Contribution to Theory and Research Award from the Society for Community Research and Action.

Sam Redding holds a doctorate in educational administration from Illinois State University and is a graduate of Harvard's Institute for Educational Management. He taught at the high school level in special education and social studies before teaching psychology and education at the college level. He served as the vice president and dean of Lincoln College until becoming the Executive Director of the Academic Development Institute in 1984. He has consulted with schools and districts throughout the country. In addition to his work with ADI, he has been a senior research associate of the Laboratory for Student Success at Temple University since 1995. He is the executive editor of the *School Community Journal*. Dr. Redding has edited two books on home–school relations and published more than 40 articles and book chapters on school improvement. In 1994, Illinois State University awarded him the Ben Hubbard Leadership Award for his service to public education. He was similarly honored by the Illinois State Board of Education in 1990.

Herbert J. Walberg is Distinguished Visiting Fellow at Stanford University and Emeritus University Scholar and Research Professor of Education and Psychology at the University of Illinois at Chicago. Holding a Ph.D. from the University of Chicago and formerly assistant professor at Harvard University, he has written and edited more than 55 books and written about 350 articles on such topics as educational effectiveness and exceptional human accomplishments. Among his latest books are *Successful Reading Instruction, International Encyclopedia of Educational Evaluation*, and *Psychology and Educational Practice*. He has given invited lectures in Australia, Belgium, China, England, France, Germany, Italy, Israel, Japan, the Netherlands, South Africa, Sweden, Taiwan, Venezuela, and the United States to educators and policymakers. He has frequently testified before U.S. congressional committees, state legislators, and federal courts.

THE CONTRIBUTORS

Amy R. Anderson is an assistant professor in the School Psychology Program at the University of South Carolina. Her research interests include student engagement; promoting successful school completion, particularly among students with mild disabilities; and Curriculum Based Measurement (CBM) and progress monitoring in the area of reading.

Nancy Feyl Chavkin is the author of three books and more than 70 publications on family/community involvement in the schools, school social work, and child welfare. Dr. Chavkin currently serves as professor of Social Work and Director of the Richter Institute of Social Work Research at Texas State University—San Marcos. She was founder and Co-Director of the university's Center for Children and Families at Texas State University—San Marcos from 1998 to 2004. Professor Chavkin is also a member of the steering committee for the National Center for Family and Community Connections with Schools. Professor Chavkin has received Texas State University's presidential awards for excellence in both scholarship and teaching, and in 2002 she was named one of the top 10 college teachers in the state of Texas by the Minnie Stevens Piper Foundation.

Sandra L. Christenson, the Birkmaier Professor of Educational Leadership at the University of Minnesota, is the Director of the School Psychology Program. Her research interests include interventions that enhance student engagement at school and with learning, and identification of home and school factors that facilitate student engagement and increase the probability for student success in school. She was the 1992 recipient of the Lightner Witmer Award from the APA for scholarship and early career contributions to the field of school psychology.

Melissa Clements is an Assistant Research Scientist with the Wisconsin Center for Education Research at the University of Wisconsin—Madison. She received her Ph.D. in developmental psychology from Wayne State University in 2001 and completed her postdoctoral training at the Waisman Center at the University of Wisconsin—Madison in 2003. Her research focuses on child social, emotional, and academic development; contextual influences (i.e., family, school); and the development and evaluation of prevention/intervention programs for high-risk populations (i.e., children with disabilities, children from economically disadvantaged backgrounds).

Pamela E. Davis-Kean is an Assistant Research Scientist at the Institute for Research on Women and Gender and the Institute for Social Research at the University of Michigan and Director of the Center for the Longitudinal Analyses of Pathways from Childhood and Adulthood. Her research focuses on the development of self-esteem over the life span; the impact of parental education attainment on children; the role that families, schools, and significant figures play in the development of children; and why gender plays a role in information technology occupations.

Jacquelynne S. Eccles is the McKeachie Collegiate Professor of Psychology at the University of Michigan. She has multiple appointments at the University of Michigan, including professor of Psychology and Women's Studies and Research Professor at the Institute for Research on Women and Gender and the Institute for Social Research. She is currently Chair of the Combined Program of Psychology. She chaired the MacArthur Foundation Network on Successful Pathways through Childhood and was a member of the MacArthur Research Network on Successful Pathways through Adolescence. She is a Fellow of the American Psychological Association, American Psychological Society, and Society for the Psychological Study of Social Issues. She has conducted research on topics ranging from gender-role socialization to classroom influences on motivation to social development in the family, school, peer, and wider cultural contexts.

Yvonne Godber is Coordinator of the Center of Excellence in Children's Mental Health at the University of Minnesota's Children, Youth, and Family Consortium. In this position, she works with university and community partners to build connections and strengthen links between research, practice, and policy to promote children's mental health and well-being. Her background as both a school psychologist in the public schools and a researcher at Columbia University's National Center for Children in Poverty continues to influence her strong interest in improving the systems that surround and affect children's development and social-emotional health. Areas of interest and expertise include prevention and intervention in schools, with a particular emphasis on family–school relationships, school climate, and evaluation.

Kathleen V. Hoover-Dempsey graduated from the University of California, Berkeley (A.B., 1964) and received her M.A. (1969) and Ph.D. (1974) from Michigan State University. She teaches at Peabody College, Vanderbilt University, and currently serves as Chair of the Department of Psychology and Human Development. Grounded in a model of the parental involvement process (Hoover-Dempsey & Sandler, 1995, 1997), her research program is focused on parents' motivations for becoming involved and the influence of involvement on students' educational and developmental outcomes. With her colleagues, she has developed and evaluated school-based interventions designed to increase the effectiveness of school invitations to parental involvement.

Luis M. Laosa obtained his Ph.D. from the University of Texas at Austin. He was the chief school psychologist of a large school district in Texas and taught at the University of California, Los Angeles. He retired as principal research scientist from the Educational Testing Service in Princeton, New Jersey. He is the author of many studies; his continuing research interests include schools and families as environments for learning and for human development.

Oliver C. Moles, Jr., a Ph.D. social psychologist, retired in 2002 from the Office of Educational Research and Improvement of the U.S. Department of Education after a long federal career. At OERI he was an Education Research Analyst and institutional monitor for OERI-sponsored R&D centers on inner-city education, students placed at risk of failure, and effective secondary schools. His specialties include family involvement and student behavior issues. He has headed research teams in both areas for the Department. He has designed several national program evaluation studies. His writings include reports for the Department's Partnership for Family Involvement in Education and a volume on student discipline strategies. He has written research syntheses and research-based articles, and edited research journal issues and two books. Before retiring he managed large comprehensive secondary school reform model development projects for OERI. He is a consultant now in social research and educational issues. He has taught at the graduate level, sits on several professional advisory boards, and is currently a Visiting Fellow at the Institute for Conflict Analysis and Resolution, George Mason University.

Arthur J. Reynolds is a professor of Social Work, Educational Psychology, and Human Development at the University of Wisconsin—Madison. He directs the Chicago Longitudinal Study, one of the largest studies of the effects of early childhood intervention. A focal point of the 19-year, ongoing project is the long-term educational and social effects of the Chicago Child-Parent Centers, a Title I preschool and early school-age program in the Chicago public schools since 1967. A cost-benefit analysis of the program has been published up to age 21. Dr. Reynolds is interested more broadly in child development and social policy, especially how research can inform public policy. He has written extensively on the implications of early childhood research. He also is affiliated with the Waisman Center and the Institute for Research on Poverty at the university.

Howard Sandler graduated from the Johns Hopkins University (B.A., 1967) and went on to complete his graduate work at Northwestern University (M.A., 1969; Ph.D., 1971). He has taught at Peabody College, Vanderbilt University, since 1970. Although his primary teaching interests are in statistics and methodology, his research has always focused on children at risk (e.g., child abuse, teen pregnancy, chronically ill children, economically disadvantaged children). For the last 10 years he has been collaborating with Professor Kathleen Hoover-Dempsey on the development of a model of understanding how parental involvement in the school translates into later achievement by children.

Pamela Sheley is a research associate at the Academic Development Institute (ADI) in Lincoln, Illinois. She is a graduate of MacMurray College with a degree in psychology and is a graduate student at the University of Illinois.

Ronald D. Taylor is associate professor in the Psychology Department at Temple University and Deputy Director of the Center for Research in Human Development and Education. Dr. Taylor is a developmental psychologist and received his Ph.D. from the University of Michigan. His research interests include family relations and social and emotional adjustment of ethnic minority adolescents. He has published numerous articles and chapters and edited several volumes, including *Social and Emotional Adjustment and Family Relations in Ethnic Minority Families* and *Resilience Across Contexts: Family, Work, Culture and Community*. Recent work has focused on the processes mediating the association of families' economic resources with adolescents' psychosocial well-being among ethnic minority families.

Joan M. T. Walker, Ph.D., served for 10 years as the director of a nationally accredited school-age child care program. Located at a public elementary school, the child care program was an integral part of the family–school community. Her experiences there were a significant influence on her thinking about contexts that support children's development. She completed her M.S. (2000) and Ph.D. (2003) in developmental psychology at Vanderbilt University. She is currently a postdoctoral Research Associate at Vanderbilt University, where she is continuing to investigate how social interactions relate to learning and development.

Index